CALCULUS OF VARIATIONS WITH APPLICATIONS

George M. Ewing

George L. Cross Research Professor
of Mathematics Emeritus,
University of Oklahoma, Norman

DOVER PUBLICATIONS, INC.
NEW YORK

Published in Canada by General Publishing Company, Ltd., 30 Lesmill Road, Don Mills, Toronto, Ontario.
Published in the United Kingdom by Constable and Company, Ltd., 10 Orange Street, London WC2H 7EG.

This Dover edition, first published in 1985, is an unabridged, corrected and slightly enlarged republication of the work first published by W. W. Norton & Co., New York, 1969. The author has supplied a new Preface and a Supplementary Bibliography for this edition.

Manufactured in the United States of America
Dover Publications, Inc., 31 East 2nd Street, Mineola, N.Y. 11501

Library of Congress Cataloging in Publication Data

Ewing, George M. (George McNaught), 1907–
  Calculus of variations with applications.

  Reprint. Originally published: New York : W. W. Norton, c1969.
With new pref. and supplemental bibliography.
  Bibliography: p.
  1. Calculus of variations. I. Title.
QA315.E9 1985   515'.64   84-18870
ISBN 0-486-64856-7

Contents

Chapter 4, *Variations and Hamilton's Principle*

Chapter 5, *The Nonparametric Problem of Bolza*

Chapter 6, *Parametric Problems*

Chapter 7, *Direct Methods*

Chapter 8, *Measure, Integrals, and Derivatives*

Chapter 9, *Variational Theory in Terms of Lebesgue Integrals*

Chapter 10, *A Miscellany of Nonclassical Problems*

Chapter 11, *Hamilton–Jacobi Theory*

Chapter 12, *Conclusion and Envoy*

Preface to the Dover Edition

This is a minor revision of the original 1969 publication. Some false statements have been changed. Clarifying remarks have been added in certain places and a short supplementary bibliography has been added. Misspellings, imperfect mathematical symbols, and numerous other errors have been corrected.

GEORGE M. EWING
Norman, Oklahoma

Preface to the Original Edition

The name calculus of variations comes from procedures of Lagrange involving an operator δ called a variation, but this restricted meaning has long been outgrown. The calculus of variations broadly interpreted includes all theory and practice concerning the existence and characterization of minima, maxima, and other critical values of a real-valued functional. To say much less would exclude works of eminent authors whose titles indicate contributions to variational theory but whose methods include no calculus in the early sense.

This book is an introduction, not a treatise. It is motivated by potential applications but is not a mere compendium of partially worked examples. It selects a path through classical conditions for an extremum and modern existence theory to problems of recent origin and with novel features. Although it begins with mild presuppositions, the intent is to expose the reader progressively to more substantial and more recent parts of the theory so as to bring him to a point where he can begin to understand specialized books and research papers. This entails compromise. Less than the traditional space is devoted to necessary conditions and sufficiency for local extrema of a succession of problems to give more attention to global extrema, to so-called direct methods, and to other twentieth-century topics. An introduction to Hamilton–Jacobi Theory makes contact with the Dynamic Programming of R. Bellman and the Maximum Principle of L. S. Pontryagin.

Chapters 1 through 6 have been used with classes including members with no special preparation beyond a course in advanced calculus. There are accordingly numerous elaborative comments and warnings against pitfalls. Certain prerequisite materials are collected in Chapter 1 for ready reference. The major objective is insight, not practice in writing Euler equations or in other techniques; hence emphasis is on conceptual and logical features of the subject. Nevertheless, the often formidable gap between theory and the analysis of particular problems is bridged by treatment of a number of examples and many exercises for the reader.

Chapters 7 through 12 require more mathematical maturity or else willingness to supplement the text as individual needs may require. The exposition is, however, largely self-contained. A brief treatment of the Lebesgue theory of integration, which is essential for important parts of modern variational theory, is in Chapter 8 for those who need it. A number of the cited books and some of the cited articles can be used in direct support of material in the text, but others begin at or beyond positions covered here and are listed in the bibliography as information on recent trends and names associated therewith. Previous experience in modern real analysis, theory of differential equations, functional analysis, or topology will be helpful, but a reader with serious intentions who lacks this advantage can still make effective use of much of the second half of the book.

Variational theory has connections with such fields as mathematical physics, differential geometry, mathematical statistics, conflict analysis, and the whole area of optimal design and performance of dynamical systems. These interrelations suggest the importance of the subject, why one book cannot be comprehensive, and why this is not an easy subject for the beginner. One never has adequate preparation for all the things with which he may be confronted under variational theory and its applications.

The author is indebted to many sources, particularly to works of G. A. Bliss, E. J. McShane, and L. Tonelli; to his association with W. T. Reid; to Marston Morse, under whose encouragement he was privileged to spend a postdoctoral year; to his teacher, W. D. A. Westfall; and to various colleagues, friends, and students.

Thanks are extended to W. T. Reid and D. K. Hughes for identifying flaws in parts of the manuscript, but this is not to suggest a shared responsibility for such flaws as may remain.

This book was sponsored in its initial stage during the summer of 1964 by the Office of Scientific Research of the Air Research and Development Command through Grant AF-AFOSR-211-63 to the University of Oklahoma Research Institute, for which the author expresses his appreciation.

GEORGE M. EWING
Norman, Oklahoma

Chapter 1

INTRODUCTION
AND ORIENTATION

1.1 PREREQUISITES

The reader is assumed to be familiar with concepts and methods usually covered by courses called advanced calculus or introduction to real analysis. Among the things presupposed are elementary set theory, real numbers, various kinds of limits and continuity, derivatives, ordinary differential equations of the first and second order, functions defined implicitly, and the Riemann integral.

A résumé of such topics is given in this chapter for review and reference and to introduce terminology, notations, and points of view to be found throughout the book. It is suggested that the chapter be read quickly for content, then returned to later for more details as needs may arise. Development of variational theory begins with Chapter 2.

1.2 FUNCTIONS

Given two nonempty sets X and Y of any nature, a function traditionally has been described as a correspondence under which to each $x \in X$ is associated $y \in Y$. This lacks the precision of a definition and is indeed circular, since the term correspondence, like mapping, operation, or transformation, is a synonym for function.

A way of avoiding this objection is to define a *function f from X to* (*or into*) *Y*, written $f: X \to Y$ to mean a set of ordered pairs (x,y), $x \in X$, $y \in Y$, such that each $x \in X$ is the first component of exactly one pair $(x,y) \in f$. Thus single-valuedness is part of the definition. Set X is the *domain* of f, the set $\{y \in Y: (x,y) \in f\}$ is the *range* of f. The function f is a subset of the *cartesian product* $X \times Y$, that is, of the set of all pairs (x,y) of respective elements from X and Y. Thus function becomes synonymous with graph of a function, by which is meant the ideal graph, not the approximation that one draws.

We follow the practice of identifying the domain X of f when we write $f: X \to Y$ but of mentioning a set Y, which in general is a superset of the range. For example, such a statement as $f(x) = 1/(x^2 + 1)$ serves to define a function $f: R \to R$, R being the set of all real numbers. The domain of f is the entire set R; the range of this f is a subset of R, namely, the half-open interval $(0,1] \equiv \{y \in R: 0 < y \leqslant 1\}$. Hence we can describe f more precisely as a function or mapping from R onto $(0,1]$, but often it suffices merely to exhibit a set of which the range is a subset.

Brackets and parentheses will be used for intervals of real numbers as follows.

$$[a,b] \equiv \{y \in R: a \leqslant y \leqslant b\},$$
$$(a,b) \equiv \{y \in R: a < y < b\},$$
$$(a,b] \equiv \{y \in R: a < y \leqslant b\},$$
$$[a,b) \equiv \{y \in R: a \leqslant y < b\}.$$

The symbol for identical equality is used here and elsewhere to indicate that the symbol on the left is being defined.

Symbol R^n denotes the cartesian product of n repetitions of the set R, that is, the set of all elements $x = (x^1, \ldots, x^n)$ of ordered n-tuples of real numbers. Elements x are alternatively spoken of as points or as vectors. For $n = 1, 2$, and 3 one visualizes either a point in the appropriate cartesian coordinate system or a directed segment with components x^i in the respective directions of the axes. Few people attempt to visualize R^n for $n > 3$, but it is often suggestive to draw planar sketches that can be thought of as crude projections into the plane of points, segments, or other objects from R^n. Our superscripts distinguish among the coordinates or components of x. We use subscripts to distinguish among different points. For example, $x_0 = (x_0^1, \ldots, x_0^n)$ and $x_1 = (x_1^1, \ldots, x_1^n)$ denote two points in R^n or two n-vectors.

A function $f: A \subset R^m \to R^n$, $m \geqslant 1$, $n \geqslant 1$, that takes each $x \in A$ to $y = f(x) \in R^n$ is often called a *vector-valued function*. For $n = 1$ it is *real-valued*.

A sequence is a function $f: N \to S$ whose domain is the set $N = \{1,2,3,\ldots\}$ of natural numbers (positive integers). The set S can be of any nature. We shall be concerned with cases in which $S = R^n$, $n \geqslant 1$, and also with cases in which S is a set of functions $f: [a,b] \to R^n$.

For many years the symbol $f(x)$ has been used either for the image of an element x in the domain of f or alternatively as a symbol for the function f, depending on the context. It is essential to distinguish between the two ideas, function and a value of a function, and it is increasingly the practice to restrict symbols f and $f(x)$ to these respective meanings. One must, however, recognize that there remain many useful books and articles even of recent date that follow the older convention. This applies in particular to the literature on variational theory and its applications.

We must deal frequently in later chapters with composite functions. For example, given $f: R \to R$ with values $f(x) = x^2$ and $g: R \to R$ with values $g(t) = e^t$, the function f composed with g, written $f \circ g: R \to R$ or $f(g): R \to R$, has values $(f \circ g)(t) \equiv f[g(t)] = e^{2t}$. Similarly, the function $g \circ f$, read "g composed with f," has values $(g \circ f)(x) \equiv g[f(x)] = \exp x^2$.

A recurring example in following chapters is a function with values $f[t,y(t),\dot{y}(t)]$. It is the composition of a function $f: [a,b] \times R \times R \to R$ with a function $g: [a,b] \to R^3$. This g is a vector-valued function with three components:

$$g^1: [a,b] \to R, \qquad g^1(t) = t,$$
$$g^2: [a,b] \to R, \qquad g^2(t) = y(t),$$
$$g^3: [a,b] \to R, \qquad g^3(t) = \dot{y}(t),$$

in which $y(t)$ and $\dot{y}(t)$ are respective values at $t \in [a,b]$ of a given function $y: [a,b] \to R$ and its derivative $\dot{y}: [a,b] \to R$.

We shall not repeat such a detailed description again but simply exhibit a typical value $f[t,y(t),\dot{y}(t)]$ of $f \circ g$ when this function is needed.

When a composition $f \circ g$ is mentioned in this book the functions f and g will always be so related that the domain of f contains the range of g.

1.3 THE EXTENDED REAL NUMBERS

The set R of real numbers, augmented by two symbols $+\infty$ (read "plus infinity" and often written without the $+$) and $-\infty$ subject to the postulates written below, constitutes the set R^* of *extended reals*. The inverted A is to be read "for all" or "for every."

$$
\begin{array}{lll}
 & \text{(i)} \quad -\infty < a < \infty, & \forall\, a \in R, \\
 & \text{(ii)} \quad a + \infty = \infty + a = \infty, & \forall\, a \in R^* \text{ except} -\infty, \\
 & \text{(iii)} \quad a + (-\infty) = -\infty + a = -\infty, & \forall\, a \in R^* \text{ except} \infty, \\
(1.1) & \text{(iv)} \quad a(\pm\infty) = (\pm\infty)a = \pm\infty & \text{if } 0 < a \in R \text{ or } a = \infty, \\
 & \text{(v)} \quad a(\pm\infty) = (\pm\infty)a = \mp\infty & \text{if } 0 > a \in R \text{ or } a = -\infty, \\
 & \text{(vi)} \quad a/\pm\infty = 0, & \forall\, a \in R.
\end{array}
$$

Such expressions as $\infty - \infty$, $-\infty + \infty$, ∞/∞, and $0(\infty)$, like $0/0$, remain meaningless. We shall later mention the special convention of assigning the value 0 to $0(\pm\infty)$ in integration theory, but this is not a useful convention in general and need not concern us at the moment.

1.4 BOUNDS, MAXIMA, AND MINIMA

That a subset S of R^* has an *upper bound* B means that $s \leqslant B$, $\forall s \in S$. A *lower bound* b of S is similarly defined. Clearly $B = \infty$ and $b = -\infty$ are an upper and a lower bound for any such set S. Only when there is a finite (that is, a real) upper or lower bound do we have something distinguishing to say. Hence, even in discussions that admit ∞ or $-\infty$, to say that a set is *bounded* means that it has finite upper and lower bounds.

We seldom have occasion in this book to extend R^n, $n > 1$, by adjoining points some of whose coordinates are ∞ or $-\infty$. That a subset S of R^n is bounded means that, for each j, the coordinates x^j of points $x \in S$ constitute a bounded subset of R. Thus, if S is bounded, it is a subset of the cartesian product of n intervals in R. Such a product is called an *n-dimensional interval* or an *n-dimensional box*. It is not difficult to verify that there exist n-dimensional balls (defined in Section 1.10) containing a given box and vice versa, hence that boundedness of S could have been defined by requiring the existence of a ball containing S.

If a set S of real numbers has a finite upper bound, it is a fundamental property of real numbers, a theorem or a postulate depending upon how the real numbers have been introduced, that S has a least upper bound or *supremum*. If S has no finite upper bound, the least upper bound or supremum is ∞. Among the symbols for the supremum of S are

$$\sup S, \qquad \sup\{x : x \in S\}, \qquad \sup x.$$

The infimum or greatest lower bound of a subset S of R^* is similarly defined and is denoted by such symbols as

$$\inf S, \qquad \inf\{x : x \in S\}, \qquad \inf x.$$

If and only if there is a largest element x^* in S, then x^* is called the *maximum* of S and its value is represented by

$$\max S, \qquad \max\{x : x \in S\}, \qquad \max x.$$

Similarly, if S includes a smallest element x_* among its elements, we call x_* the *minimum* of S and use such symbols as

$$\min S, \qquad \min\{x : x \in S\}, \qquad \min x.$$

Given any subset S of R^* there is always an answer to the question: What are inf S and sup S? If it is meaningful to speak of max S, that is, if there is a largest extended real number in S, then max $S = $ sup S and under similar circumstances min $S = $ inf S. If, for example, S is the open interval $(0,1)$, then inf $S = 0$ and sup $S = 1$ but neither max S nor min S exists.

If S is a finite nonempty set, min S and max S always exist; if S is an infinite set either or both of the two can fail to exist. If S is the empty set $\emptyset$, then every extended real number is both an upper bound and a lower bound of S. It follows that inf $S = \infty$ and sup $S = -\infty$. Clearly min $\emptyset$ and max $\emptyset$ are meaningless.

1.5   LIMITS

By the δ-*neighborhood* $U(\delta,x_0)$ of $x_0 \in R$ we shall mean the open interval $(x_0 - \delta, x_0 + \delta), \delta > 0$. Alternatively stated,

(1.2) $$U(\delta,x_0) \equiv \{x \in R: |x - x_0| < \delta\}.$$

We shall use $V(\delta,x_0)$ for the *deleted* δ-neighborhood.

(1.3) $$V(\delta,x_0) \equiv U(\delta,x_0) - \{x_0\} = \{x \in R: 0 < |x - x_0| < \delta\}.$$

Neighborhoods and deleted neighborhoods of ∞ and $-\infty$ are defined as follows:

(1.4) $$U(\delta,\infty) \equiv \{x \in R^*: x > 1/\delta\},$$

(1.5) $$V(\delta,\infty) \equiv U(\delta,\infty) - \{\infty\} = \{x \in R: x > 1/\delta\},$$

(1.6) $$U(\delta,-\infty) \equiv \{x \in R^*: x < -1/\delta\},$$

(1.7) $$V(\delta,-\infty) \equiv U(\delta,-\infty) - \{-\infty\} = \{x \in R: x < -1/\delta\}.$$

In stating various definitions and theorems it is convenient to use *symbols* $\forall$ and $\equiv$ introduced in Sections 1.2 and 1.3 and also

$\exists$, read "there exists,"
$\Rightarrow$, read "implies,"
$\Leftrightarrow$, read "is logically equivalent to."

The reader is assumed to be familiar with symbols $\cup$ and $\cap$ for the union and intersection of sets and with the meaning of these operations. We have already followed the practice of using braces either enclosing symbols for typical elements or enclosing conditions that serve to select the elements as a symbol for a set. We use the ordinary minus sign for a difference between sets and the simple inclusion symbol ⊂ (without the bar underneath preferred by some writers). Thus $A \subset B$ means that A is a subset of B, possibly a proper subset and possibly equal to B. We agree that $A = B \Leftrightarrow A \subset B$ and $B \subset A$.

That $a \in R^*$ is an *accumulation* (cluster) *point* of $S \subset R^*$ means that

$$S \cap V(\delta, a) \neq \phi, \qquad \forall \, \delta > 0.$$

A function $f \colon S \subset R \to R^*$ is said to have the *limit* $\lambda \in R^*$ *at* a, an accumulation point of S, if

$$(1.8) \quad \forall \epsilon > 0, \, \exists \delta_{a,\epsilon} > 0 \quad \text{such that} \quad x \in S \cap V(\delta, a) \Rightarrow f(x) \in U(\epsilon, \lambda).$$

This formulation includes various familiar cases that have often been treated separately, for example, a function $f \colon R \to R^*$ or a function (sequence) $f \colon N \to R^*$. Either λ or a can be ∞ or $-\infty$ as well as finite.

Subscripts a, ϵ on δ in (1.8) signify that, in general, δ depends on the choice of both a and ϵ even though, for certain functions f, δ may be independent of a or ϵ. We avoid using $\delta(a, \epsilon)$, since this might be mistaken to mean that δ was a function having $\delta(a, \epsilon)$ as a value and we prefer not to say this. Observe that $\delta_{a,\epsilon}$ is not unique. Given any one such value for which (1.8) is true, every positive real number smaller than the first one also serves. Although there is often a largest such δ, we seldom need to identify this value when using (1.8). Similar remarks apply to other definitions that involve an ϵ.

The limit defined by (1.8) is a *deleted limit*. Some writers also define a *nondeleted limit* by replacing the deleted neighborhood $V(\delta, a)$ with the nondeleted neighborhood $U(\delta, a)$. All limits mentioned in this book will be deleted limits unless there is explicit statement to the contrary.

There may or may not exist an element λ of R^* with the property stated in (1.8). Since a is an accumulation point of S, there necessarily exists a sequence $\{x_n \in S \colon n \in N\}$ having a as limit, but the corresponding sequence $\{f(x_n) \in R^* \colon n \in N\}$ need have no limit. However, every sequence of extended reals has at least one subsequence with a finite or infinite limit. That a bounded sequence necessarily has such a subsequence is a classic theorem. If the given sequence has no finite upper (or lower) bound, it is easy to verify the existence of a subsequence having, in accord with (1.8), the limit ∞ (or $-\infty$), respectively. Thus there exists a sequence $\{x_n\}$ with the limit a such that the sequence $\{f(x_n)\}$ of functional values has a limit, finite, ∞, or $-\infty$ as the case may be.

Given $f \colon S \subset R \to R^*$ and the accumulation point a of S in the broad sense stated above, consider the class of all sequences $\{x_n\}$ such that x_n has the limit a and such that $f(x_n)$ has some limit $\lambda \in R^*$. Denote by $\{\lambda\}$ the subset of R^* consisting of all such limits λ. In the special case where f has a limit under definition (1.8), set $\{\lambda\}$ is a singleton set. It is easy to construct sequences such as 0, 1, 0, 1, 0, 1, ... for which $\{\lambda\}$ consists of exactly two elements or other examples for which $\{\lambda\}$ has a preassigned finite number of elements. For the example $f: R - \{0\} \to R$, $f(x) = \sin(1/x)$, and with $a = 0$, the set $\{\lambda\}$ is the closed interval $[-1,1]$, whereas, if $f(x) = (1/x^2)\sin(1/x)$, then $\{\lambda\} = R^*$.

The *limit inferior* and *limit superior* of a function f at an accumulation point of its domain can be defined by the statement that

(1.9) $$\lim{}_a \inf f \equiv \inf\{\lambda\},$$

(1.10) $$\lim{}_a \sup f \equiv \sup\{\lambda\}.$$

Clearly $\lim \inf f \leqslant \lim \sup f$. Iff (read "if and only if") equality holds, $\{\lambda\}$ is a singleton set and f has a limit at a as defined by (1.8).

In the event that there is a sequence $\{x_n\}$ in S converging to a from the left (or right) we can define left and right limits inferior by using, respectively, the subsets of $\{\lambda\}$ obtained by considering only sequences $\{x_n\}$ such that $x_n < a$ (or $x_n > a$) for all n. One-sided limits superior are similarly defined. These four limits are denoted by

(1.11) $$\lim{}_{a-} \inf f, \qquad \lim{}_{a+} \inf f,$$

(1.12) $$\lim{}_{a-} \sup f, \qquad \lim{}_{a+} \sup f.$$

Iff the left (or right) limits inferior and superior are equal, the common value can be used as the definition of the left (right) limit of f at a. Iff the left and right limits of f at a both exist and are equal, the common value has the property of λ in (1.8).

Exercise 1.1

1. Given $f: \{x \in R: x \neq 0\} \to R$, $f(x) = 1 + 2\sin(1/x)$ or $-2 + \sin(1/x)$ according as $x < 0$ or > 0, identify the relevant subsets of $\{\lambda\}$ and determine the four limits (1.11) and (1.12).
2. Discuss (1.11) and (1.12) for the special case in which f is a sequence; that is, the domain of f is the set N of positive integers and $a = \infty$. Construct an example of a sequence of real numbers such that the set $\{\lambda\}$ consists of exactly three different numbers. What is the set $\{\lambda\}$ if $f(n) = (1/r_n)^2 \sin(1/r_n)$, where $\{r_n: n \in N\}$ is a sequentialization of all rational real numbers?
3. If f and g both have finite limits at a, use (1.8) in proving that $f + g$ has a limit at a. Point out by an example why this conclusion would not in general be correct without the restriction to finite limits. If one of the given limits is finite and the other is ∞ or $-\infty$, is it or is it not true that $f + g$ has a limit and why?

1.6   CONTINUITY AND SEMI-CONTINUITY

We shall be mainly interested in functions $f: I \to R$, where I is an interval of the reals of positive length. However, the definition can just as easily be stated for the more general case in which the domain of f is any non-empty subset S of the reals.

That f is *continuous at* $c \in S$ (c may or may not be an accumulation point of S) means that

$$(1.13) \quad \forall \epsilon > 0, \exists \delta_{c,\epsilon} > 0, \quad \text{such that} \quad x \in S \cap V(\delta_{c,\epsilon}, c)$$
$$\Rightarrow |f(x) - f(c)| < \epsilon.$$

The final inequality has been used rather than to require that $f(x) \in U(\epsilon, f(c))$ in order to make finiteness of $f(c)$ part of the definition of continuity at c. Although it is useful to our purposes to include ∞ or $-\infty$ in the ranges of certain functions, we prefer never to say that a function f is continuous at a point c if $f(c) = \infty$ or $-\infty$.

The function f is called *continuous on* S if it is continuous at each $c \in S$. It is called *uniformly continuous on* S if there is a δ_ϵ free of c such that (1.13) holds (with δ_ϵ in place of $\delta_{c,\epsilon}$) for every $c \in S$. The following equivalent definition of uniform continuity is often convenient.

$$(1.14) \quad \forall \epsilon > 0, \exists \delta_\epsilon > 0 \quad \text{such that} \quad x, x' \in S$$
$$\text{and} \quad |x - x'| < \delta_\epsilon \Rightarrow |f(x) - f(x')| < \epsilon.$$

One defines left (right) continuity of f at c by adding to (1.13) the restriction $x < c$ ($x > c$). If the domain S of f is a closed interval $[a,b]$, then there is no x below a in S and the definition of left continuity applies vacuously at a. A similar remark applies to right continuity at b.

If the final inequality in (1.13) is replaced by $f(x) > f(c) - \epsilon$ or by $f(x) < f(c) + \epsilon$, we have respective definitions of *lower* and *upper semi-continuity at* c and with the restriction built into (1.13) that $f(c)$ be finite. In contrast with the concept of continuity, it is convenient to speak of semi-continuity even when $f(c) = \infty$ or $-\infty$. The respective statements that

$$(1.15) \qquad\qquad \lim_c \inf f \geq f(c)$$

and

$$(1.16) \qquad\qquad \lim_c \sup f \leq f(c)$$

serve as definitions of lower and upper semi-continuity without the restriction that $f(c)$ be finite. Function f is continuous at $c \in S$ iff (1.15) and (1.16) both hold and $f(c)$ is finite.

The simplest examples of functions that are semi-continuous but not continuous at c are obtained by starting with a point c that is both a point and an accumulation point of the domain of a function f continuous at c and then altering the value $f(c)$.

Exercise 1.2

1. Given that $f(x) = 1 - x^2$ or 0 according as $x \neq 0$ or $= 0$, verify that f is lower semi-continuous at every real c.
2. Investigate the semi-continuity of the Dirichlet function with values $f(x) = 1$ or 0 according as x is irrational or rational.
3. Construct an example of a function $f: R \to R$ that is uniformly continuous on R.
4. Construct your own proof of the classic theorem that, if $f: [a,b] \to R$ is continuous on $[a,b]$, then f is uniformly continuous on $[a,b]$.
5. Construct an example of a function $f: S \subset R \to R$ that is continuous on S but not uniformly continuous on S.
6. Define left and right lower and upper semi-continuity. Construct an example of a function $f: R \to R$ right continuous at 0, not left continuous at 0, but left lower semi-continuous at 0.

1.7 DERIVATIVES

We restrict attention to a function $f: [a,b] \to R$, $b > a$. Given x and $c \in [a,b]$, define the difference quotient

$$(1.17) \qquad Q(x,c) = \frac{f(x) - f(c)}{x - c}, \qquad x \neq c.$$

With c fixed and with reference to (1.11) and (1.12), define

$$(1.18) \qquad (D^- f)(c) \equiv \lim_{c-} \sup Q,$$

$$(1.19) \qquad (D_- f)(c) \equiv \lim_{c-} \inf Q,$$

$$(1.20) \qquad (D^+ f)(c) \equiv \lim_{c+} \sup Q,$$

$$(1.21) \qquad (D_+ f)(c) \equiv \lim_{c+} \inf Q.$$

Consistently with Section 1.5, each right member may be finite, ∞, or $-\infty$. Statements (1.18) through (1.21) define four functions

$$(1.22) \qquad \begin{aligned} D^- f &: (a,b] \to R^*, \\ D_- f &: (a,b] \to R^*, \\ D^+ f &: [a,b) \to R^*, \\ D_+ f &: [a,b) \to R^*, \end{aligned}$$

called the *upper left, lower left, upper right,* and *lower right, derivate of f,* respectively. Parentheses around D^-f, etc., in (1.18) through (1.21) are omitted by some writers. They are included here to emphasize that we are stating the values at c of the respective functions (1.22).

Clearly $(D_-f)(c) \leq (D^-f)(c)$. Iff equality holds, the common extended real value is by definition the *left derivative of f at c* and is denoted by $f'^-(c)$. The *right derivative of f at c* is similarly defined and is denoted by $f'^+(c)$. Iff $f'^-(c)$ and $f'^+(c)$ exist and are equal, the common extended real value is called the *derivative of f at c* and is denoted by the familiar symbol $f'(c)$.

The domains of the function f'^-, f'^+ are, in general, proper subsets of the respective half-open intervals $(a,b]$ and $[a,b)$, and that of f' is a subset of the open interval (a,b). It is, however, convenient, in the event that $f'^+(a)$ exists, to extend the domain of f' to include $x = a$ by the additional definition $f'(a) \equiv f'^+(a)$. Similarly, if $f'^-(b)$ exists, we define $f'(b) \equiv f'^-(b)$. Alternatively stated, $f'(c)$ is now defined at $c \in [a,b]$ iff all derivates that are defined at c have a common extended real value and then $f'(c)$ is this common value.

The function f is said to be *differentiable* at c iff $f'(c)$ exists. Some authors make finiteness part of the definition of the derivative, as we have not. Given, for example, the function $f: R \to R, f(x) = x^{1/3}$, we say, from the viewpoint of this section, that f is everywhere differentiable. Specifically,

$$(1.23) \qquad f'(x) = \begin{cases} \frac{1}{3} x^{-2/3}, & x \neq 0, \\ \infty, & x = 0. \end{cases}$$

The discussion of this section applies after appropriate modifications when the domain of f is not the closed interval $[a,b]$ but is an interval of some other type, possibly of infinite length, as in the preceding example, or when the domain of f is some other subset S of R.

Exercise 1.3

1. Given $f: R \to R, f(x) = |x|$, identify the domains of f'^-, f'^+, and state the values of these functions in form (1.23).

2. Given $f: R \to R, f(x) = x \sin(1/x)$ if $x \neq 0$ and $f(0) = 0$, determine the values of the four derivates at 0.

3. Given that $f(x) = x^2$ or $-x^2$ according as x is rational or irrational, identify the domain of f' and determine $f'(x)$ at each point x of that domain.

4. If $f(x) = x^{2/3}$ for all real x, demonstrate that f does or does not have a derivative at 0 under our definition, whichever is correct.

5. State and prove Rolle's theorem for a function $f: [a,b] \to R$ that is continuous on $[a,b]$ and differentiable on (a,b).

6. Explain the difference between $f'^-(c)$, the left derivative at c, and $f'(c-)$, the left limit at c of f'. Explain similarly the conceptual distinction between $f'^+(c)$ and $f'(c+)$. Finally, illustrate these distinctions by the following function $f: R \to R$ with $c = 0$:

$$f(x) = \begin{cases} x^2 \sin(1/x), & x < 0, \\ 0, & x = 0, \\ x^2, & x > 0. \end{cases}$$

7. Given the absolute value function $| \cdot |: R \to R$, $|x|$ being the absolute value of x, identify the domain of the derivative $| \cdot |'$ of this function. Explain why

$$\int_{-1}^{2} |x|' \, dx = |2| - |-1| = 2 - 1 = 1.$$

8. Given the *unit step-function* $u: R \to R$,

$$u(x) = \begin{cases} 0, & x < 0, \\ \frac{1}{2}, & x = 0, \\ 1, & x > 0, \end{cases}$$

state the values for $u'(x)$ for $x < 0, = 0$, and > 0. Explain why u' is not Riemann integrable over $[-2,1]$ or any other interval of which 0 is a point.

1.8   PIECEWISE CONTINUOUS FUNCTIONS

Let S denote a closed interval $[a,b]$ less a finite, possibly empty, set of interior points of $[a,b]$. Visualize S as either $[a,b]$ or as $[a,b]$ with a possible finite number of interior points removed. A function $\phi: S \to R$ is called *piecewise continuous* (abbreviated PWC) on $[a,b]$ if

(i) ϕ *is bounded on* S,
(ii) *the right limit* $\phi(x+)$ *exists and is finite on* $[a,b)$,
(iii) *the left limit* $\phi(x-)$ *exists and is finite on* $(a,b]$,
(iv) $\phi(x-) = \phi(x+)$ *on* $S \cap (a,b)$.

The domain S of such a function is either the whole of $[a,b]$, the union $[a,c) \cup (c,b]$ of two half-open subintervals, or a union $[a,c_1) \cup (c_1,c_2) \cdots \cup (c_{n-1},c_n) \cup (c_n,b]$. The *restriction* of ϕ to any interval I of this decomposition of S into intervals is continuous on I.

The following are examples of PWC function on $[-1,1]$.

(1.25) $\phi, \quad \phi(x) = 1, \qquad x \in [-1,0) \cup (0,1]$,

(1.26) $\phi, \quad \phi(x) = \begin{cases} x^2, & x \in [-1,0), \\ x+1, & x \in (0,1]. \end{cases}$

We have left the function ϕ undefined at points of the finite set $[a,b] - S$ because this is precisely the type of function with which we must deal in Section 2.5. If we had different purposes in mind we might require as part of the definition of piecewise continuity on $[a,b]$ that the domain of ϕ be the whole of $[a,b]$.

Exercise 1.4

1. Extend the function ϕ of (1.25) to $[-1,1]$ by assigning the arbitrary real value r as $\phi(0)$. Explain why this extension of ϕ is Riemann integrable over $[-1,1]$ and that $\int_{-1}^{1} \phi(x)\,dx = 2$ independently of r. Because of these facts, one often says that the original ϕ is integrable over $[-1,1]$ even though its domain lacks point 0 of $[-1,1]$.

2. Extend the function ϕ of (1.26) to $[-1,1]$ and evaluate $\int_{-1}^{1} \phi(x)\,dx$.

3. Formulate a general theorem on the Riemann integrability over $[a,b]$ of a function $\phi: [a,b] \to R$ that is PWC on $[a,b]$ but otherwise arbitrary. Describe a procedure for evaluating $\int_{a}^{b} \phi(x)\,dx$ and illustrate by an example in which ϕ is continuous on $[a,c)$, (c,d), and $(d,b]$.

1.9 CONTINUOUS PIECEWISE SMOOTH FUNCTIONS

A function $\psi: [a,b] \to R$ will be called *smooth* on a subinterval I of its domain if its restriction to I has no discontinuities in direction, that is, if its derivative ψ' exists and is continuous and is hence finite on I. Usages vary among mathematicians. The word smooth may be assigned a different meaning elsewhere.

In this book a function $\psi: [a,b] \to R$ will be called *piecewise smooth* (abbreviated PWS) *on* $[a,b]$ if its value $\psi(x)$ is an indefinite integral of a function ϕ that is PWC on $[a,b]$. More explicitly stated, that ψ is PWS on $[a,b]$ will mean that

$$(1.27) \qquad \psi(x) = \int_{a}^{x} \phi(\xi)\,d\xi + \psi(a), \qquad \forall x \in [a,b].$$

Since an integral with a variable upper limit x is continuous in x we have made continuity on $[a,b]$ part of the meaning of PWS on $[a,b]$. Even though the term piecewise smooth does not include any reference to continuity, it is nevertheless partially descriptive of a function ψ satisfying (1.27) and short enough to be convenient. It is used in this book as it has been in this same sense by Akhiezer (see the Bibliography, reference I, first page of chapter 1) and others. If one goes to other subjects, he will find the same term used for essentially a larger class of functions ψ that are not necessarily continuous on $[a,b]$.

A function ψ that is PWS in the sense of (1.27) is said to be of class D' by many writers, the symbol being suggested by the partially descriptive term, discontinuous derivative. Others use a variant, D^1, of the preceding symbol. Similarly, a function ψ that is smooth on $[a,b]$, that is, which has a continuous first derivative on $[a,b]$, is said to be of class C' or of class C^1.

If a function ψ is smooth (is of class C', is of class C^1) on $[a,b]$, then it is PWS (is of class D', is of class D^1). The PWS function ψ on $[a,b]$ is smooth on $[a,b]$ iff the PWC function ϕ appearing in (1.27) is continuous on $[a,b]$.

One verifies from (1.27) and the definition of a derivative that if ϕ is continuous at x, that is, if x is in the set S of Section 1.8, then

$$(1.28) \qquad \psi'(x) = \phi(x), \qquad \forall x \in S.$$

Given $t \in [a,b] - S$, we can use (1.27) and the definitions of left and right derivatives to find that

$$\psi'^-(t) = \phi(t-) \qquad and \qquad \psi'^+(t) = \phi(t+).$$

Exercise 1.5

1. If $\phi: [a,b] \to R$ is Riemann integrable over $[a,b]$, prove that there exists $\mu \in R$ such that

$$\int_a^b \phi(x)\, dx = \mu(b-a).$$

What can be said of the value of μ if ϕ is continuous on $[a,b]$? If ϕ is not continuous on $[a,b]$?

2. Given (1.27), (1.28), and the mean value theorem of problem 1, prove that $\psi'^-(x) = \psi'(x-)$ on $(a,b]$ and that $\psi'^+(x) = \psi'(x+)$ on $[a,b)$.

3. Given that $\psi: [a,b] \to R$ is PWS on $[a,b]$, prove that its derivative ψ' is integrable over $[a,b]$ in the sense of problem 1, Exercise 1.4, and that

$$\int_a^x \psi'(u)\, du = \psi(x) - \psi(a), \qquad \forall x \in [a,b].$$

1.10　METRIC SPACES

Let S be a nonempty set with elements $x, y, z, \ldots$ of an arbitrary nature. Any function $d: S \times S \to R$ that satisfies the following postulates (axioms) is called a *distance* or a *metric*.

(1.29)
$$\begin{array}{l} \text{(i)} \ \ d(x,y) \geqslant 0, \\ \text{(ii)} \ \ d(x,y) = 0 \qquad \textit{iff} \ x = y, \\ \text{(iii)} \ \ d(x,y) = d(y,x), \\ \text{(iv)} \ \ d(x,y) + d(y,z) \geqslant d(x,z). \end{array}$$

The given set S together with a distance d with properties (1.29) constitutes a *metric space*. Alternatively stated, a metric space is an ordered pair (S,d), of which S is a nonempty set and d is a distance.

Of major importance among metric spaces are the *euclidean spaces* E_n, $n = 1, 2, \ldots, E_n \equiv (R^n, d)$, where R^n is the n-fold cartesian product of the set R of reals and $d(x,y)$ is the *euclidean distance* $|x - y|$,

(1.30)
$$|x - y| \equiv \left[\sum_1^n (x^i - y^i)^2 \right]^{1/2}.$$

For $n = 1$, $|x - y|$ is an ordinary absolute value. For a general n, we read the symbol $|x - y|$ as "the *norm* (*length* or *modulus*) of the difference vector" or " the distance between x and y," whichever seems more convenient at a particular time.

If we understand symbols $|x - y|$ and $|f(x) - f(c)|$ in the sense of (1.30), then much of Sections 1.5 and 1.6 automatically covers higher-dimensional cases. The set denoted by $U(\delta, x_0)$ in (1.2) is by definition the n-dimensional *open ball* and definition (1.13) of continuity at c can now be taken as the definition of continuity at $c \in R^p$ of a function $f: R^p \to R^q$.

Euclidean distance (1.30) clearly has properties (1.29)(i), (ii), and (iii). That it also has the *triangle property*, (1.29)(iv), is intuitively evident when $n = 1, 2, 3$. A proof of this property for a general n is not very difficult.

If S is a suitable set of functions and d is a distance, then the metric space is also a function-space. For example, given the fixed interval $[a,b]$, let S be the set of all functions $x: [a,b] \to R^n$ each continuous on $[a,b]$ and define

(1.31)
$$d(x,y) \equiv \sup_t |x(t) - y(t)|.$$

One sees easily that d has properties (1.29) (i), (ii), and (iii). To see that it has the triangle property, observe first that the distance $|x(t) - z(t)|$ for E_n is continuous in t; consequently, there exists, by a classic theorem on the existence of a greatest value, $t_1 \in [a,b]$ such that

$$|x(t_1) - z(t_1)| = d(x,z).$$

By the triangle inequality for E_n,

(1.32)
$$|x(t_1) - y(t_1)| + |y(t_1) - z(t_1)| \geqslant d(x,z),$$

while, as a consequence of definition (1.31), $d(x,y)$ and $d(y,z)$ dominate the respective terms on the left in (1.32).

Exercise 1.6

1. Prove algebraically that euclidean distance between points of R^2 has the triangle property.
2. Given the cartesian plane R^2, define $d(x,y)$ as 0 or 1 according as $x = y$ or $x \neq y$ and verify that (R^2,d) is a metric space.
3. Define $d(x,y) = |x-y|^{1/2}$ for real numbers x, y and show that (R,d) is a metric space.
4. Let S be the set of all smooth functions $x: [0,1] \to R$, define $d(x,y)$ by (1.31), and define $r(x,y)$ as $d(x',y')$ for the derivatives x' and y' of x and y. Show that $(S, d+r)$ is a metric space but that (S,r) is not.

1.11 FUNCTIONS DEFINED IMPLICITLY

Given a function f whose domain A and range B are subsets of $R^p \times R^q$ and R^q, respectively, and given an equation $f(x,y) = 0$ known to hold at $(x_0,y_0) \in R^p \times R^q$, it is often crucial to know whether there is a function ϕ from R^p to R^q, such that

(1.33) $f[x, \phi(x)] = 0$ *for all x in some open subset of R^p containing x_0.*

A function ϕ satisfying (1.33) is said to be *implicitly defined* by the equation $f(x,y) = 0$.

Few equations $f(x,y) = 0$ are simple enough so that by a sequence of elementary operations one can find an equivalent equation with y or x on the left and with a right member free of y or x, respectively. As a substitute we need theorems on the existence of a function ϕ with property (1.33), traditionally called *implicit function theorems.*

A variety of such theorems is to be found in books on advanced calculus, real analysis, or functions of real variables. The following is a typical example for the case $p = q = 1$.

Theorem 1.1

Given a function $f: A \subset (R \times R) \to R$ that is continuous on A and has continuous first-order partial derivatives f_x and f_y on A, if $f(x_0,y_0) = 0$ at an interior point (x_0,y_0) of A and if $f_y(x_0,y_0) \neq 0$, then there exists a function ϕ from an open interval $I = (x_0 - \delta, x_0 + \delta)$ to R such that

(i) $f[x,\phi(x)] = 0$, $\forall x \in I$, *with ϕ unique,*
(ii) $\phi'(x)$ *exists and is continuous on I, hence ϕ is continuous on I, and*
(iii) $\phi'(x) = -f_x[x,\phi(x)]/f_y[x,\phi(x)]$, $\forall x \in I$.

This theorem and its extension to a general p and q are *local* existence theorems. When we examine a proof we find that δ is positive but that this is all we can say about it. It may turn out to be large for certain particular functions f. However, given an arbitrarily small positive real number e, there are functions f satisfying the hypotheses of the theorem and for which $\delta < e$. Consequently, in the usual absence of a determination of δ, one views the conclusions conservatively and realizes that δ is possibly quite small.

It is possible to relax the hypotheses and still obtain conclusion (i) without (ii) and (iii). With suitably strengthened hypotheses we can prove conclusion (i) for an interval I given at the outset and not merely for an interval of undetermined possibly small length that appears in the proof. The last type is a *global* implicit function theorem.

A simple example of a global implicit function theorem is that in which $f(x,y)$ is of the form $x - g(y) = 0$, with g as a monotone function from R to R. There then exists a function $\phi: R \rightarrow R$ such that

$$x - g[\phi(x)] = 0, \qquad \forall x \in R.$$

1.12   ORDINARY DIFFERENTIAL EQUATIONS

In later chapters we shall meet a number of first- and second-order equations

(1.34)
$$y' = g(x,y)$$

and

(1.35)
$$y'' = h(x,y,y').$$

A *solution* of (1.34) means traditionally a function y on some interval I such that
$$y'(x) = g[x,y(x)], \qquad \forall x \in I.$$

For right members $g(x,y)$ of certain particular forms, there are devices found in elementary books that yield expressions for $y(x)$ involving some finite combination or other of known functions and/or integrals of such functions. These cases are the exceptions. In general, the analyst must use existence theory in order to proceed. If a solution exists, there are numerical methods for approximating it with the aid of modern computers.

Experience with elementary examples leads one to anticipate that, under reasonable hypotheses on the right member of (1.34), the equation should have a unique solution satisfying a preassigned condition

(1.36)
$$y(\xi) = \eta.$$

The following is a local existence theorem that speaks to this point.

Theorem 1.2

Given a function $g: A \subset (R \times R) \to R$ *that is continuous on* A *and satisfies a so-called Lipschitz condition*

$$(1.37) \qquad |g(x,y) - g(x,z)| \leq k\,|y - z|, \qquad \forall\, (x,y),\, (x,z) \in A,$$

then, given any point (ξ,η) *interior to* A, *there exists a positive* δ *and a function* $y: I \equiv (\xi - \delta, \xi + \delta) \to R$ *such that*

        (i) $y(\xi) = \eta$,
        (ii) $y'(x) = g[x, y(x)], \forall\, x \in I$, *and*
        (iii) *this solution* y *is unique.*

Examination of a proof will reveal that the solution y involves the value η and might be written $y(\eta,\,\cdot\,)$, the dot indicating that the value of the function at x will be $y(\eta,x)$. We thus have a *general solution*, that is, a one-parameter family of solutions with η as the parameter or arbitrary constant.

To investigate a second-order equation (1.35), it is convenient to set $y' = z$ and study the system

$$y' = z,$$
$$z' = h(x,y,z),$$

of first-order equations. This is a special case of the more general system $y' = g(x,y,z)$ and $z' = h(x,y,z)$.

If for a general n we shift to vector notation and to a dot in place of the prime we can reinterpret (1.34) as meaning the system

$$\dot{y}^1 = g^1(x,y),$$
$$\vdots$$
$$\dot{y}^n = g^n(x,y).$$

Both primes and dots are used in this book to denote derivatives without any explicit restriction in the case of the dot that the differentiation is with respect to time.

An extension of Theorem 1.2 to this case is obtained by taking as g a function from $A \subset (R \times R^n)$ to R^n, interpreting (1.37) as a condition in terms of euclidean norms (1.30) and understanding (ξ,η) to mean $(\xi, \eta^1, \ldots, \eta^n)$. There exists under this revision of hypotheses a unique function $y(\eta,\cdot)$ depending on n parameters, the components of η, such that $y(\eta,\xi) = \eta$ and the vector equation

$$\dot{y}(\eta,x) = g[x, y(\eta,x)]$$

holds for each fixed η such that (ξ, η) is interior to A and for all x on an interval I_η containing ξ and generally depending on η.

1.13   THE RIEMANN INTEGRAL

The reader is assumed to be familiar with some definition of the Riemann integral, either as a limit of Riemann sums

$$\Sigma \phi(\xi_i)(x_i - x_{i-1})$$

or by way of upper and lower sums.

Given the bounded function ϕ: $[a,b] \to R$, a sufficient condition for ϕ to be Riemann integrable over $[a,b]$ is that ϕ have at most a finite number of discontinuities. This condition is not necessary. We mention in passing that a necessary and sufficient condition for Riemann integrability of the bounded function ϕ is that the set of points in $[a,b]$ at which ϕ is discontinuous be a set of Lebesgue measure zero, a term to be defined in Chapter 8.

We shall often have occasion to differentiate an integral with respect to a parameter in accord with the following classic theorem.

Theorem 1.3

   Given two intervals I_1 and I_2 of the reals each of positive length and three functions f: $I_1 \times I_2 \to R$, g: $I_2 \to I_1$ and h: $I_2 \to I_1$ with values denoted by $f(x,\alpha)$, $g(\alpha)$, and $h(\alpha)$, suppose that f and its partial derivative f_α are both continuous on $I_1 \times I_2$ and that g and h have finite derivatives at a point β of I_2. Then $g(\beta)$ and $h(\beta)$ are values in I_1, the function F: $I_2 \to R$ with values

$$F(\alpha) \equiv \int_{g(\alpha)}^{h(\alpha)} f(x,\alpha) \, dx$$

has a finite derivative at β, and

$$F'(\beta) = \int_{g(\beta)}^{h(\beta)} f_\alpha(x,\beta) \, dx + f[h(\beta),\beta]h'(\beta) - f[g(\beta),\beta]g'(\beta).$$

Let f be the function mentioned near the end of Section 1.2 and require that f be continuous on its domain. If y: $[a,b] \to R$ is PWS, it follows from the appropriate definitions that the composite function of Section 1.2 with values $f[t,y(t),\dot{y}(t)]$ has at most a finite number of discontinuities on $[a,b]$, namely, discontinuities at those t at which y has corners. Therefore, this composite function is Riemann integrable over $[a,b]$ in the sense of problem 1, Exercise 1.4, a fact needed frequently in later chapters.

In Section 2.6 we shall need the derivative of a function F, where

$$F(\epsilon) = \int_{t_0}^{t_1} f[t, y_0 + \epsilon\eta, \dot{y}_0 + \epsilon\dot{\eta}] \, dt$$

with suitable hypotheses on f and with $y_0 + \epsilon\eta$ known to be PWS on $[t_0, t_1]$. The composite integrand satisfies the hypotheses of Theorem 1.3 iff $y_0 + \epsilon\eta$ happens to be smooth on $[t_0, t_1]$. In general $\dot{y}_0 + \epsilon\dot{\eta}$ has discontinuities at one or more interior points $c_1, \ldots, c_n$ of the interval. Set $c_0 \equiv t_0$, $c_{n+1} \equiv t_1$, and express the integral as the sum of integrals over intervals $[c_{i-1}, c_i]$, $i = 1, \ldots, n+1$. Hypotheses of Theorem 1.3, although not satisfied on the interval $[t_0, t_1]$, are satisfied on the subinterval $[c_{i-1}, c_i]$ provided that we interpret $\dot{y}(c_{i-1})$ as $\dot{y}^+(c_{i-1})$ and $\dot{y}(c_i)$ as $\dot{y}^-(c_i)$. The $n+1$ integrals obtained by Theorem 1.3 as derivatives of the separate integrals can then be combined into an integral over $[t_0, t_1]$ of the derivative of the integrand, exactly what we would have obtained if we had applied the conclusion of Theorem 1.3 to the original integral without justification. Upper and lower limits on the various integrals are all independent of ϵ; consequently, the boundary terms in the general expression for the derivative are both zero.

Similar procedure can be used to justify the expression we shall exhibit for $F''(\epsilon)$ in Section 2.12 and the derivatives of integrals in other sections.

We shall apply integration by parts to a number of definite integrals. That the familiar technique is meaningful and valid under a variety of circumstances is attested by a number of theorems, of which we give three.

Theorem 1.4

If $u: [a,b] \rightarrow R$ and $v: [a,b] \rightarrow R$ are both smooth on $[a,b]$, then
(i) uv', vu', and $(uv)' = uv' + vu'$ are all Riemann integrable over $[a,b]$ and
(ii) $\int_a^b u(t)v'(t) \, dt = u(b)v(b) - u(a)v(a) - \int_a^b v(t)u'(t) \, dt.$

PROOF

That u and v are smooth on $[a,b]$ means (Section 1.9) that u, v, u' and v' are all continuous on $[a,b]$. It follows that uv', vu', and $(uv)' = uv' + vu'$ are continuous on $[a,b]$ and hence Riemann integrable over $[a,b]$. Conclusion (ii) is then obtained by integration of $(uv)'$ over $[a,b]$.

In Section 2.6 and elsewhere, an extension of Theorem 1.4 to functions u and v that are PWS in the sense of Section 1.9 is needed.

Theorem 1.5

If $u: [a,b] \rightarrow R$ and $v: [a,b] \rightarrow R$ are both PWS on $[a,b]$, then conclusions (i) and (ii) of the Theorem 1.4 remain valid.

PROOF

Since u and v are PWS on $[a,b]$ they are both continuous on that interval. It follows that uv', vu', and $(uv)' = uv' + vu'$ are PWC on $[a,b]$ and hence Riemann integrable over $[a,b]$. Recall with possible reference to problem 1, Exercise 1.4, that when a function such as uv' is PWC it can be assigned an arbitrary real value at each point t where it is discontinuous. Let $c_1, \ldots, c_n$ be all points of $[a,b]$ at each of which u' or v' or both have a discontinuity and set $c_0 \equiv t_0$, $c_{n+1} \equiv t_1$. The hypotheses of Theorem 1.4 are satisfied on each closed subinterval $[c_{i-1}, c_i]$ provided that we take $u'(c_{i-1})$ and $u'(c_i)$ to mean the respective right and left derivatives of u at these points and similarly for v. Consider the $n+1$ equations (ii) expressing the result of integration by parts over $[c_{i-1}, c_i]$, $i = 1, \ldots, n+1$. The sum of the left members is the left member of (ii) as written in Theorem 1.4 and similarly for the sum of the right members.

The next and final theorem of this section is for possible reference in connection with Chapters 8 and 9. Readers not already familiar with the Lebesgue integral and absolutely continuous functions need not concern themselves with the theorem for the time being.

Theorem 1.6

If $u: [a,b] \to R$ and $v: [a,b] \to R$ are both absolutely continuous on $[a,b]$, then

(i) *uv', vu', and $(uv)' = uv' + vu'$ are all Lebesgue integrable over $[a,b]$, and*

(ii) *$\int_a^b u(t)v'(t)\, dt = u(b)v(b) - u(a)v(a) - \int_a^b v(t)u'(t)\, dt$,*

with the integrals now understood as Lebesgue integrals.

PROOF

Since u is absolutely continuous, it is of bounded variation and hence is expressible as the difference between two monotone functions by standard theorems (Theorems 8.24 and 8.23). The derivative $u'(t)$ then exists and is finite at each point t of a set $[a,b] - Z_1$, where Z_1 is of measure zero (Theorem 8.32). Similarly, $v'(t)$ exists and is finite on $[a,b] - Z_2$ with Z_2 of measure zero. The product uv is necessarily absolutely continuous; hence its derivative $(uv)'(t)$ also exists and is finite on $[a,b]$ except for a set Z_3 of measure zero. One next verifies by essentially the treatment of the derivative of a product in elementary calculus that the equation

$$(uv)'(t) = u(t)v'(t) + v(t)u'(t)$$

is meaningful and valid for all $t \in [a,b] - Z$, $Z \equiv Z_1 \cup Z_2 \cup Z_3$. The set Z has measure zero.

If we reinterpret the symbol $u'(t)$ by arbitrarily assigning it the value zero on Z but retain its original meaning as the derivative of u on $[a,b] - Z$ and do the same for $v'(t)$ and $(uv)'(t)$, then the equation above holds everywhere on $[a,b]$.

By the Fundamental Theorem of the Integral Calculus (Theorem 8.38),

$$\int_a^b (uv)'(t)\, dt = (uv)(b) - (uv)(a) = u(b)v(b) - u(a)v(a).$$

The function u being absolutely continuous is bounded and measurable on $[a,b]$ and v' is Lebesque integrable over $[a,b]$, hence measurable on $[a,b]$. The product uv' of functions measurable on $[a,b]$ is measurable on $[a,b]$ [Theorem 8.11(v)]. Let M denote an upper bound for $|u|$. Then $0 \leq |uv'| \leq M|v'|$. The function 0 is clearly integrable over $[a,b]$ and $|v'|$ is integrable (problem 9, Exercise 8.4); hence $M|v'|$ is integrable. It follows from the double inequality above that $|uv'|$ is integrable (problem 7, Exercise 8.4) and, since uv' is measurable, it is then integrable (problem 9, Exercise 8.4). We see in the same manner that vu' is integrable over $[a,b]$. Consequently,

$$\int_a^b (uv)'(t)\, dt = \int_a^b u(t)v'(t)\, dt + \int_a^b v(t)u'(t)\, dt,$$

and the proof is complete.

1.14   WHAT IS THE CALCULUS OF VARIATIONS?

Let $\mathcal{Y}$ denote a given class of functions $y: [a,b] \to R^n$ and consider a function $J: \mathcal{Y} \to R$, often called a functional, the suffix "al" serving as a reminder that a value $J(y)$ depends not upon the choice of a point y in some subset of R^n but upon the choice of a function y in a set of $\mathcal{Y}$ of functions. Since, however, the definition of a function as stated in Section 1.2 covers the case $J: \mathcal{Y} \to R$ as well as the more familiar types studied in elementary calculus, we shall generally omit the suffix.

If, for example, $[a,b]$ is a fixed interval and $\mathcal{Y}$ denotes the class of all functions y that are Riemann integrable over $[a,b]$, then

$$(1.38) \qquad\qquad J(y) \equiv \int_a^b y(x)\, dx$$

is a value of a function $J: \mathcal{Y} \to R$.

If $\mathcal{Y}$ is a suitable class of functions $y: R \to R$, then $J(y) = y(c)$ is a value of a function $J: \mathcal{Y} \to R$.

Variational theory is concerned with the existence and determination of $y_0 \in \mathcal{Y}$ such that $J(y_0)$ is a minimum or a maximum value of $J(y)$ and also with so-called stationary values to be defined later that may or may not be extreme values. To say that $J(y_0)$ is an *extremum* means that it is a minimum or a maximum.

The two examples mentioned above are overly simple. We close this chapter with some more complex examples drawn from the rich and constantly growing supply. We are interested for the moment only in the forms of typical problems, not in how to deal with them.

EXAMPLE 1.1

Given a particle of fixed mass m that is required to move from position $y(0) = 0$ to $y(T) = Y$ during the fixed time T. Does there exist, in the class $\mathscr{Y}$ of all PWS functions $y: [0,T] \to R$ having the stated initial and terminal values, a particular y such that the average kinetic energy

$$J(y) \equiv \int_0^T \tfrac{1}{2} m \, \dot{y}^2(t) \, dt$$

has a minimum value? If so, what function y_0 minimizes $J(y)$, and is it unique? On the basis of Chapter 3 there is a unique such y_0, namely, the linear function with the assigned initial and terminal values.

EXAMPLES 1.2 and 1.3

Enlarge the class $\mathscr{Y}$ of Example 1.1 by assigning no value for $y(T)$. We can then investigate the existence and possible nature of $y_0 \in \mathscr{Y}$ minimizing

$$J(y) \equiv y^2(T)$$

or

$$J(y) \equiv y^2(T) + \int_0^T \tfrac{1}{2} m \, \dot{y}^2(t) \, dt.$$

These minimum problems are simple examples of the Problem of Mayer and the Problem of Bolza to be treated in Chapter 5.

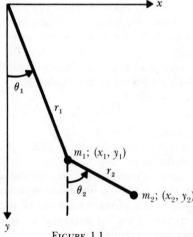

FIGURE 1.1

EXAMPLE 1.4

An idealized double pendulum consists of masses m_1 and m_2 attached to weightless inextensible cords of lengths r_1, r_2, as shown in Fig. 1.1. What system of differential equations in the signed angular displacements θ_1, θ_2 and the time t will govern the motions of the two particles?

These equations are found by variational methods in Example 4.5. If one is sufficiently adept at identifying all static and dynamic forces affecting the two particles, the desired equations can be derived directly from Newton's laws of motion. The calculus of variations, although not essential for this problem, offers certain advantages.

EXAMPLE 1.5

The mathematical model for a certain industrial process involving consecutive chemical reactions is the pair of equations

$$(1.39) \qquad \dot{x} = -Axp^m, \qquad \dot{y} + \dot{x} = Byp^n,$$

subject to the boundary conditions

$$(1.40) \qquad t_0 = 0, \quad x(0) = a, \quad y(0) = b, \quad x(T) = h, \quad y(T) = k.$$

Symbols A, B, m, n, a, b, h, and k denote constants; x y, and p are functions on $[t_0, T]$ to R, but terminal time T is not fixed. Indeed, T depends upon the choice of a triple (x, y, p) of functions from $[0, T]$ to R satisfying (1.39) and (1.40). If the constants a, b, h, and k have been chosen arbitrarily, we do not know whether there exists such a triple (x, y, p). The terminal state $[x(T), y(T)] = (h, k)$ may not be attainable by a system satisfying the other conditions (1.40) together with the dynamical equations (1.39).

If, however, we suppose given a positive time T and regard $p(t)$ as given, $0 \leqslant t \leqslant T$, with p restricted to be continuous on $[0, T]$ for simplicity, then equations (1.39) are linear in x and y with continuous coefficients. There are theorems for such systems of linear equations that ensure the existence of a pair $x: [0, T] \to R$ and $y: [0, T] \to R$ having the initial values $x(0) = a$, $y(0) = b$ and satisfying (1.39) with the given p on the given interval $[0, T]$. The set of *attainable terminal states* can be defined as the set of all pairs $[x(T), y(T)]$ of terminal values corresponding to all choices of a positive time T and of a continuous function p.

Since our purpose at the moment is to exhibit a typical problem and not to investigate the questions to which it gives rise, let us simply suppose that the values h and k mentioned under (1.40) constitute an attainable terminal state. The time T at which the state (h, k) is attained then depends upon the choice of a triple (x, y, p) satisfying (1.39) and (1.40) and we regard T as a value $J(x, y, p)$ of a function J, the domain of which consists of all triples (x,y,p) satisfying (1.39) and (1.40). We wish to know whether there is a particular such triple minimizing

$$J(x,y,p) \equiv T,$$

and, if so, how to determine the minimizing triple.

This more complex problem of Mayer will be discussed further in Examples 5.2 and 5.9. In the terminology of control theory, it is a time-optimal control problem in two state variables x, y representing concentrations of reactants and one control variable p representing a controlled pressure.

EXAMPLE 1.6

Consider the problem of maximizing

$$J(y,v,m) \equiv \int_{t_0}^{t_1} \dot{y}(t)\ dt$$

on the class of all PWS triples $(y,v,m)\colon [t_0,t_1] \to R^3$ satisfying the side-conditions

(1.41) $\qquad m\dot{v} + c\dot{m} + mg = 0, \qquad \dot{y} - v = 0, \qquad \dot{m} \leqslant 0$

and the end-conditions

(1.42) $\quad t_0 = 0, \quad y(t_0) = 0, \quad v(t_0) = 0, \quad m(t_0) = M_0, \quad m(t_1) = M_1 < M_0.$

Equations (1.41) are an overly idealized but much used mathematical model for the vertical motion of a rocket-propelled vehicle of decreasing mass $m(t)$. End values (1.42) leave the time of flight t_1 free but fix the initial and terminal values of the mass. Since the value of the integral is $y(t_1) - y(t_0)$, with $y(t_0) = 0$ by (1.42), and also $v(t_0) = 0$, we see that the idealized missile is to start from rest at the origin at the time 0 and that with a given mass $M_0 - M_1$ of propellant we wish to achieve a maximum terminal height $y(t_1)$ by selecting, if possible, from among all PWS triples (y, v, m) satisfying (1.41) and (1.42) a particular such triple corresponding to which the terminal height $y(t_1)$ has a maximum value.

This is a reasonable-looking mathematical problem even if one doesn't know yet at which end to attack it. One is inclined to guess from the nature of the physical problem behind the mathematical formulation that there ought to exist a maximizing triple in the given class of triples and that it only remains to find a way of identifying it. In actuality, however, there exists no such triple for this example. Everyone prefers problems with solutions, but intuition alone is not sufficient to identify them and we often need theorems on the existence of solutions.

A dynamical system from any branch of science or engineering is a source of variational problems. We are always interested in optimal design, optimal control, and optimal performance. When these vague desires are translated into the language of mathematics we sometimes must maximize or minimize a real-valued function of types met in the ordinary calculus, but often it is a real-valued functional.

Variational theory also has potential applications to conflict analysis and this brings it into contact with the social and behavioral sciences, including mathematical economics and the dynamics of military operations.

Although problems of the types described in this section bear a superficial resemblance to maximum and minimum problems from the first course in calculus, it is well for a beginner to anticipate that the former are generally much more complicated. Particular examples tend to be of two kinds. First are artificial ones deliberately constructed to illustrate some aspect of the theory and yet simple enough so that one can see exactly what is going on. In contrast are those that arise in a natural manner as the formulations of optimization questions that are important per se.

A complete analysis of a variational problem of the second type can demand considerable knowledge and ingenuity, absorb hundreds of man hours, and entail all the hazards of a doctoral dissertation or of any research project. Relatively few particular examples of the difficult kind have ever been definitively treated. One does what he can, and often even a partial analysis yields a certain amount of firm information that can be quite useful.

Chapter 2

NECESSARY CONDITIONS FOR AN EXTREMUM

2.1 INTRODUCTION

This chapter is mainly devoted to the simplest problem of the calculus of variations, or, in more technical language, the classic fixed-endpoint nonparametric problem in the plane. Since all problems appearing in Chapters 2 through 5 are of the type called nonparametric, it will be convenient to understand this without repeating that qualification. Most variational problems arising from questions in engineering and science are of this type. A second type, called parametric, is introduced in Chapter 6.

Various references from the list near the end of the book will be cited in this and other chapters using such notations as (VI, pp. 73–74) for a book or (7b, p. 598) for an article.

2.2   THE FIXED-ENDPOINT PROBLEM IN THE PLANE

Let $[t_0,t_1]$ be a nondegenerate compact interval, that is, a closed interval of positive finite length. Let $\mathscr{Y}$ denote the class of all PWS functions $y: [t_0,t_1] \to R$ satisfying end conditions

$$(2.1) \qquad y(t_0) = h_0, \qquad y(t_1) = h_1.$$

Let $f: [t_0,t_1] \times R^2 \to R$ be a function with values $f(t,y,r)$, $t \in [t_0,t_1]$, $(y,r) \in R^2$. We wish to express various conclusions in terms of partial derivatives of f; hence the discussion must be restricted at each stage to functions f having the needed derivatives. Appropriate hypotheses on f could be stated with each theorem, but it is convenient to agree at the outset that f together with any partial derivatives f_r, f_{rr}, f_{yt}, etc., that may appear will be understood to exist and be continuous at all triples (t,y,r) of the domain of f that are appropriate to a particular theorem or discussion. This agreement will be referred to as the *blanket hypothesis*.

Given the class $\mathscr{Y}$ described above, consider the functional $J: \mathscr{Y} \to R$ with values

$$(2.2) \qquad J(y) \equiv \int_{t_0}^{t_1} f(t,y,\dot{y})\, dt, \qquad y \in \mathscr{Y}.$$

Functions $y \in \mathscr{Y}$ are called *admissible*; they and only they are admitted to competition. By *the problem $J(y) = minimum$*, we shall mean the following combination of questions:

(i) *Does there exist $y_0 \in \mathscr{Y}$ such that $J(y_0) \leqslant J(y)$, $\forall y \in \mathscr{Y}$?*
(ii) *If so, is the minimizing admissible y_0 unique?*
(iii) *How can all such y_0 be characterized?*

A function $y_0 \in \mathscr{Y}$ such that $J(y_0) \leqslant J(y)$, $\forall y \in \mathscr{Y}$ is said to furnish the *global* or *absolute minimum* of $J(y)$. This, as remarked by Bolza (X,p.10), is the ultimate objective. Local or relative minima, which are also important, are defined in Section 2.4.

To *characterize* y_0 means to determine it. Ideally, this means to be able to say that $y_0(t) = t^2$ or sin t or some combination of familiar special functions. This can be done only for exceptionally simple examples. It can occur, for instance, that a unique minimizing y_0 is known to exist, to have no corners, and to satisfy a certain known differential equation. This could be a sufficient indirect characterization for the purposes of further mathematical development.

By an *analysis* of the problem $J(y) = minimum$ we mean either a substantiated negative answer to question (i) or a positive answer together with a characterization of all minimizing functions y_0. This describes an ideal that is seldom attained. One often achieves a partial analysis and for lack of time or because of difficulties can go no further.

We have avoided the term "solution of the problem," because this tends to be used with several different meanings. An analysis as defined above may be called a solution, or more often a published partial analysis of a special problem is cited as that author's solution of the problem. A minimizing function y_0 is frequently called a solution of the problem. In many instances a function y_0 satisfying the first necessary condition for a minimum is referred to as a solution, when all that is known for certain of y_0 is that it is a solution of an Euler equation.

An analysis need not and often does not take up questions (i), (ii), and (iii) in that order.

We have followed the common convention of phrasing the discussion in terms of minima and shall continue to do so. Since $J(y_0)$ is a maximum of $J(y)$ iff $-J(y_0)$ is a minimum of $-J(y)$, any statements or results for minima immediately translate into statements for maxima; hence it is unnecessary to give a separate complete discussion of maxima. For instance, the inequality $-J(y_0) \le -J(y)$ is equivalent to $J(y_0) \ge J(y)$ and that this hold for all admissible y defines $J(y_0)$ as the *global maximum*.

2.3 MINIMA OF ORDINARY POINT-FUNCTIONS

It is helpful to be reminded of some of the facts concerning minima of an ordinary function $\phi: [a,b] \to R$ in preparation for analogous concepts and results for the function J of Section 2.2.

If ϕ is continuous on $[a,b]$, we have the classic theorem that there exists $x_0 \in [a,b]$, not necessarily unique, such that

$$(2.3) \qquad \phi(x_0) \le \phi(x), \qquad \forall x \in [a,b].$$

Such a value $\phi(x_0)$ is the *global* or *absolute minimum* of $\phi(x)$ on $[a,b]$.

If there is a positive δ such that

$$(2.4) \qquad \phi(x_0) \le \phi(x), \qquad \forall x \in [a,b] \cap U(\delta, x_0),$$

then $\phi(x_0)$ is a *local* or *relative minimum*. Clearly (2.3) is a stronger statement than (2.4). If $\phi(x_0)$ is the global minimum, then it is also a local minimum but not conversely, as one sees from simple examples.

Since a real-valued function ϕ can be very complicated, there is no general method for locating values x_0 for which (2.3) or (2.4) holds. If, however, ϕ has first and second derivatives on the open interval (a,b) and has a minimum of either kind at $x_0 \in (a,b)$, then it is necessary that

$$(2.5) \qquad \phi'(x_0) = 0, \qquad \phi''(x_0) \ge 0.$$

That conditions (2.5) are not sufficient for a minimum is shown by such examples as $\phi(x) = x^3$ or $-x^4$ with $x_0 = 0$. If, however, $x_0 \in (a,b)$ and

$$(2.6) \qquad \phi'(x_0) = 0, \qquad \phi''(x_0) > 0,$$

these are sufficient to guarantee that $\phi(x_0)$ is a local minimum. This does not exclude the possibility that $\phi(x_0)$ is actually the global minimum. The question simply remains open.

The combination of conditions

$$(2.7) \qquad \phi'(x_0) = 0, \qquad \phi''(x) \geq 0, \qquad \forall x \in [a,b],$$

is sufficient for $\phi(x_0)$ to be the global minimum. Conditions (2.6) interpreted descriptively say that ϕ has a horizontal tangent and is convex at x_0. The second condition as strengthened in (2.7) says that ϕ is convex everywhere on $[a,b]$. Contemplate the difference.

Necessary conditions for $\phi(a)$ or $\phi(b)$ to be a minimum are, respectively, that

$$(2.8) \qquad \phi'(a) \geq 0 \qquad or \qquad \phi'(b) \leq 0.$$

Consideration of some simple examples will show that $\phi''(a)$ can be of either sign when $\phi(a)$ is a minimum, and similarly for $\phi''(b)$.

Exercise 2.1

Construct an elementary example of a function $\phi: I \to R$ to illustrate each of the following possibilities.
1. I is an open interval and ϕ has a relative but no global minimum on I.
2. I is any kind of interval and ϕ has infinitely many local minima on I. Does such a ϕ necessarily have a global minimum on I, and why?
3. $\phi(x_0)$ is a minimum of some kind and $\phi'(x)$, $\phi''(x)$ both exist and are finite on $[a,b]$ but (2.6) does not hold.
4. ϕ is not continuous on $[a,b]$ but (2.3) holds.
5. $x_0 \in (a,b)$ and $\phi(x_0)$ is a global minimum but $\phi'(x_0)$ does not exist.

2.4 DIFFERENT KINDS OF MINIMA OF $J(y)$

The global minimum of $J(y)$ has been defined in Section 2.2. With reference to Section 1.10, we now denote the distance (1.31) by $d_0(x,y)$ and call it a *distance of order zero*. We also define a *distance of order one*

$$(2.9) \qquad d_1(x,y) \equiv d_0(x,y) + \sup\{|\dot{x}(t) - \dot{y}(t)| : t \in [t_0,t_1]^*\},$$

where $[t_0,t_1]^*$ denotes the closed interval $[t_0,t_1]$ less those t, if any, where the derivatives $\dot{x}(t)$ or $\dot{y}(t)$ fail to exist. Functions x and y are understood to be admissible, as defined in Section 2.2, and hence are PWS in the sense of Section 1.9; therefore, right derivatives exist everywhere on the half-open interval $[t_0,t_1)$ and left derivatives everywhere on $(t_0,t_1]$. We can omit the star in stating (2.9) without change in the meaning if we look ahead to convention (2.17) in Section 2.6.

Observe that if $d_1(x,y) < \delta$, then $|x(t)-y(t)| < \delta$, $\forall t \in [t_0,t_1]$, and $|\dot{x}(t)-\dot{y}(t)| < \delta$, $\forall t \in [t_0,t_1]^*$. Moreover, $d_1(x,y) < \delta \Rightarrow |x(t)-y(t)| + |\dot{x}(t)-\dot{y}(t)| < \delta$, $\forall t \in [t_0,t_1]^*$.

A *neighborhood of order zero* $U_0(\delta,x) \equiv \{y \in \mathscr{Y}: d_0(x,y) < \delta\}$ is easily visualized. A function $y \in U_0(\delta,x)$ is in the strip of the (t,y) plane bounded by $x(t) + \delta$ and $x(t) - \delta$.

A *neighborhood of order one* $U_1(\delta,x) \equiv \{y \in \mathscr{Y}: d_1(x,y) < \delta\}$ is not so easily visualized. In order that y be in $U_1(\delta,x)$, it is necessary that y and $\dot{y}$ be in the respective zero-order neighborhoods $U_0(\delta,x)$, $U_0(\delta,\dot{x})$, but this is not sufficient. A sufficient condition is that y and $\dot{y}$ be in zero-order neighborhoods $U_0(\alpha,x)$, $U_0(\beta,\dot{x})$ such that $\alpha+\beta \le \delta$. In descriptive language $y \in U_1(\delta,x)$ means that $y(t)$ and $\dot{y}(t)$ are, respectively, near to $x(t)$ and $\dot{x}(t)$, $\forall t \in [t_0,t_1]^*$.

Observe that $d_1(x,y) < \delta$ implies that $d_0(x,y) < \delta$ but not conversely. Alternatively stated, $y \in U_1(\delta,x)$ implies $y \in U_0(\delta,x)$, but not conversely.

If there is a function $y_0 \in \mathscr{Y}$ and a positive δ such that

$$(2.10) \qquad J(y_0) \le J(y), \qquad \forall y \in \mathscr{Y} \cap U_0(\delta,y_0),$$

then $J(y_0)$ is called a *strong local* (or *strong relative*) *minimum*.

Similarly, if

$$(2.11) \qquad J(y_0) \le J(y), \qquad \forall y \in \mathscr{Y} \cap U_1(\delta,y_0),$$

then $J(y_0)$ is called a *weak local* (or *weak relative*) minimum.

The global minimum defined in Section 2.2, the strong local minimum defined by (2.10), and the weak local minimum defined by (2.11) compare y_0 with successively smaller classes of functions $y \in \mathscr{Y}$. If $J(y_0)$ is a global minimum, then it is necessarily a strong local minimum but not conversely. If $J(y_0)$ is a strong local minimum, then it is also a weak local minimum but not conversely. This ordering of the three types of minima is conveniently described by saying that one type is *stronger* (*weaker*) than another, respectively, if the inequality $J(y_0) \le J(y)$ holds for a larger (smaller) class of functions y than it does for the other.

Any property that $y_0 \in \mathscr{Y}$ must have to furnish a minimum of one of the three types must then also obtain if y_0 furnishes a stronger type of minimum. Any conditions that may be sufficient to ensure that $J(y_0)$ is a minimum of one type automatically ensure that $J(y_0)$ is a minimum of weaker type.

These distinctions and relations among three types of minima have been important ingredients in the calculus of variations since the time of Weierstrass.

The reader may notice that the terms zero-order and first-order for distances and neighborhoods seem to be counter to the ordering of types of minima described above. This use of the words first (or one) and zero is consistent with that of McShane (33f) and of Akhiezer (I,p.5). One might prefer to use the adjective weak for neighborhoods and distances associated with the weak minimum and similarly for the word strong. However, in view of the fact that $U_1(\delta,y) \subset U_0(\delta,y)$, to say weak and strong in place of one and zero is counter to a dominant usage in comparing neighborhood topologies.

Exercise 2.2

1. Formulate definitions for global, strong local, and weak local maxima. Discuss relationships among the three.
2. Given $y(t) = t^2$ and $z(t) = 1$, $t \in [-1,1]$, find the values of $d_0(y,z)$ and $d_1(y,z)$.
3. Given the sequence $\{y_n: [0,2\pi] \to R: n \in N\}$, $y_n(t) \equiv (\sin nt)/n$ together with $y_0: [0,2\pi] \to R$, $y_0(t) \equiv 0$, show whether or not $d_0(y_n,y_0) \to 0$ and $d_1(y_n,y_0) \to 0$.
4. Given a sequence of functions of general term $y_n: [a,b] \to R$, if there is a function $y_0: [a,b] \to R$ such that $d_0(y_n,y_0) \to 0$, verify that y_n converges uniformly to y_0 and conversely. What can be said of the convergence of y_n and $\dot{y}_n$ if d_0 is replaced by d_1?

2.5   THE LEMMA OF DU BOIS REYMOND

Theorem 2.1

If $m: S \to R$ with S defined as in Section 1.8 is a fixed PWC function and if the relation

$$(2.12) \qquad \int_{t_0}^{t_1} m\dot{\eta}\, dt = 0$$

holds for every PWS function $\eta: [t_0,t_1] \to R$ such that $\eta(t_0) = \eta(t_1) = 0$, then $m(t)$ is constant on $[t_0,t_1]$ except possibly for a finite number of points in (t_0,t_1), at which $m(t)$ remains undefined.

PROOF

If $m(t)$ is to be a constant c in the sense stated in the theorem, then it must satisfy the equation

$$(2.13) \qquad c = \frac{1}{t_1 - t_0} \int_{t_0}^{t_1} m\, dt.$$

Since $\eta(t)$ is required to vanish at t_0 and t_1, then

$$\int_{t_0}^{t_1} c\,\dot{\eta}(t)\,dt = c\,[\eta(t_1) - \eta(t_0)] = 0,$$

and we see with the aid of hypothesis (2.12) that

(2.14) $$\int_{t_0}^{t_1} (m-c)\dot{\eta}\,dt = 0.$$

The particular function η with values

(2.15) $$\eta(t) = \int_{t_0}^{t} [m(\tau) - c]\,d\tau$$

is PWS under definition (1.27). Clearly $\eta(t_0) = 0$, while $\eta(t_1) = 0$ by (2.13); hence this function η has the properties required in the theorem. Moreover, $\dot{\eta}(t) = m(t) - c$ except for a possible finite subset of $[t_0, t_1]$ by (1.28) or the discussion following Theorem 1.3 in Section 1.13. After substituting $\dot{\eta} = m - c$ in (2.14) we have that

(2.16) $$\int_{t_0}^{t_1} (m-c)^2\,dt = 0.$$

Let t_2 be any point of $[t_0, t_1]$ at which m is continuous. If $m(t_2) \neq c$, then $m(t)$ must differ from c on some subinterval of $[t_0, t_1]$ of positive length. Integral (2.16) cannot then vanish as stated and, from this contradiction, we infer that $m(t_2) = c$, hence that $m(t) = c$ wherever m is continuous.

The proof is complete but we remark further of m that what has been shown is that m consist of a horizontal segment in the (t,m) plane, possibly punctured by removal of a finite number of points.

2.6   THE EULER NECESSARY CONDITION

In stating the next theorem and others to follow we use the phrase "if y_0 minimizes $J(y)$" as an abbreviation for "if y_0 furnishes at least a weak local minimum for $J(y)$."

It is also convenient in stating the Euler condition and other conditions later to adopt the special convention that

(2.17)   *when a condition involves $[t_0, t_1]$ and the symbol $\dot{y}(t)$, then at any interior point where the derivative of the PWS function y fails to exist, the stated condition is understood to hold with $\dot{y}(t)$ interpreted as either $\dot{y}^-(t)$ or $\dot{y}^+(t)$.*

Here and elsewhere in this chapter when we mention a problem it is always that of Section 2.2 unless there is explicit statement to the contrary.

Theorem 2.2

If y_0 minimizes J (y) on $\mathscr{Y}$, then there exists a constant c such that

$$(2.18) \qquad f_r[t,y_0(t),\dot{y}_0(t)] = \int_{t_0}^{t} f_y[\tau,y_0(\tau),\dot{y}_0(\tau)] \, d\tau + c, \qquad \forall t \in [t_0,t_1],$$

subject to convention (2.17).

PROOF

Let η: $[t_0,t_1] \to R$ be PWS with $\eta(t_0) = \eta(t_1) = 0$ but otherwise arbitrary. Then $y_0 + \epsilon\eta \in \mathscr{Y}$ for each real ϵ. Moreover, $y_0 + \epsilon\eta$ and $\dot{y}_0 + \epsilon\dot{\eta}$ converge, respectively, to y_0 and $\dot{y}_0$ as $\epsilon \to 0$; consequently, given $\delta > 0$, then if $|\epsilon|$ is sufficiently small, $y_0 + \epsilon\eta$ is in the neighborhood $U_1(\delta,y_0)$ appearing in definition (2.11). Thus if $J(y_0)$ is a weak local minimum or if it is one of the stronger types of minima discussed in Section 2.4, we can proceed as follows.

The function $F: R \to R$,

$$(2.19) \qquad F(\epsilon) \equiv \int_{t_0}^{t_1} f(t, y_0 + \epsilon\eta, \dot{y}_0 + \epsilon\dot{\eta}) \, dt,$$

necessarily has a local minimum at $\epsilon = 0$, an interior point of the domain of F. The blanket hypothesis on f permits us to differentiate F and we have as a necessary condition on y_0 that

$$(2.20) \qquad\qquad F'(0) = 0.$$

By applying Theorem 1.3 in the manner described in Section 1.13,

$$(2.21) \qquad\qquad F'(\epsilon) = \int_{t_0}^{t_1} (f_y\eta + f_r\dot{\eta}) \, dt,$$

with the arguments $t, y_0 + \epsilon\eta$, $\dot{y}_0 + \epsilon\dot{\eta}$ of f_y and f_r suppressed. After setting $\epsilon = 0$, integration by parts of the first term with the aid of Theorem 1.5 yields the relation

$$(2.22) \qquad \int_{t_0}^{t_1} f_y\eta \, dt = \eta(t) \int_{t_0}^{t} f_y \, d\tau \Big]_{t_0}^{t_1} - \int_{t_0}^{t_1} \dot{\eta} \int_{t_0}^{t} f_y \, d\tau \, dt.$$

The first term on the right vanishes as a result of our choice of η; hence, by (2.20), (2.21), and (2.22),

$$(2.23) \qquad \int_{t_0}^{t_1} \{f_r[t,y_0(t),\dot{y}_0(t)] - \int_{t_0}^{t} f_y[\tau,y_0(\tau),\dot{y}_0(\tau)] \, d\tau\}\dot{\eta} \, dt = 0.$$

The Lemma of du Bois Reymond was designed for the next step. The expression in braces fails to be defined and continuous only at those possible points of $[t_0,t_1]$ corresponding to corners of the minimizing function y_0; therefore, the expression is PWC on $[t_0,t_1]$ and plays the role of m in Theorem 2.1. That theorem implies (2.18) except for those t, if any, that correspond to corners of y_0.

To establish that (2.18) holds for all $t \in [t_0,t_1]$ subject to convention (2.17), it remains to investigate corners. If t_2 corresponds to a corner, observe that the integral in (2.18) is continuous in t on $[t_0,t_1]$ and hence at t_2. If $t \to t_2$ from the left and then from the right $\dot{y}(t) \to \dot{y}_0(t_2-)$ and $\dot{y}_0(t_2+)$, respectively. But the left and right limits of $\dot{y}(t)$ at t_2 are equal to the left and right derivatives $\dot{y}^-(t_2)$ and $\dot{y}^+(t_2)$, respectively, as pointed out by problem 2, Exercise 1.5. Consequently (2.18) holds at t_2 in the sense of (2.17).

The preceding proof applies without change if $J(y_0)$ is a maximum of any of the types considered in Section 2.4.

Theorem 2.3

If y_0 minimizes $J(y)$ on $\mathscr{Y}$ and if t is any point of $[t_0,t_1]$ where the derivative $\dot{y}_0(t)$ exists, then, at such a value t, $\dfrac{d}{dt}f_r[t,y_0(t),\dot{y}_0(t)]$ exists and

$$(2.24) \qquad f_y[t,y_0(t),\dot{y}_0(t)] = \frac{d}{dt}f_r[t,y_0(t),\dot{y}_0(t)].$$

Moreover, (2.24) holds everywhere on $[t_0,t_1]$ in the sense of convention (2.17).

PROOF

We can differentiate (2.18) by Theorem 1.3 at any t where the integrand is continuous. If $\dot{y}_0$ is discontinuous at $t \in (t_0,t_1)$, the respective right and left derivatives of the integral are given by the integrand with $\dot{y}(t+)$ and $\dot{y}(t-)$ as respective values of the third argument.

Theorem 2.4 (Hilbert)

If y_0 extremizes $J(y)$ on $\mathscr{Y}$, if t_2 is a point of $[t_0,t_1]$ such that $\dot{y}_0(t_2)$ exists, and if $f_{rr}[t_2,y_0(t_2),\dot{y}_0(t_2)] \neq 0$, then (i) $\ddot{y}_0(t_2)$ exists and is finite, (ii) there exists a subinterval I of $[t_0,t_1]$ containing t_2 such that $\ddot{y}_0$ is continuous on I, and (iii)

$$(2.25) \qquad f_y = f_{rt} + f_{ry}\dot{y}_0(t) + f_{rr}\ddot{y}_0(t), \qquad \forall t \in I,$$

the arguments of f_y, f_{rt}, f_{ry}, and f_{rr} being $t,y_0(t),\dot{y}_0(t)$. If the hypotheses hold everywhere on $[t_0,t_1]$, then so also do the conclusions.

PROOF

Abbreviate $y_0(t_2)$, $\dot{y}_0(t_2)$ by y_{02}, $\dot{y}_{02}$. Set $\Delta y_0 \equiv y_0(t_2+\Delta t) - y_{02}$ and $\Delta \dot{y}_0 \equiv \dot{y}_0(t_2+\Delta t) - \dot{y}_{02}$. Restrict Δt to be positive or negative, respectively, if t_2 coincides with t_0 or t_1; otherwise, Δt may have either sign but not the value zero.

By the mean value theorem for a function of three arguments,

$$(2.26) \qquad \frac{\Delta f_r}{\Delta t} = f_{rt} + f_{ry} \frac{\Delta y_0}{\Delta t} + f_{rr} \frac{\Delta \dot{y}_0}{\Delta t},$$

arguments $t_2 + \theta \Delta t$, $y_{02} + \theta \Delta y_0$, $\dot{y}_{02} + \theta \Delta \dot{y}_0$ of f_{rt}, f_{ry}, and f_{rr} being suppressed.

Function y_0 being admissible, is PWS. Since $\dot{y}_0(t_2)$ exists by hypothesis, t_2 is not the abscissa of a corner, and therefore $\dot{y}_0$ exists and is continuous on some interval I_1 to which t_2 is interior. It follows that Δy_0 and $\Delta \dot{y}_0$ converge to 0 with Δt. We observe that $\Delta y_0/\Delta t$ has a finite limit, namely, $\dot{y}_0(t_2)$. The limit of the left member of (2.26) exists and equals the left member of (2.24) by Theorem 2.3.

It follows from (2.26) and our hypothesis on f_{rr} that $\Delta \dot{y}_0/\Delta t$ is expressible as a quotient with f_{rr} in the denominator. Each term in the numerator has a finite limit as a result of the blanket hypothesis on f and conclusions in the preceding paragraph. The denominator has a limit, which is not zero by hypothesis. Consequently $\Delta \dot{y}_0/\Delta t$ has a finite limit, which by definition is $\ddot{y}_0(t_2)$ and (2.25) holds at t_2.

Under the blanket hypothesis, f_{rr} is continuous and, since $f_{rr}[t_2, y_0(t_2), \dot{y}_0(t_2)] \neq 0$ and $\dot{y}_0$ is continuous on the interval I_1, there must exist an interval I_2, to which t_2 is interior, such that $f_{rr}[t, y_0(t), \dot{y}_0(t)]$ does not vanish on I_2. For the same reasons stated in the preceding paragraph, $\ddot{y}_0(t)$ exists and is finite, (2.26) is satisfied on $I_1 \cap I_2$, and this is the interval I mentioned in the theorem.

Observe that the three forms of the Euler condition apply under increasingly more restrictive conditions as we pass from the du Bois Reymond form (2.18) to (2.24) to (2.25).

The most widely known result from the calculus of variations appears to be form (2.24) of the Euler condition, and there is a tendency in applying the calculus of variations to problems in the sciences to ignore the hypotheses of Theorem 2.3 and to treat (2.24) as a universal solvent. If a smooth solution y_0 of (2.24) satisfying the given end-conditions can be found, then $J(y_0)$ is often supposed to be the desired minimum or maximum value of $J(y)$, whichever of the two may be desired.

Such a step is based on a combination of assumptions: (i) that the nature of the intended application is sufficiently clear to the analyst so that he knows there must be a minimum or maximum as the case may be, (ii) that the mathematical model for the original extremum problem represented by the variational problem is a sufficiently realistic approximation to the original so that the analyst's intuition can safely be transfered to the mathematical problem, and (iii) that the solution of the latter, now assumed to exist, is necessarily smooth. Any or all of such assumptions can turn out to be false.

All that has been proved thus far is that, if the integral $J(y)$ of Section 2.2 has a minimum or a maximum value $J(y_0)$, then y_0 satisfies (2.18), whereas if y_0 has certain other properties, it satisfies (2.24) or (2.25). A function $y_0 \in \mathcal{Y}$ satisfying one of these equations is a candidate. It may minimize or maximize or it may not. To determine whether it actually furnishes an extreme value for $J(y)$ and of which type requires further investigation.

We mention in this connection that the term *extremal* is variously used in the literature, sometimes for any smooth function y satisfying the Euler condition and again with other meanings. We prefer to avoid the term as misleading, even though it is traditional and will be found in various of the books that we cite. An extremal may or may not yield an extreme value of $J(y)$.

2.7 EXAMPLES

It frequently happens that a minimizing PWS function y_0 is actually smooth and even has a second derivative on $[t_0, t_1]$. We may not know of a particular example whether or not this will be so or even whether a minimizing y_0 exists. Nevertheless, in practice we often begin by examining (2.24) or (2.25).

The second-order differential equation (2.25) is in general nonlinear, and very few such equations have elementary solutions. We can turn to numerical methods for an example that is important enough to justify this investment but, as has been remarked, even if we solve an Euler equation this is only one step in the analysis of a problem.

We shall restrict ourselves at present to examples having simple Euler equations.

EXAMPLE 2.1

$f(t,y,r)$ *is free of* t *and* y. Equation (2.25) is then of the form

$$(2.27) \qquad\qquad f_{rr}(\dot{y})\ddot{y} = 0.$$

The equation $\ddot{y} = 0$ has the general solution $y = at + b$. If the first factor has a real zero $\dot{y} = m$, then $y = mt + b$ is a solution of (2.27), but it is already included in the result for the other factor. An extremizing function with a corner must satisfy (2.18) on the given interval $[t_0, t_1]$ and (2.24) on each subinterval that is free of corners. The reader should verify that $y = at + b$ also satisfies (2.24) for the present problem. Verify that an admissible y_0 satisfying (2.27) need not be linear in t if $f(t,y,r) = (r + 1)^4$, 0, or $(r - 1)^4$ when $r < -1$, $-1 \le r \le 1$, or $r > 1$, respectively.

EXAMPLE 2.2

$f(t,y,r) = r^2$. *The fixed endpoints are* $(0,0)$ *and* $(1,1)$. This is a special case of Example 2.1; hence y_0, $y_0(t) = t$, may possibly minimize or maximize

$J(y)$ or it may not if all we have to go on is Section 2.6. Theorems in Chapter 3 will tell us that this function y_0 furnishes the global minimum for the present example and that it is unique.

We can often get negative information by simply trying out some particular admissible functions. Any function y with values $y(t) = at^2 + (1-a)t$ is admissible for this problem. By elementary calculation

$$J(y) = \int_0^1 (2at+1-a)^2 \, dt = \frac{a^2}{3}+1,$$

and $J(y) \to \infty$ as $a \to \infty$ or $-\infty$. Therefore, $J(y)$ has no global maximum. Conceivably some function y_1 might furnish a strong or weak relative maximum, but the necessary condition of Legendre in Section 2.11 will eliminate this possibility.

EXAMPLE 2.3

$f(t,y,r) = r^{3/2}$, *with the fixed endpoints* $(0,0)$ *and* $(1,0)$. The global minimum of the given point-function f is clearly zero, corresponding to $r = 0$. Therefore, the example

$$J(y) = \int_0^1 \dot{y}^{3/2} = global\ minimum\ on\ \mathcal{Y}$$

has the unique minimizing function y_0, $y_0(t) = 0$. Moreover, we have the strict inequality $J(y_0) < J(y)$ for all $y \in \mathcal{Y}$ distinct from y_0, and we say that $J(y_0)$ is a *proper* global minimum. This is an exceptional example, so simple that all the facts constituting a complete analysis in the sense of Section 2.2 are available by inspection.

We know from Section 2.6 that this function y_0, since it has no corners, must satisfy the Euler condition in form (2.24), namely, the equation

(2.28) $$\frac{d}{dt}\left[\frac{3}{2}\dot{y}_0^{1/2}(t)\right] = \frac{d0}{dt} = 0.$$

If we consider (2.25), which for this example is the equation

$$3\ddot{y}/4\dot{y}^{1/2} = 0,$$

we see that it reduces to the meaningless form $0/0$ if $y = y_0$. Under the blanket hypothesis stated in Section 2.2, Theorem 2.4 includes the tacit hypothesis that all derivatives of f appearing in the theorem exist and are continuous. Therefore, the theorem does not apply to the present example because of the fact that

$$f_{rr}(t,y,r) = \begin{cases} \frac{3}{4}r^{-1/2}, & r > 0, \\ \infty, & r = 0, \end{cases}$$

so that f_{rr} is not continuous at $r = 0$. We remark, with reference to the first paragraph of this section, that although (2.25) is often the first thing we examine in approaching an example, we must be prepared to go back to (2.24) or (2.18).

EXAMPLE 2.4

$f(t,y,r) = r^{1/2}$ *with the fixed endpoints* (0,0) *and* (1,0). It is again clear by inspection that y_0, $y_0(t) = 0$, furnishes a proper global minimum, and yet not even (2.18) is applicable. The left member is meaningless. The tacit hypotheses in Theorem 2.2 are, in accord with the blanket hypothesis, that f, f_r, and f_y exist and are continuous for all $t \in [t_0,t_1]$ and all real values of y and r.

The present function f is not defined if $r < 0$ and f_r is not continuous at $r = 0$.

EXAMPLE 2.5

$f(t,y,r) = r^4$ *with the fixed endpoints* (0,0) *and* (1,0). The smooth admissible function y_0, $y_0(t) = 0$, again furnishes a proper global minimum. Equation (2.25) is $12\dot{y}^2\ddot{y} = 0$, and clearly the above function y_0 is a solution of this equation. Observe, however, that $f_{rr}[t,y(t),y_0(t)] = 12\dot{y}_0^2(t) \equiv 0$; consequently, a hypothesis of Theorem 2.4 is not satisfied. The hypotheses of this theorem are sufficient to imply that a minimizing function y_0 will satisfy (2.25) but not necessary to that end, as shown by this example.

Exercise 2.3

In problems 1 through 4 find a function y_0 satisfying (2.25) and through the given endpoints.

1. $\int (\dot{y}^2 + 2y) \, dt$, (0,0) and (1,1).
2. $\int (\dot{y}^2 + 2y\dot{y}) \, dt$, (−1,1) and (2,0).
3. $\int (\dot{y}^2 + 2t\dot{y} + t^2) \, dt$, (0,0) and (1,0).
4. $\int (\dot{y}^2 + 2y\dot{y} + y^2) \, dt$, (0,0) and (2,1).

5. To what does (2.18) reduce in the special case where $f(t,y,r)$ is free of y? Construct an example with both t and r present such that (2.18) has an elementary solution.
6. Discuss the Euler condition for the case in which $f(t,y,r)$ is free of t with the help if necessary of (X,p.27), (XI,p.32), (XXXII,pp.42–43), or some other reference.
7. Discuss the degenerate case in which $f(t,y,r)$ is free of r, illustrating your conclusions by examples.
8. Discuss the special case $f(t,y,r) = [r - g(t,y)]^2$. Give an example for which you can demonstrate that $J(y_0)$ is an extremum.
9. Given that $f(t,y,r) = g^2(t,y,r)h(t,y,r)$ with g and h subject to the blanket hypothesis, what relation does the first-order equation $g(t,y,\dot{y}) = 0$

have to the second-order equation (2.25)? Construct an example for which (2.25) has an elementary general solution.

10. Discuss the so-called inverse problem of the problem of Section 2.2 with reference to (X,pp.30–32), (XI,pp.37–39), or (I,pp.164–166). Identify the class of integrands f for which the general solution of (2.25) is $y = at + b$.

11. Given $\phi(t,r) = r^4 - 8t^2r^2$, regard t as a parameter and find values $r(t)$ that minimize or maximize $\phi(t,r)$. Then given $J(y) = \int (\dot{y}^4 - 8t^2\dot{y}^2)\, dt$ and the initial point $(0,0)$, find a terminal point $(1,h)$ and a function y_0 through these points and satisfying the first-order equation $\dot{y} = r(t)$. Is $J(y_0)$ a maximum or a minimum and of which type?

12. Given $J(y) = \int_{t_1}^{t_2} f(t,y,\dot{y})\, dt$, where y is now a PWS vector-valued function $y = (y^1, \ldots, y^n)$ through fixed points $(t,y^1, \ldots, y^n)$ in R^{n+1}, the procedure used in proving Theorem 2.2 can be applied to any one component y^i of y. State and prove such a theorem.

2.8   THE WEIERSTRASS NECESSARY CONDITION

The function $E\colon [t_0,t_1] \times R^3 \to R$ defined by the equation

$$(2.29) \qquad E(t,y,r,q) \equiv f(t,y,q) - f(t,y,r) - (q-r)f_r(t,y,r)$$

is called the Weierstrass *excess-function* (for a reason that will appear in Section 3.4) or simply the E-function. It is easily remembered by observing that, with t and y fixed, it is the difference between a term which we now write as $f(q)$ and the first two terms of a Taylor expansion for $f(q)$ in powers of $q - r$.

Theorem 2.5

If $y_0 \in \mathscr{Y}$ furnishes either a strong local or a global minimum for $J(y)$, then

$$(2.30) \qquad E[t,y_0(t),\dot{y}_0(t),q] \geqslant 0, \qquad \forall t \in [t_0,t_1] \text{ and } \forall q \in R,$$

with symbol $\dot{y}_0(t)$ understood in the sense (2.17).

PROOF

Given $\tau \in [t_0,t_1)$ and not the abscissa of a corner of y_0, select a number $a \in (\tau,t_1]$ that is so near to τ that the interval $[\tau,a]$ does not include the abscissa of a corner of y_0. Let Y denote the linear function with values

$$(2.31) \qquad\qquad Y(t) \equiv y_0(\tau) + q(t-\tau)$$

in which q denotes an arbitrary real number, different from $\dot{y}_0(\tau)$.

Given $u \in [\tau,a)$, define

(2.32)
$$y(t) \equiv \begin{cases} y_0(t), & t \in [t_0,\tau] \cup [a,t_1], \\ Y(t), & t \in [\tau,u], \\ \phi(t,u), & t \in [u,a], \end{cases}$$

with

(2.33)
$$\phi(t,u) \equiv y_0(t) + \frac{Y(u)-y_0(u)}{a-u}(a-t).$$

One verifies that the function $y: [t_0,t_1] \to R$ with values (2.32) is admissible, that it coincides with y_0 except generally on the interval (τ,a), and that if $q > \dot{y}_0(\tau)$, then y has the nature suggested by the path $0\,2\,3\,4\,1$ in Fig. 2.1. If $q < \dot{y}_0(\tau)$, the point labeled 3 would fall below y_0. When $u = \tau$ we see from (2.32) and (2.33) that $y(t)$ reduces to $y_0(t)$ on the entire interval $[t_0,t_1]$.

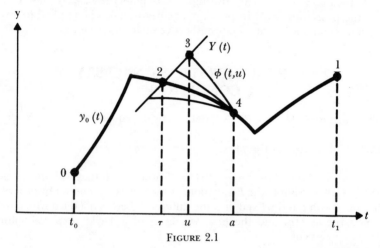

FIGURE 2.1

Define $\Phi(u) \equiv J(y) - J(y_0)$. The function $\Phi: [\tau,a) \to R$ is differentiable and, under the hypothesis that $J(y_0)$ is at least a strong local minimum, we have as a necessary condition on y_0 from (2.8) that

(2.34)
$$\Phi'(\tau) \geqslant 0.$$

Since y and y_0 coincide along paths 02 and 41 in Fig. 2.1,

$$\Phi(u) = \int_\tau^u [f(t,Y,\dot{Y}) - f(t,y_0,\dot{y}_0)]\, dt$$
$$+ \int_u^a \{f[t,\phi(t,u)\phi_t(t,u)] - f(t,y_0,\dot{y}_0)\}\, dt,$$

or, after rearranging the right member,

$$\Phi(u) = \int_\tau^u f(t,Y,\dot{Y})\, dt + \int_u^a f[t,\phi(t,u),\phi_t(t,u)]\, dt - \int_\tau^a f(t,y_0,\dot{y}_0)\, dt.$$

The last integral is free of u. Theorem 1.3 on differentiation of an integral with respect to a parameter applies to each of the other integrals and we find that

(2.35)
$$\Phi'(u) = f[u,Y(u),\dot{Y}(u)]$$
$$-f[u,\phi(u,u),\phi_t(u,u)] + \int_u^a \{f_y[\ ]\phi_u + f_r[\ ]\phi_{tu}\}\, dt.$$

The integrand in (2.35) has the same form as that in (2.21). After setting $u = \tau$, integrating the second term by parts, and using property (2.24) of y_0, the integral in (2.35) is found to have the value

$$-f_r[\tau,y_0(\tau),\dot{y}_0(\tau)]\phi_u(\tau,\tau).$$

By elementary differentiation of (2.33) with respect to u and then setting $u = \tau$, we find that

$$\phi_u(\tau,\tau) = \dot{Y}(\tau) - \dot{y}_0(\tau).$$

Therefore, the necessary condition (2.34) on y_0 is expressible in the form

$$E[\tau,y_0(\tau),\dot{y}_0(\tau),\dot{Y}(\tau)] \geq 0.$$

That $\dot{Y}(\tau)$ is the derivative at τ of the auxiliary function Y can now be forgotten. The important thing is that $\dot{Y}(\tau)$ is an arbitrary real number q. We have thus proved the necessity of the condition

(2.36)
$$E[\tau,y_0(\tau),\dot{y}_0(\tau),q] \geq 0, \qquad \forall q \in R,$$

subject thus far to the restrictions introduced at the beginning of the proof that $t_0 \leq \tau < t_1$ and that τ is not the abscissa of a corner.

Under our blanket hypothesis on f, the function E is continuous on its domain. Let $\tau \to t_1$ from below. It follows that (2.36) must also hold with $\tau = t_1$. If $y_0(t_2)$ is a corner, let $\tau \to t_2$ in (2.36) once from below and once from above. Then $\dot{y}_0(\tau) \to \dot{y}_0^-(t_2)$ and $\dot{y}_0^+(t_2)$, respectively, as a result of problem 2, Exercise 1.5, so that (2.36) must hold in the sense of convention (2.17) for all $t \in [t_0,t_1]$ and all real q as stated in the theorem.

Exercise 2.4

1. Identify features of the proof of Theorem 2.5 that depend upon the restrictions (i) $\tau \in [t_0,t_1]$ and (ii) τ is not the abscissa of a corner.
2. Prove as a corollary to Theorem 2.5 the Weierstrass necessary condition for a strong local or a global maximum by applying Theorem 2.5 to $-J$.

3. Given the integral $J\ (y) = \int\ (1+\dot{y}^2)^{1/2}\ dt$ with the fixed endpoints (0,0) and (1,1), find the function y_0 satisfying the Euler condition and show that y_0 also satisfies the Weierstrass necessary condition for a minimum.

4. Point out in the statement and proof of Theorem 2.5 those features that exclude the weak local minimum. Then modify the proof by suitably bounding the value of $|\dot{Y}(\tau) - \dot{y}_0(\tau)|$ so as to obtain a Weierstrass necessary condition for $J(y_0)$ to be a weak local minimum. State carefully the theorem that goes with your proof.

2.9   THE ERDMANN CORNER CONDITIONS

Theorem 2.6

If $J(y_0)$ is a weak local extremum of $J(y)$ on $\mathscr{Y}$ and $[t,y_0(t)]$ is a corner, then

$$(2.37) \qquad f_r[t,y_0(t),\dot{y}_0^-(t)] = f_r[t,y_0(t),\dot{y}_0^+(t)].$$

PROOF

The proof has already been given in the second paragraph following (2.23).

Theorem 2.7

*If $J(y_0)$ is a **strong** local extremum of $J(y)$ on $\mathscr{Y}$ and $[t,y_0(t)]$ is a corner, then*

$$(2.38) \quad f[t,y_0(t),\dot{y}_0^-(t)] - \dot{y}_0^-(t) f_r[t,y_0(t),\dot{y}_0^-(t)]$$

$$= f[t,y_0(t),\dot{y}_0^+(t)] - \dot{y}_0^+(t) f_r[t,y_0(t),\dot{y}_0^+(t)].$$

PROOF

Suppose that $J(y_0)$ is a strong local minimum. Apply the Weierstrass necessary condition (2.30) first with $\dot{y}_0(t) = \dot{y}_0^-(t)$ and $q = \dot{y}_0^+(t)$ and again with $\dot{y}_0(t) = \dot{y}_0^+(t)$ and $q = \dot{y}_0^-(t)$. From these two inequalities and (2.37), relation (2.38) follows. If $J(y_0)$ is a strong local maximum, proceed similarly with reference to problem 2, Exercise 2.4.

EXAMPLE 2.6

$f(t,y,r) = (r^2-1)^2$. Write p and q for the respective left and right derivatives. Equations (2.37) and (2.38) for this example are

$$p(p^2-1) = q(q^2-1),$$

$$(p^2 - 1)(3p^2 + 1) = (q^2 - 1)(3q^2 + 1).$$

This system of equations has infinitely many trivial solutions with $p = q$. The only nontrivial solutions are found to be $(p,q) = (1,-1)$ and $(-1,1)$. The algebra, although elementary, is a little tedious.

Given an example for which the integrand involves t and/or y, the pair of equations (2.37), (2.38) is unlikely to be solvable by elementary algebra and one must be prepared if necessary to select or devise an approximation procedure.

Theorems 2.6 and 2.7 and the conclusion for Example 2.6 ensure that the only slopes that can occur at a corner of a minimizing function are 1 and −1. The integral

$$J(y) = \int (\dot{y}^2 - 1)^2 \, dt$$

clearly assumes its infimum zero if $\dot{y}(t) = 1$ or -1 for all values of t. Consequently, given as the fixed endpoints any two that can be joined by a piecewise linear function consisting of alternate line segments of slopes 1 and −1, such a function furnishes a global minimum for $J(y)$. Such points as $(0,0)$ and $(1,2)$ cannot be joined by such a function and $J(y_0) \neq 0$ for the linear function y_0, $y_0(t) = 2t$, satisfying the Euler equation. Whether $J(y_0)$ is some kind of extremum remains open until some of the results in Chapter 3 become available.

We have remarked that an extremizing function in the class of PWS functions often turns out to be smooth. The next theorem gives a sufficient condition for this to occur.

Theorem 2.8

If $f_{rr}(t,y,r) \neq 0$, $\forall\, t \in (t_0,t_1)$, $\forall\, (y,r) \in R^2$, then no function y_0 having a corner can minimize or maximize $J(y)$ on $\mathcal{Y}$.

PROOF

Suppose that $J(y_0)$ is an extremum of any type and that $[t,y_0(t)]$ is a corner. Then by Theorem 2.6 with p and q for the two derivatives,

(2.39) $f_r[t,y_0(t),p] - f_r[t,y_0(t),q] = 0.$

By the Mean Value Theorem of the Differential Calculus, the left member can be expressed as

$$(p-q)f_{rr}[t, y_0(t), q + \theta(p-q)], \qquad 0 < \theta < 1,$$

and this must vanish by (2.39). From this contradiction with the given hypothesis, we infer the truth of the theorem.

2.10 THE FIGURATIVE

With (t,y) fixed, consider f as a function from R to R and set $u = f(r)$ with t, y suppressed. This function f is called the *figurative* (or *indicatrix*) *at* (t,y). Except in cases where f is free of (t,y), we have a family of figuratives with (t,y) as parameter.

One verifies by elementary calculus that the line tangent to the figurative at $(r,u) = [\dot{y}, f(t,y,\dot{y})]$ is

$$(2.40) \qquad u = f(t,y,\dot{y}) + (r - \dot{y}) f_r(t,y,\dot{y}).$$

The difference between the ordinate $u = f(t,y,r)$ to the figurative and that to this tangent line is precisely $E(t,y,\dot{y},r)$.

It is a further exercise in the calculus to verify that the two corner conditions (2.37), (2.38) are satisfied at (t,y) iff the figurative has a double tangent, more precisely, iff the tangent lines at $(r,u) = (p,u)$ and (q,u) coincide.

These observations are an aid in verifying for a particular example whether or not the Weierstrass necessary condition (2.30) or the Erdmann corner conditions (2.37), (2.38) are satisfied. Consequently they are an aid in constructing examples.

Example 2.6 is a stock example. One readily visualizes the graph of the quartic $u = (r^2 - 1)^2 = (r-1)^2(r+1)^2$, tangent to the r axis at each of two minimum points $(-1,0)$, $(1,0)$ and having a relative maximum $(0,1)$. Clearly any solution $y = mt + b$ of (2.25) with slope $m = 1$ or -1 corresponds to a tangent line (2.40) that is below the figurative; hence each such function y satisfies (2.30). Moreover, identification of the double tangent property is easier for this example than the details of solving the simultaneous equations discussed in Section 2.9.

A minimizing or maximizing function y_0 with corners traditionally has been called a *discontinuous solution* of the given extremum problem. Of course, it is the derivative $\dot{y}$ that has discontinuities and not y.

One can ask whether it might be worthwhile to extend the domain $\mathscr{Y}$ of J by admitting functions y with discontinuities. This requires a new formulation of the problem, since the Riemann integral with the integrand f affected by $\dot{y}$ is no longer adequate. Although such a move seems a bit artificial unless one is able to point to an important example that depends on it, there were unsuccessful attempts in the 1920s mentioned in (18b) and Lawden discusses such an extension in (27a). Example 1.6, Section 1.14, is such an example.

2.11 THE LEGENDRE NECESSARY CONDITION

This condition was published by Legendre in 1786 and, although his proof was faulty, it can be derived (X,p.47) independently of the Weierstrass condition, which appeared circa 1879. Granted the Weierstrass condition, however, then that of Legendre is an easy corollary. It continues to be useful even though it is weaker than the Weierstrass condition. Each of these conditions is an analogue of the necessary condition $\phi''(a) \geqslant 0$ for a minimum of a point-function.

Theorem 2.9

If $y_0 \in \mathscr{Y}$ minimizes $J(y)$ on $\mathscr{Y}$, then, in the sense (2.17),

$$(2.41) \qquad f_{rr}[t,y_0(t),\dot{y}_0(t)] \geqslant 0, \qquad \forall\, t \in [t_0,t_1].$$

PROOF

The hypothesis that y_0 minimizes is understood to mean that $J(y_0)$ is at least a weak local minimum but may be a strong local minimum or the global minimum.

Apply Taylor's Formula with second-order remainder to $f(t,y,q)$ as a function of q in powers of $q-\dot{y}$. With t, y suppressed, we find that

$$f(q) = f(\dot{y}) + (q-\dot{y})f_r(\dot{y}) + \frac{(q-\dot{y})^2}{2} f_{rr}[\dot{y}+\theta(q-\dot{y})];$$

hence, by definition (2.29) of the E-function,

$$(2.42) \qquad E(t,y,\dot{y},q) = \frac{(q-\dot{y})^2}{2} f_{rr}[t,y,\dot{y}+\theta(q-\dot{y})], \qquad 0 < \theta < 1.$$

Now $J(y_0)$ is a minimum by hypothesis and, by the companion theorem to Theorem 2.5 called for in problem 4, Exercise 2.4, inequality (2.30) must hold at each t for all q such that $|q-\dot{y}_0(t)|$ is sufficiently small. Suppose that for some t_2, not the abscissa of a corner, $f_{rr}[t_2,y_0(t_2),\dot{y}_0(t_2)]$ < 0. It then follows that

$$f_{rr}\{t_2,y_0(t_2),\dot{y}_0(t_2) + \theta[q-\dot{y}_0(t_2)]\} < 0$$

for all q sufficiently near to $\dot{y}_0(t_2)$. The right member of (2.42) is then negative for all such q, in contradiction to the Weierstrass necessary condition for a weak local minimum. If t_2 is the abscissa of a corner, repeat the last steps once with $\dot{y}_0^-(t_2)$ and once with $\dot{y}_0^+(t_2)$ in place of $\dot{y}_0(t_2)$.

From the contradiction we infer the truth of the theorem as stated.

Exercise 2.5

1. Obtain, as a corollary to Theorem 2.9, the Legendre necessary condition for a maximum. Given $f(t,y,r) = (r^2 - 1)^2$ with general solution $y = mt + b$ to the Euler equation (2.25), apply the Legendre condition to the functions $y_1, y_1(t) = 1$ and $y_2, y_2(t) = 2t$.

2. Given $f(t,y,r) = (1 + r^2)^{1/2}$ find a solution y_0 of (2.25) through $(0,0)$ and $(1,1)$. Show that the integral based on this integrand can have no extremizing function with a corner. Show that y_0 satisfies both the Weierstrass and Legendre necessary conditions.

3. Demonstrate the equivalence of the two corner conditions with the double tangent property of the figurative.

4. Discuss the possible location of corners and determine left and right slopes at such corners for the problem $\int (\dot{y}^2 - 4t^2 y^2)^2 \, dt = minimum$ with endpoints $(-1,1)$ and $(2,e^3)$.

5. The same for $\int \sin \dot{y} \, dt$ with endpoints $(0,0)$ and $(1,1)$.

6. Point out why the problem $\int |(\dot{y} - 1)(\dot{y} + 1)| \, dt = minimum$ with endpoints $(0,0)$ and $(1,1)$ is pathological. What can be said by inspection concerning the class of piecewise linear admissible functions y having alternate slopes of 1 and -1?

7. Given $f(t,y,r) = (r^2 - 1)^2 (4 - r^2)$, examine the figurative. Depending on the choice of the two fixed endpoints, what can be said about possible weak (strong) minima and maxima of $J(y)$?

8. Obtain all the information available from the theory, as developed thus far, concerning possible minima or maxima of

$$J(y) = \int_{(0,2)}^{(1,0)} \dot{y}^2 e^{\dot{y}} \, dt.$$

9. Define $F(t,y,p,q) \equiv f_r(t,y,p) - f_r(t,y,q)$ and $G(t,y,p,q) \equiv f(t,y,p) - pf_r(t,y,p) - [f(t,y,q) - qf_r(t,y,q)]$. With such aid as may be needed from the literature on implicit functions, formulate a theorem giving conditions sufficient for the pair of equations $F(t,y,p,q) = 0$, $G(t,y,p,q) = 0$ to determine p and q. Then interpret this result in terms of possible extremizing functions y with corners for an integral with integrand f.

2.12   THE JACOBI NECESSARY CONDITION

We have seen that an extremizing PWS function y_0 is often smooth. In this section we consider only the smooth case.

If y_0 is smooth and $J(y_0)$ is a minimum of any one of the types that have been discussed in Section 2.4, then the function F of (2.19) must satisfy both of the conditions

$$(2.43) \qquad\qquad F'(0) = 0, \qquad F''(0) \geqslant 0.$$

By differentiation of expression (2.21) for $F'(\epsilon)$ we find that

$$F''(\epsilon) = \int_{t_0}^{t_1} (f_{yy}\eta^2 + 2f_{yr}\eta\dot{\eta} + f_{rr}\dot{\eta}^2)\, dt,$$

arguments of f_{yy}, etc., being t, $y_0 + \epsilon\eta$, $\dot{y}_0 + \epsilon\dot{\eta}$; therefore,

$$(2.44) \qquad F''(0) = \int_{t_0}^{t_1} [f_{yy}(t,y_0,\dot{y}_0)\eta^2 + 2f_{yr}(\ )\eta\dot{\eta} + f_{rr}(\ )\dot{\eta}^2]\, dt.$$

With y_0 understood as a fixed minimizing function, let $\mathscr{H}$ be the class of all PWS $\eta: [t_0,t_1] \to R$ such that $\eta(t_0) = \eta(t_1) = 0$. We now denote the integral (2.44) by $\mathscr{J}(\eta)$ and the integrand by $2\omega(t,\eta,\dot{\eta})$. Thus

$$(2.45) \qquad \mathscr{J}(\eta) = \int_{t_0}^{t_1} 2\omega(t,\eta,\dot{\eta})\, dt,$$

and, since $J(y_0)$ is a minimum by hypothesis, it follows from (2.43_2) that $\mathscr{J}(\eta) \geq 0$ on $\mathscr{H}$. By inspection of (2.44), $\mathscr{J}(\eta) = 0$ if $\eta(t) \equiv 0$; consequently, $\mathscr{J}(\eta)$ has the minimum value zero on $\mathscr{H}$.

The problem $\mathscr{J}(\eta) = \text{minimum}$ on $\mathscr{H}$, called the *accessory minimum problem*, has the following Euler equation in form (2.25):

$$(2.46) \qquad \omega_\eta = \omega_{\dot{\eta}t} + \omega_{\dot{\eta}\eta}\dot{\eta} + \omega_{\dot{\eta}\dot{\eta}}\ddot{\eta}.$$

In order that this be a necessary condition for the accessory problem by application of Theorem 2.4 to that problem, we proceed subject to the further hypothesis that condition (2.41) for the original problem holds with the strict inequality, namely, that

$$(2.47) \qquad f_{rr}[t,y_0(t),\dot{y}_0(t)] > 0, \qquad \forall\, t \in [t_0,t_1].$$

We find by differentiation of ω that

$$\omega_{\dot{\eta}\dot{\eta}}(t,\eta,\dot{\eta}) = f_{rr}[t,y_0(t),\dot{y}_0(t)],$$

hence by (2.47) and Theorem 2.4 that form (2.46) of the Euler equation applies and moreover by Theorem 2.8 that no minimizing $\eta \in \mathscr{H}$ for $\mathscr{J}(\eta)$ can have a corner.

If $u: [t_0,t_2] \to R$ satisfies (2.46) on $[t_0,t_2]$ together with the condition $u(t_0) = u(t_2) = 0$ but $u(t)$ does not vanish if $t_0 < t < t_2$, then t_2 is said to be a *conjugate value* to t_0 and $[t_2,y_0(t_2)]$ is called a *conjugate point* to the initial point $[t_0,y_0(t_0)]$ of y_0.

Theorem 2.10

If y_0 is a smooth function minimizing $J(y)$ on the class $\mathscr{Y}$ of PWS functions and if (2.47) holds, then there cannot exist a value t_2 conjugate to t_0 with $t_2 < t_1$.

PROOF

Suppose the contrary—that there is a value t_2 conjugate to t_0 and that $t_2 < t_1$. Then there is a solution $u(t)$ of the Jacobi equation (2.46) vanishing at t_0 and t_2 but nowhere between. Define $\eta_0 \in \mathcal{H}$ by the relation

$$(2.48) \qquad \eta_0(t) \equiv \begin{cases} u(t), & t_0 \le t \le t_2, \\ 0, & t_2 < t \le t_1. \end{cases}$$

We find, by differentiation of 2ω, or alternatively by Euler's Theorem for homogeneous functions, that

$$(2.49) \qquad 2\omega = \eta\omega_\eta + \dot{\eta}\omega_{\dot{\eta}}.$$

Observe that η_0 defined by (2.48) satisfies the Euler equation

$$(2.50) \qquad \omega_\eta[t,\eta_0(t),\dot{\eta}_0(t)] = \frac{d}{dt}\omega_{\dot{\eta}}[t,\eta_0(t),\dot{\eta}_0(t)]$$

on each of the half-open intervals $[t_0,t_2)$, (t_2,t_1). After substituting the right member of (2.50) for ω_η in (2.49) we see that

$$2\omega = \eta\frac{d}{dt}\omega_{\dot{\eta}} + \dot{\eta}\omega_{\dot{\eta}} = \frac{d}{dt}(\eta\omega_{\dot{\eta}}), \qquad t \in [t_0,t_2) \cup (t_2,t_1].$$

Therefore,

$$\mathscr{J}(\eta_0) = \int_{t_0}^{t_2} d[\eta_0\omega_{\dot{\eta}}(t,\eta_0,\dot{\eta}_0)] + \int_{t_2}^{t_1} 2\omega(t,\eta_0,\dot{\eta}_0)\,dt.$$

The second integral vanishes by the definition of η_0, the first from the fact that $\eta_0(t_0) = \eta_0(t_2) = 0$. Thus $\mathscr{J}(\eta_0) = 0$, the infimum of $\mathscr{J}(\eta)$ on $\mathcal{H}$, so that η_0 is a minimizing function.

However, by standard existence theorems, the second-order equation (2.46) with nonvanishing coefficient of $\ddot{\eta}$ has a unique integral curve through a given point in a given direction so that the solution u appearing in (2.48) cannot have the derivative $\dot{u}(t_2) = 0$. For if $\dot{u}(t_2) = 0$, then $u(t)$ would vanish identically on $[t_0,t_1]$, which it does not under our hypothesis that t_2 is conjugate to t_0. Therefore, the minimizing function η_0 for $\mathscr{J}(\eta)$ has a corner contrary to our observation that this is inconsistent with (2.47) and Theorem 2.8. From this contradiction we infer the truth of the theorem.

2.13   OTHER FORMS OF THE JACOBI CONDITION

We suppose given $y_0 \in \mathcal{Y}$ satisfying the hypotheses of Theorem 2.10. As a consequence of (2.47), the blanket hypothesis on f, and existence theory for differential equations, Euler equation (2.25) for f has a two-parameter family of solutions $y(t,a,b)$ that includes $y_0(t)$. Moreover, the partial derivatives y_a, y_b, y_t, y_{at}, y_{bt} all exist and are continuous for $t \in [t_0,t_1]$, and for a and b, respectively, near the particular values a_0, b_0 such that

$$(2.51) \qquad y(t,a_0,b_0) = y_0(t), \qquad \forall\, t \in [t_0,t_1].$$

The Euler equation in form (2.24), namely,

$$(2.52) \qquad f_y[t,y(t,a,b),y_t(t,a,b)] = \frac{d}{dt}f_r[t,y(t,a,b),y_t(t,a,b)],$$

holds for all t, a, b mentioned above. With t held fixed, we differentiate with respect to a and find that

$$(2.53) \qquad f_{yy}y_a + f_{yr}y_{at} = \frac{d}{dt}[f_{ry}y_a + f_{rr}y_{at}].$$

We also find by differentiation of ω that

$$(2.54) \qquad \omega_\eta = f_{yy}\eta + f_{yr}\dot{\eta} \qquad \text{and} \qquad \omega_{\dot{\eta}} = f_{yr}\eta + f_{rr}\dot{\eta};$$

consequently (2.53) is the Jacobi equation (2.50) with y_a, y_{at} replacing η, $\dot{\eta}$ and consequently $y_a(t,a_0,b_0)$ is a solution of the Jacobi equation. We find in the same way that $y_b(t,a_0,b_0)$ is a solution.

Situations arise in which one wishes to use various one-parameter subfamilies of $y(t,a,b)$. For example, let a and b be differentiable functions of a parameter α with $a(\alpha_0) = a_0$ and $b(\alpha_0) = b_0$ and set

$$\phi(t,\alpha) \equiv y[t,a(\alpha),b(\alpha)].$$

Then $\phi_\alpha(t,\alpha_0) = y_a(t,a_0,b_0)a'(\alpha_0) + y_b(t,a_0,b_0)b'(\alpha_0)$ is a linear combination of solutions of the Jacobi equation. Equation (2.46) is seen to be linear in η with the aid of relations (2.54), and hence $\phi_\alpha(t,\alpha_0)$ is another solution of (2.46).

If $\phi(t,\alpha)$ is the particular one-parameter family consisting of all solutions of (2.25) through the initial point $[t_0,y_0(t_0)]$, the Jacobi condition has the following geometric interpretation [(VII,p.131), (X,p.60), and (I,p.81)]. The family $\phi(t,\alpha)$ either has no envelope or, if it does have, the minimizing function y_0 for $J(y)$ cannot touch the envelope at a point with abscissa smaller than t_1.

Since a conjugate point has been defined preceding Theorem 2.10 by way of a solution u of (2.46), conceivably that point is dependent on the choice of u. That this fortunately is not the case is shown by the next theorem.

Theorem 2.11

The conjugate value t_2 is independent of the particular choice of a solution of the Jacobi equation. Any two solutions vanishing at t_0 differ by a nonzero constant factor.

PROOF

Let $\eta_1(t)$, $\eta_2(t)$ be distinct solutions of (2.46), both vanishing at t_0 and neither of which vanishes identically on $[t_0,t_1]$. Then $\dot{\eta}_1(t_0) \neq 0$ and $\dot{\eta}_2(t_0) \neq 0$ and there is a constant k such that

$$\dot{\eta}_1(t_0) - k\,\dot{\eta}_2(t_0) = 0.$$

The linear combination $\eta_1(t) - k\eta_2(t)$ is a solution, which together with its derivative has the value 0 at t_0; consequently, this solution must be identically 0 on $[t_0,t_1]$. It follows that $\eta_1(t) \equiv k\eta_2(t)$, $k \neq 0$, and therefore that a conjugate value based on either η_1 or η_2 is also determined by the other.

Exercise 2.6

1. State a companion theorem to Theorem 2.10 for the case of a maximum and prove it as a corollary to Theorem 2.10.
2. Demonstrate that the Jacobi equation (2.46) is linear in η. Then show that if η_1 and η_2 are any solutions, so also is $a\eta_1 + b\eta_2$ for any constants a, b.
3. Given the problem $\int (\dot{y}^2 + \dot{y})\, dt = minimum$ with endpoints $(0,0)$ and $(1,1)$, find a function y_0 satisfying the Euler condition and through these points. Then show that y_0 satisfies the Weierstrass, Legendre, and Jacobi conditions.
4. If $f(t,y,\dot{y}) = \dot{y}^4 + 2\dot{y}^2 + ty^2$ and the fixed endpoints are $(0,0)$ and $(1,0)$, verify that $y_0(t) = 0$ is an admissible function satisfying the Euler condition without finding a general solution of that equation. Then show that y_0 satisfies the Jacobi condition.
5. Given relation (2.51) and that $y_a(t_0,a_0,b_0)$, $y_b(t_0,a_0,b_0)$ are not both zero, state reasons why

$$(2.56) \qquad \Delta(t,t_0) \equiv \begin{vmatrix} y_a(t,a_0,b_0) & y_b(t,a_0,b_0) \\ y_a(t_0,a_0,b_0) & y_b(t_0,a_0,b_0) \end{vmatrix}$$

is a solution of the Jacobi equation that can be used to locate possible conjugate points.
6. If $J(y) = \int (\dot{y}^2 - y^2)\, dt$ and the endpoints are $(0,0)$ and $(4,0)$, what definite statement can be made about the problem $J(y) = extremum$ with the aid of (2.56)?

7. Point out why the proof of Jacobi's condition breaks down for the integrand $f(t, y, \dot{y}) = \dot{y}^4$ when the endvalues $y(t_0), y(t_1)$ are equal.

8. Investigate all the necessary conditions discussed in this chapter for $\int (\dot{y}^2 + y\dot{y} + y^2) \, dt$ with endpoints $(0,0)$ and $(1,1)$.

9. Given the fixed interval $[t_0, t_1]$ and the class $\mathcal{Y}$ of all functions $y: [t_0, t_1] \to R$ with fixed endvalues each of which has a continuous derivative $\dot{y}$ and a PWC second derivative $\ddot{y}$, derive an Euler necessary condition for the problem $J(y) = $ extremum on $\mathcal{Y}$, where $J(y) = \int f(t, y, \dot{y}, \ddot{y}) \, dt$, first in an integral form analogous to (2.18) and then in other forms.

10. With the interval $[t_0, t_1]$ fixed, let $\mathcal{Y}$ be the class of all PWS functions $y: [t_0, t_1] \to R$ such that $y(t_0) = h_0$ but with $y(t_1)$ free. Given that $J(y_0)$ is a minimum on $\mathcal{Y}$, point out why, as a corollary to Theorem 2.2, relation (2.18) remains necessary and show that we have the further necessary condition $f_r[t_1, y_0(t_1), \dot{y}_0(t_1)] = 0$.

11. With t_0 fixed but not t_1, let $\mathcal{Y}$ be the class of all PWS functions $y: [t_0, t_1] \to R$ such that $y(t_0) = h_0$ and $y(t_1) = Y(t_1)$, where $Y: R \to R$ is a given differentiable function. Show, as a corollary to Theorem 2.2, that (2.18) is necessary and show that the classical *transversality condition*

$$(2.57) \qquad f[t_1, y_0(t_1), \dot{y}_0(t_1)] + [\dot{Y}(t_1) - \dot{y}_0(t_1)] f_r[t_1, y_0(t_1), \dot{y}_0(t_1)] = 0$$

is then also a necessary condition for an extremum.

2.14   CONCLUDING REMARKS

It is easy to construct problems such as Examples 2.3 and 2.4 of Section 2.7 where the integrand f fails to meet the hypotheses of some or even all of the theorems in this chapter. Sometimes parts but not all of the existing theory are applicable to an example, and beyond that the analyst is dependent upon his own ingenuity.

The basic problem is not always formulated in precisely the pattern of our Section 2.2. We have admitted all PWS functions $y: [t_0, t_1] \to R$ with the given endvalues and the domain of f is $[t_0, t_1] \times R^2$. However, examples not covered by this formulation can occur. The domain of f might be some smaller set. Sometimes there is a constraint (restriction) on the values $y(t)$ or $\dot{y}(t)$ associated with an admissible function. We prefer to regard such modifications as separate problems and have not included them in this chapter.

A reader meeting the calculus of variations for the first time is likely to feel at this stage that a great many things are not completely clear. This is normal. We are just getting started. To feel otherwise is either a mark of genius or indication of real trouble.

One may wonder, for example, why admissible functions y in Section 2.2 were taken to be PWS. This has been a common practice for about a century. Prior to that time admissible functions were usually required to be smooth in the sense of Section 1.9 or even to have higher derivatives, but it is desirable to admit the largest class of functions for which the theory can be developed. Why not then enlarge the class still further to some proper superset of the PWS functions satisfying end-conditions (2.1)? A need for such a move was indeed recognized by Weierstrass around 1879. Work in this direction, which has played an important role in modern variational theory, requires other integrals than that of Riemann. In keeping with the introductory character of this book, only the Riemann integral is presupposed and all integrals used in Chapters 1 through 6 are to be understood in the sense of Riemann. A sufficient (although not necessary) condition for (2.2) and similar integrals to be meaningful under the Riemann definition is that y be PWS. Essential steps in many of the proofs in the first six chapters depend upon properties of integrals, which we suppose to be known for those of Riemann but not necessarily known for other integrals. Our basic problem in Section 2.2 has not been formulated with caprice but with historical, pedagogical, and other considerations in mind.

Chapter 3

SUFFICIENT CONDITIONS FOR AN EXTREMUM

3.1 INTRODUCTION

Recall with reference to (2.5) and (2.6), Section 2.3, that, to obtain sufficient conditions for a local minimum of a twice-differentiable point-function ϕ, one strengthens a combination of necessary conditions. After further strengthening in (2.7), we have conditions sufficient for a global minimum.

It is then reasonable to hope that, by suitably strengthening certain combinations of the necessary conditions found in Chapter 2, we can generate conditions sufficient to ensure that $J(y_0)$ is a weak local, strong local, or global minimum, as the case may be.

This chapter presents such a development of sufficient conditions for the problem of Section 2.2. Also included are some sufficient conditions for a global extremum of the general fixed-endpoint nonparametric problem in $(n+1)$-space, $n \geqslant 1$.

Although the class $\mathscr{Y}$ of Chapter 2 included all PWS functions y satisfying the given end-conditions, often a function y_0 without corners has the desired extremizing property. We confine attention to smooth functions y_0 in Sections 3.2 through 3.7.

It is convenient to follow Bliss in referring to the respective necessary conditions of Euler, Weierstrass, Legendre, and Jacobi by the Roman numerals I, II, III, and IV and to recall that the theory given in Chapter 2 for the Jacobi condition IV was only for the case of a smooth y_0 and was subject to the strengthened form (2.47) of the Legendre condition.

Since Bolza's books continue to be useful, it should be noticed that his II, III, and IV must be translated into III, IV, and II of Bliss. We also remark in passing that there is a fifth necessary condition (XI,pp.117–119) due to Bolza but that it is seldom used and does not appear among the usual sets of sufficient conditions. Although there possibly remain still other undiscovered necessary conditions, searching for them seems unlikely to be a profitable endeavor.

Various strengthened forms of conditions II, III, and IV that are to appear in theorems are listed in this section for reference and comparison. An accent on II or III signifies exclusion of the equality in the corresponding necessary condition; an accent on IV indicates the inclusion of an equality. Subscript N on II or III means that the condition must hold for triples (x,y,p) in a neighborhood of *line elements* $[t,y_0(t),\dot{y}_0(t)]$ of y_0. We use subscript R to suggest a further strengthening to all real values of certain arguments.

In the various conditions I, II, III, and IV and their strengthened forms it is important to take notice of precisely what arguments appear in each of the last two positions of symbol $f_{rr}(t, \cdot , \cdot)$ and the last three positions for $E(t, \cdot , \cdot , \cdot)$.

II′ $E[t,y_0(t),\dot{y}_0(t),q] > 0, \forall\ (t,q)$ *such that* $t \in [t_0,t_1]$ *and* $q \neq \dot{y}_0(t)$.

II_N $E[t,y,p,q] \geq 0, \forall\ (t,y,p,q)$ *such that* $t \in [t_0,t_1]$, $q \in R$ *and such that* $|y-y_0(t)|$ *and* $|p-\dot{y}_0(t)|$ *are, respectively, below some pair* δ *and* δ' *of positive real numbers.*

II_R $E[t,y,p,q] \geq 0, \forall\ (t,y,p,q)$ *such that* $t \in [t_0,t_1]$ *and* $y,p,q \in R$.

II_N' *Condition* II_N *with the strict inequality wherever* $q \neq p$.

II_R' *Condition* II_R *with the strict inequality wherever* $q \neq p$.

III′ $f_{rr}[t,y_0(t),\dot{y}_0(t)] > 0, \forall\ t \in [t_0,t_1]$.

III_N $f_{rr}(t,y,p) \geq 0, \forall\ (t,y,p)$ *such that* $t \in [t_0,t_1]$ *and such that* $|y-y_0(t)|$ *and* $|p-\dot{y}_0(t)|$ *are, respectively, below some pair* δ *and* δ' *of positive real numbers.*

III_N' *Condition* III_N *with the strict inequality.*

III_R $f_{rr}(t,y,p) \geq 0, \forall\ (t,y,p)$ *such that* $t \in [t_0,t_1]$ *and* $y,p \in R$.

III_R' *Condition* III_R *with the strict inequality.*

IV′ *No value* t_2 *conjugate to* t_0 *is on the half-open interval* $(t_0,t_1]$.

Since sufficient conditions (2.6) for a local minimum of a point-function are obtained from necessary conditions (2.5) by replacing a weak inequality with a strict inequality, it is plausible that perhaps the combination of conditions I, II′, III′, and IV′ may be sufficient for a minimum of $J(y)$. That this is not sufficient for a strong local minimum is pointed out by Bolza with an example (XI,pp.116–117). The conditions do turn out to suffice for a weak local minimum and indeed with II' omitted.

3.2 FIELDS

Suppose given a particular solution y_0: $[t_0,t_1] \rightarrow R$ of the second-order Euler equation (2.25) together with a one-parameter sub-family $\phi(\cdot,\alpha)$ of the two-parameter totality of solutions of (2.25) with the following properties:

(3.1) (i) $\phi(t,\alpha_0) = y_0(t)$, $\forall\, t \in [t_0,t_1]$.
This means that y_0 is included for $\alpha = \alpha_0$ in the family $\phi(\cdot,\alpha)$ of functions satisfying (2.25).

(3.1) (ii) The relation $y - \phi(t,\alpha) = 0$ defines implicitly a function $\alpha: S \rightarrow R$ where S is a subset of the (t,y) plane containing a set of the form $\{(t,y): t_0 \leqslant t \leqslant t_1, y_0(t) - k < y < y_0(t) + k, 0 < k \leqslant \infty\}$. Moreover, the partial derivatives α_t and α_y are required to exist and to be continuous on S.

(3.1) (iii) The function ϕ together with its partial derivatives ϕ_t, ϕ_α, $\phi_{t\alpha}$, ϕ_{tt} exists and is continuous in (t,α) for t in $[t_0,t_1]$ and α in some interval I to which α_0 is interior.

Existence theory for solutions of differential equations and for implicit functions will supply conditions on the integrand f that will imply the three properties (3.1) but we shall proceed on the basis of (3.1) rather than to digress into a development of these important related matters.

We visualize such a family $\phi(\cdot,\alpha)$ as a sheaf of integral curves of (2.25) covering a subset S of the strip $\{(t,y): t_0 \leqslant t \leqslant t_1\}$ of the (t,y) plane with exactly one such curve through each point of S.

Define a function $p: S \rightarrow R$ called the *slope-function*, by the equation

(3.2) $$p(t,y) \equiv \phi_t[t,\alpha(t,y)].$$

This is the slope at $(t,y) \in S$ of the unique function $\phi(\cdot,\alpha)$ of the given family through (t,y). It follows that

(3.3) $$p[t,\phi(t,\alpha)] = \phi_t(t,\alpha),$$

and, as a consequence of (3.1)(iii), that p has finite partial derivatives p_t, p_y. By differentiation of (3.3),

(3.4) $\qquad p_t[t,\phi(t,\alpha)] + p_y[t,\phi(t,\alpha)]\phi_t(t,\alpha) = \phi_{tt}(t,\alpha),$

or, with the aid of (3.3),

(3.5) $\qquad p_t(t,y) + p(t,y)p_y(t,y) = \phi_{tt}[t,\alpha(t,y)].$

The ordered pair (S,p) consisting of the domain S of the function α of (3.1)(ii) and the slope function p is called a *field* $\mathscr{F}$ about y_0 and y_0 is said to be *embedded in* the field $\mathscr{F}$. The word field has a similar meaning in mathematical physics, for example, a force field.

The simplest case is that of any integrand f (Bolza, XI,p.39) corresponding to which $y = at + b$ is a general solution of the Euler equation (2.25). If, for example, the fixed endpoints are $(0,0)$ and $(1,1)$, the family of parallel lines $y = \phi(t,\alpha) \equiv t + \alpha$ or any pencil of lines $y - k = \alpha(t - k)$ with fixed $k < 0$ or > 1 defines a field (S,p) such that S is the entire strip $\{(t,y): 0 \leq t \leq 1\}$ of the (t,y) plane. Of course, the maximal set S on which p is defined by any one of these families is larger than this strip, but the latter is all that is needed. The family $y = \alpha t$ fails to define a field (S,p) because of the defect that p is not defined at the point $(0,0)$, and a similar remark applies to each $k \in [0,1]$ in the above example. It is essential that S be *simply covered* by the family, that is, that through each $(t,y) \in S$ pass exactly one function $\phi(\cdot,\alpha)$ of the family which generates the field.

As soon as we go to more complicated examples, the possibilities become chaotic. For instance, if the general solution of the Euler equation is $y = a \cos t + b \sin t$ and the end values t_1 and t_0 differ by more than π, there exists no field about y_0. If the fixed endpoints are $(0,0)$ and $(2\pi,0)$, then $y_0(t) = \sin t$ is a member of the family $\alpha \sin t$, but this family does not define a field with property (3.1)(ii) on any set S. Neither does any other one-parameter family of solutions of (2.25).

Few integrands f yield an Euler equation for which there is an elementary general solution. When there is one we can investigate the existence and extent of fields directly from this general solution. To proceed with a general theory we need the following existence theorem.

In the proof of this theorem and hence in others to follow that make use of it, we need to strengthen the formulation of Section 2.2 by supposing the domain of the integrand f to be $[t_0 - e, t_1] \times R^2$, where e is some positive number. The blanket continuity and differentiability hypothesis on f is now understood to apply to this enlarged domain.

Theorem 3.1

If $y_0: [t_0,t_1] \to R$ is a smooth function in $\mathscr{Y}$ satisfying conditions I, III', and IV', then there exists a field $\mathscr{F}$ about y_0.

PROOF

The details depend upon properties of solutions of differential equations, for which we refer to Bliss (VII,pp.154–157), Bolza (X,pp.78–83), and (XI,pp.100–102) and to books on the theory of differential equations. We shall give only a descriptive outline.

Condition IV' ensures that the one-parameter family of solutions of (2.25) through $[t_0,y_0(t_0)]$ either has no envelope or, if there be one, that no point $[t,y_0(t)]$, $t_0 < t \leqslant t_1$, is on the envelope. This family fails to define a field because of the fact that all members include the initial point of y_0. However, under our hypotheses, y_0 has an extension from $[t_0,t_1]$ to an interval $[t_0 - e_0,t_1]$ satisfying (2.25) on the larger interval with e_0 between 0 and the e mentioned preceding the theorem. The one-parameter family of solutions of the Euler equation (2.25) through $[t_0 - e_0,y_0(t_0 - e_0)]$ simply covers a subset of the (t,y) plane of the type described in condition (3.1)(ii), and this family determines a field (S,p). This theorem provides no information on the extent of the field, that is, on the possible value of k in (3.1)(ii). Given an example that meets the hypotheses of Theorem 3.1, it may be possible to verify from features of the example that $k = \infty$, alternatively stated, that the common domain S of α and p is the infinite strip $\{(t,y) : t_0 \leqslant t \leqslant t_1\}$, in which event we have a *global field* or *field in the large*. There is no simple criterion that identifies such cases. All that the theorem guarantees is that there is a field (S,p), with set S possibly being a thin band about y_0. Such a *field in the small* or *local field* suffices for sufficiency theorems on local extrema.

3.3   THE HILBERT INTEGRAL

Given a field $\mathscr{F} = (S,p)$, consider the functional $J^* \colon \mathscr{Y} \to R$ defined by the equation

$$(3.6) \qquad J^*(y) \equiv \int \{ f[t,y,p(t,y)] + [\dot{y} - p(t,y)] f_r[t,y,p(t,y)] \} \, dt$$

$$= \int \{ f[\ \ ] - p f_r[\ \ ] \} \, dt + f_r[\ \ ] \, dy.$$

This integral, due to Hilbert, is a curvilinear integral of the form $\int P \, dt + Q \, dy$, which, by a classic theorem, is independent of the choice of a PWS function y in the set S through fixed endpoints iff $P_y(t,y) \equiv Q_t(t,y)$ on S. This condition applied to (3.6) can be expressed in the form

$$(3.7) \qquad f_y = f_{rt} + f_{ry} p + f_{rr}(p_t + p p_y),$$

the arguments of p, p_t, p_y being (t,y) and those of f_y, etc., being $[t,y,p(t,y)]$.

Let $\phi(\cdot,\alpha)$ be the one-parameter family that generates the given field $\mathscr{F}$. It follows from (3.2) through (3.5) and the blanket hypothesis on f that (3.7) is valid, and therefore we have proved the following theorem.

Theorem 3.2

Given any PWS functions y_1 and y_2 in the set S associated with a field $\mathscr{F}$ such that $y_1(t)$ and $y_2(t)$ have common end values, then $J^*(y_1) = J^*(y_2)$.

Hilbert's integral is commonly referred to as invariant, which is another way of saying that it is independent of path in the sense of Theorem 3.2.

3.4 THE FUNDAMENTAL SUFFICIENCY THEOREM

The Weierstrass E-function or *excess-function* is so named because its average value represented by the integral in the next theorem is a measure of the (positive or negative) excess of $J(y)$ over $J(y_0)$.

Theorem 3.3 *(Weierstrass–Hilbert)*

If y_0 is a smooth admissible function embedded in a field $\mathscr{F} = (S, p)$ and y is any PWS function in S through the same endpoints as y_0, then

$$(3.8) \qquad J(y) - J(y_0) = \int_{t_0}^{t_1} E\{t,y(t),p[t,y(t)],\dot{y}(t)\}\, dt.$$

PROOF

That $J^*(y_0) = J(y_0)$ is immediate from the definition of the E-function and the fact that $\dot{y}_0(t) = p[t,y_0(t)]$, $\forall\, t \in [t_0,t_1]$. By Theorem 3.2, $J^*(y) = J^*(y_0)$. It follows that

$$J(y) - J(y_0) = J(y) - J^*(y_0) = J(y) - J^*(y),$$

and the last difference is precisely the right member of (3.8).

Theorem 3.4 *(Fundamental Sufficiency Theorem)*

If y_0 is a smooth admissible function embedded in a field $\mathscr{F} = (S,p)$, if y is any PWS function in S through the same endpoints as y_0, and if

$$E\{t,y(t),p[t,y(t)],\dot{y}(t)\} \geq 0\ (>0\ provided\ \dot{y}(t) \neq p[t,y(t)]),$$

then

$$(3.9) \qquad J(y_0) \leq J(y) \qquad \left(<J(y)\right).$$

PROOF

Use (3.8).

There is no restriction on slopes $\overset{\bullet}{y}(t)$ in Theorems 3.3 and 3.4; hence, if either alternative inequality involving the E-function holds for all admissible functions y in some strip,

$$\{(t,y) \in R^2 : t \in [t_0, t_1], y_0(t) - k < y < y_0(t) + k, k > 0\}$$

about y_0, it follows from Theorem 3.4 that $J(y_0)$ is a strong local minimum, called *improper* or *proper* according as equality can or cannot hold in (3.9).

This conclusion by itself does not exclude the possibility that $J(y_0)$ is actually a global minimum. The question simply remains open unless some way to settle it can be found. Sometimes for a particular example it can be verified that there is a field (S,p) for which S is the infinite strip bounded by $t = t_0$ and $t = t_1$ and that the inequality on the E-function holds for all admissible functions y in this strip. Theorem 3.4 then shows that $J(y_0)$ is a global minimum, improper or proper as the case may be.

Certain examples not satisfying the hypotheses of Theorem 3.4 fall under the following modification of that theorem.

Theorem 3.5

If $y_0 \in \mathcal{Y}$ is smooth and embedded in a field $\mathscr{F} = (S,p)$, if $y \in \mathcal{Y}$ is in S, if, for some positive δ', $|\overset{\bullet}{y}(t) - \overset{\bullet}{y}_0(t)| < \delta'$, $\forall t \in [t_0, t_1]$, and if for all such y

$$E\{t, y(t), p[t, y(t)], \overset{\bullet}{y}(t)\} \geq 0 \qquad \left(> 0 \text{ provided } \overset{\bullet}{y}(t) \neq p[t, y(t)]\right),$$

then

$$(3.10) \qquad J(y_0) \leq J(y) \qquad or \qquad \left(< J(y)\right).$$

When this theorem applies it establishes that $J(y_0)$ is a weak local minimum, proper or improper according as $<$ or $\leq$ holds. For comparison functions y with corners, the restriction $|\overset{\bullet}{y}(t) - \overset{\bullet}{y}_0(t)| < \delta'$ is to be understood in the sense of convention (2.17).

Companion theorems for maxima are obtained by reversing the inequalities on E in Theorem 3.4 and 3.5. This yields a reversal of inequalities in (3.9) and (3.10).

3.5 EXAMPLES

EXAMPLE 3.1

$$f(t,y,r) = ar^2 + br + c, a \neq 0.$$

Discussion

Whatever the endpoints (t_0,h_0), (t_1,h_1) may be, so long as $t_0 < t_1$, a function y_0, $y_0(t) = \alpha t + \beta$ satisfies condition I. The figurative is a parabola that is convex or concave according as $a > 0$ or $a < 0$; hence y_0 satisfies II' and III' or the corresponding conditions for a maximum. Since we know the general linear solution of (2.25), the most convenient form of condition IV is in terms of the determinant Δ of relation (2.56) attached to problem 5, Exercise 2.6. We find that there is no value t_2 conjugate to t_0, therefore that IV' is satisfied. Thus there exists a field in the small by Theorem 3.1 but, by the earlier discussion in Section 3.2, we actually have a field in the large. It therefore follows from Theorem 3.4 that y_0 is the unique admissible function such that $J(y_0)$ is the global minimum or maximum of $J(y)$ on $\mathscr{Y}$ according as $a > 0$ or $a < 0$, respectively.

This is a complete analysis of the problem for this special type of integrand f. A positive answer to the existence problem [question (i), Section 2.2] comes out as a by-product.

EXAMPLE 3.2

$f(t,y,r) = r^4 - 2r^2t/(y^2 + 1)$ *with the fixed endpoints* (1,0) *and* (2,0).

Discussion

The reader is asked to examine the Euler equation in form (2.24). To obtain an elementary general solution appears hopeless but we have picked the example so that y_0, $y_0(t) = 0$ is easily seen to satisfy the equation and to have the required end values. We find that

$$f_{rr}(t,y,\dot{y}) = 12\dot{y}^2 - 4t/(y^2 + 1),$$

hence that $f_{rr}[t,y_0(t),\dot{y}_0(t)] = -4t$ and condition III' for a maximum is satisfied. To investigate the Jacobi condition without a general solution of (2.25) at our disposal we return to the integrand of (2.45) for the accessory problem and find that, in the present instance,

$$2\omega(t,\eta,\dot{\eta}) = -4t\dot{\eta}^2.$$

The Jacobi differential equation, which is the Euler equation (2.24) for this integrand, is of the form $d(t\dot{\eta})/dt = 0$, hence $t\dot{\eta} = a$ and $\eta = a \ln t + b$. A particular solution vanishing at $t = 1$ but not identically is $\eta = a \ln t$, $a \neq 0$. Clearly $\eta(t)$ vanishes nowhere except at $t_0 = 1$; therefore, there exists no value conjugate to t_0 and condition IV' holds. Our function y_0 satisfies I, III', and IV', and hence a local field exists by Theorem 3.1.

Finally, to examine the E-function, think of the family of figuratives. Regardless of the choice of $t \in [1,2]$ and y, the figurative is a quartic polynomial in r with a relative maximum at $r = 0$ and a global minimum at $r = \pm[t/(y^2+1)]^{1/2}$. The companion theorem for maxima to Theorem 3.5 applies and $J(y_0)$ is weak local maximum. The reader is asked to investigate the solutions $y_1(t)$ and $y_2(t)$ of the two differential equations $\dot{y}(t) = r$ corresponding to the two values of r that minimize the given integrand, verifying that a solution $y_1(t)$ through $(1,0)$ of one of the equations intersects a solution $y_2(t)$ through $(2,0)$ of the other. Then observe that the composite function $y(t) = y_1(t)$ or $y_2(t)$ according as $1 \leq t \leq (0.5+\sqrt{2})^{2/3}$ or $(0.5+\sqrt{2})^{2/3} < x \leq 2$ furnishes a global minimum for the given integral.

Exercise 3.1

1. If $f(t,y,r) = r^3$ and the endpoints are $(0,0)$ and $(1,1)$, show that $J(y)$ has a weak local minimum. Then show with the aid of admissible functions of the form

$$y(t) = \begin{cases} -t/\epsilon, & 0 \leq t \leq \epsilon^2, \\ 1+(1+\epsilon)(t-1)/(1-\epsilon^2), & \epsilon^2 < t \leq 1, \end{cases}$$

   and by letting $\epsilon \to 0$ through positive values that the function y_0 that furnishes a weak local minimum does not furnish a strong local minimum.

2. If $f(t,y,r) = r^2 - y^2$ and the endpoints are $(0,0)$ and $(\pi/2,1)$, establish that there exists a field in the large and that there is an admissible function y_0 furnishing a proper global minimum.

For each of problems 3 through 6, find a function y_0 satisfying the Euler equation and having the given end values. Then either show that Theorem 3.1 applies or establish that there is a field about y_0 by direct use of the Euler equation.

3. Investigate $J(y) = \int [1+\dot{y}^2]^{1/2}\,dt$ with the endpoints $(1,1)$ and $(3,0)$ for possible extreme values.

4. Given $J(y) = \int (\dot{y}^2 + ty^2)\,dt$ with the endpoints $(0,0)$ and $(1,0)$, observe that $y_0(t) = 0$ satisfies the Euler equation and end-conditions. Then investigate the minimizing or maximizing nature of y_0. A general solution of the Euler equation can be found in the form $y = aS_1(t) + btS_2(t)$, where S_1 and S_2 are power series. Consider other choices of endpoints.

5. If $f(t,y,r)$ is free of y and there exists a smooth function y_0 satisfying the Euler equation and having the required end values, demonstrate that $y_0 + \alpha$ is a solution for all real α and that there exists a field in the large. Construct a particular example with both t and r present for which you can show with the aid of Theorem 3.4 that $J(y_0)$ is the global minimum and supply the details of a complete analysis of the problem.

6. $J(y) = \int (e^{-\dot{y}} + t y \dot{y}) \, dt$ with endpoints $(0,0)$ and $(1,0)$. Say the most that you can about the minimizing or maximizing nature of $y_0(t) = 0$.

The next three problems have integrands that fail to satisfy the blanket hypothesis, and yet parts of the theory apply.

7. Investigate $J(y) = \int \dot{y}^2 [1 - 100 y^2]^{1/2} \, dt$ with the fixed endpoints $(0,0)$ and $(1,0)$.
8. Investigate $J(y) = \int y^2 [1 - 100 \dot{y}^2]^{1/2} \, dt$ with the fixed endpoints $(0,0)$ and $(1,0)$.
9. Investigate $J(y) = \int \dot{y}^{2/3} \, dt$ with the fixed endpoints $(0,0)$ and $(1,1)$; also with $(0,0)$ and $(1,0)$.

3.6   SUFFICIENT COMBINATIONS OF CONDITIONS

This section depends upon Sections 3.2, 3.3, and 3.4 and hence is restricted to smooth functions y_0. Our methods, in particular those of Section 3.3, continue to admit general PWS comparison functions.

Theorem 3.6

If the smooth function $y_0 \in \mathcal{Y}$ satisfies conditions I, II$_N$ *(or* II$'_N$*),* III$'$, *and* IV$'$, *then* $J(y_0)$ *is a strong (or proper strong) local minimum.*

PROOF

A local field about y_0 exists by Theorem 3.1. Recall that $p(t,y)$ is the slope at $(t,y) \in S$ of the unique function through (t,y) from the family that generates the field (S,p). As a consequence of property (3.1)(iii) in the definition of a field, the function $p: S \to R$ is continuous on S. Consider relation (3.8) with hypothesis II$_N$ above. It follows with the aid of the continuity of p that the integrand on the right in (3.8) is non-negative provided that $|y(t) - y_0(t)|$ is below some positive constant for all $t \in [t_0, t_1]$. If the alternative hypothesis II$'_N$ holds, then similarly the integrand in (3.8) remains positive. The alternative conclusions of the theorem then follow from Theorem 3.4.

It can be verified that conditions III$'$ and II$_N$ taken together imply that II$'_N$ holds on a smaller neighborhood of y_0 than that for which the given inequality II$_N$ holds. It follows that I, II$_N$, III$'$, and IV$'$ actually suffice for a proper strong local minimum.

Theorem 3.7

If the smooth function $y_0 \in \mathcal{Y}$ satisfies conditions I, III$'$, III$_N$, *and* IV$'$ *(or* I, III$'_N$, *and* IV$'$*), then* $J(y_0)$ *is a strong (or proper strong) local minimum.*

PROOF

Condition III'_N implies III'; hence, under either of the alternative hypotheses, we have I, III', and IV' and consequently have a local field by Theorem 3.1. We then use relation (2.42) in the form

$$(3.11) \quad E[t,y,p(t,y),q] = \tfrac{1}{2}[q-p(t,y)]^2 f_{rr}\{t,y,p(t,y) + \theta[q-p(t,y)]\},$$

to see that III_N and III'_N respectively, imply II_N and II'_N. The theorem is then a corollary to Theorem 3.6.

Theorem 3.8

If the smooth function $y_0 \in \mathscr{Y}$ satisfies I, III', and IV', then $J(y_0)$ is a proper weak local minimum.

PROOF

The present hypotheses are identical with those of Theorem 3.1; consequently a local field exists. It follows from III', the continuity of p, and the continuity of f_{rr}, under our blanket hypothesis, that $f_{rr}[t,y,p(t,y)] > 0$, provided that $|y - y_0(t)|$ is sufficiently small for all $t \in [t_0, t_1]$. The last factor on the right in (3.11) is positive provided that y and q are uniformly near $y_0(t)$ and $\dot{y}_0(t)$ for all $t \in [t_0, t_1]$. The stated conclusion is then a consequence of Theorem 3.5. This also can be proved (X, pp. 68–71) without any use of the notion of a field.

Theorem 3.9

If the smooth function $y_0 \in \mathscr{Y}$ is embedded in a field (S,p) in the large, that is, a field such that S is the infinite strip $\{(t,y): t_0 \le t \le t_1\}$, then II_R or III_R is sufficient for $J(y_0)$ to be a global minimum, and either II'_R or III'_R is sufficient for a proper global minimum. Moreover, under either of the last two alternatives, y_0 is the unique function furnishing the proper global minimum.

PROOF

The various conclusions follow from Theorem 3.4 and relation (3.11).

To apply any of these theorems to a problem $J(y) = \text{maximum}$, consider the equivalent problem, $-J(y) = \text{minimum}$.

It can require considerable ingenuity to apply the theorems to particular examples. Even in those exceptional cases for which the Euler equation (2.25) has an elementary general solution, it can be difficult to determine whether there are values of the two parameters consistent with the given end values. If such values have been found, it still requires ingenuity to verify whether or not an expression in terms of E or f_{rr} is nonnegative or whether the strengthened Jacobi condition IV' holds.

Given a particular example, the various strengthened forms of condition II and III are related to the convexity of f in r and hence to the family of figuratives. If the hypotheses of Theorem 3.8 hold for a certain y_0, one should not be content with the weak conclusion of that theorem until he has looked into the possibility that y_0 may satisfy the hypotheses of one of the stronger theorems. For some examples it is best to ignore the theorems of this section and work directly with Theorem 3.4 or 3.5.

3.7 PROBLEMS FOR WHICH CONDITION III' FAILS

There is a gap between the set of necessary conditions I, II, and III and hypotheses of any of the sufficiency theorems of Sections 3.4 and 3.6. The theory of the Jacobi condition depends upon the hypothesis that the function y_0 being examined is *nonsingular (regular)*, that is, that y_0 satisfies condition III'. If it does, then condition IV is also necessary.

It is easy to find examples for which $f_{rr}[t,y_0(t),\dot{y}_0(t)]$ vanishes at one or more points of $[t_0,t_1]$ or even for which this expression vanishes identically on $[t_0,t_1]$. [See Mancill (31a) or Miele (36b).] The function y_0 is then called *singular (nonregular)*. Hypotheses of Theorems 3.1, 3.6, 3.7, and 3.8 include III' and hence exclude the singular case. Theorems 3.4, 3.5, and 3.9 are still available provided we can establish directly from the properties of a particular example that there exists a field in the small or in the large as the case may be.

We can partially close the gap between necessary conditions and sufficient conditions by an elementary device discussed in reference (12a) of adding a *penalty term*.

Let $J(y)$ be the usual integral and let y_0 be an admissible function satisfying the Euler condition. With k as a real-valued parameter, define

$$(3.12) \qquad J_k(y) \equiv \int_{t_0}^{t_1} \{f(t,y,\dot{y}) + k^2[\dot{y} - \dot{y}_0(t)]^2\}\, dt.$$

Since $J_0(y) = J(y)$, it is clear from the form of (3.12) that if $J(y_0)$ is a minimum of $J(y)$ of any of the types discussed, then $J_k(y_0)$ is some type of minimum for $J_k(y)$. By Chapter 2, if $J(y_0)$ is a minimum, y_0 must satisfy conditions I, II, and III for J. An effect of the penalty term is to ensure that y_0 will then satisfy III' for J_k; consequently, y_0 must satisfy the Jacobi condition IV for J_k. If y_0 satisfies the strengthened Jacobi condition IV' for the integral $J_k(y)$, then y_0 satisfies all of the conditions I, III', and IV', and this ensures by Theorem 3.8 that $J_k(y_0)$ is a weak local minimum of J_k. If y_0 happens to satisfy the condition II_N (or II'_N) for J_k, then $J_k(y_0)$ is a strong (or proper strong) local minimum for J_k.

EXAMPLE 3.3

$J(y) = \int \dot{y}^4\, dt$ *with endpoints* $(0,0)$ *and* $(1,0)$.

Discussion

We have remarked under Example 2.5 that $y_0(t) = 0$ satisfies Euler equation (2.25). Since 0 is the global minimum of the integrand, it is clear that $J(y_0) = 0$ is the global minimum of $J(y)$. However, we wish to consider the integral $J_k(y) = \int (\dot{y}^4 + k^2 \dot{y}^2)\, dt$ for this example. It is easy to verify that if $k \neq 0$, then y_0 satisfies I, III′, and IV′ for J_k and, using the family $y = \alpha$ to define a field (S,p), where S is the infinite strip bounded by $t = 0$ and 1 and $p(t,y) = 0$, we verify that y_0 satisfies II'_R. Therefore, by Theorem 3.9,

(3.13) $J_k(y_0) < J_k(y)$ *if* $y_0 \neq y \in \mathscr{Y}$ *and* $k \neq 0$.

Since this is true for all nonzero k and since it is clear from the form of (3.12) that $J_k(y) \to J(y)$ as $k \to 0$, it then follows from (3.13) that $J(y_0) \leqslant J(y)$. From this argument based on Theorem 3.9 and J_k we can only infer the weak inequality in the limit, even though we had the strict inequality in (3.13). That actually $J(y_0) < J(y)$ if $y \neq y_0$ was clear by inspection, and this conclusion also would be obtainable by applying Theorem 3.4 directly to J.

For more complex integrands, we would generally have a field (S_k,p_k) depending on k with a set S_k being a proper subset of the strip bounded by t_0 and t_1 and (3.13) would hold iff y is in S_k. Hence we could only conclude that $J_k(y_0)$ is a local minimum. Moreover, we could only establish that y_0 furnishes a local minimum, for the original integral J, if we could verify for the example that S_k does not collapse onto y_0 as $k \to 0$.

Although the use of J_k sheds light on the theory, it provides no panacea for attacking particular examples. Indeed there are no panaceas!

EXAMPLE 3.4

$f(t,y,\dot{y}) = y^2 + t^2\dot{y}.$

Discussion

This falls under the degenerate case of an integrand that is linear in $\dot{y}$ (XI, pp. 35–37). The Euler equation (2.25) is $2t = 2y$; hence the only possibility for an extremizing function is $y_0, y_0(t) = t$. Given a pair of endpoints not both on this line, there can exist neither a minimizing nor a maximizing function. Given a pair of endpoints such as $(0,0)$ and $(1,1)$, whether $J(y_0)$ is an extremum of some type remains an open question. There is no use in looking for a field, hence none of the Theorems 3.1 through 3.9 applies to this integral.

After adding the penalty term $k^2(\dot{y}-1)^2$, we have an integral J_k, for which the Euler equation is

$$k^2\ddot{y} - y = -t.$$

If $k \neq 0$, the general solution is $y = a \exp(t/k) + b \exp(-t/k) + t$, and with $b = a = \alpha/2$ we have the one-parameter family $y = \alpha \cosh(t/k) + t$, which generates a field on the infinite strip $\{(t,y): 0 \leqslant t \leqslant 1\}$ provided $k \neq 0$. Moreover, y_0 satisfies condition III'_R; hence from Theorem 3.9 we again reach conclusion (3.13). Letting $k \to 0$, we see that $J(y_0) \leqslant J(y)$ for every admissible y distinct from y_0.

If we change the sign of the penalty term the Euler equation becomes

$$k^2\ddot{y} + y = t,$$

and the general solution is $y = a \cos(t/k) + b \sin(t/k) + t$. The integral $-J_k$ now satisfies III'_R if $k \neq 0$; hence if y_0 is embedded in a field we would have the inequality $-J_k(y_0) < -J(y)$ or $J_k(y_0) > J_k(y)$ if $y \neq y_0$ and are tempted to conclude that $J(y_0) \geqslant J(y)$. This with the complementary inequality above would imply that $J(y_0) = J(y)$, therefore that J is independent of the choice of an admissible y. This is manifestly false in view of the fact that J is not of the special form described following (3.6). We must infer that either our supposition of the existence of a field about y_0 was false for values of k near zero or, if there is a field (S_k, p_k) as $k \to 0$, then the set S_k must converge to $y_0 = \{(t,y) | y = t, 0 \leqslant t \leqslant 1\}$ as $k \to 0$. Consider the Jacobi condition in terms of the solution $\Delta(t,t_0)$ of (2.56). We verify that $\Delta(t,0) = \sin(t/k)$, hence that, if $k\pi < 1$, then $\Delta(t,0)$ vanishes for $t = k\pi < 1$. Thus y_0 does not satisfy the Jacobi necessary condition and $J_k(y_0)$ cannot be a maximum for values of k near zero. Our conclusion for the problem remains as stated at the end of the preceding paragraph.

EXAMPLE 3.5

$J(y) = \int (\dot{y}^4 - y^2) \, dt$ with endpoints $(0,0)$ and $(1,0)$.

Discussion

This is not a very formidable-looking integral. The figuratives corresponding to fixed values of y are all convex; hence one is likely to guess that there will be a function y_0 satisfying II'_N and enough other conditions to ensure that $J(y_0)$ is some sort of minimum. If the last term of the integrand were $+y^2$, then $y_0(t) = 0$ would clearly furnish the infimum 0 of possible values of $J(y)$. However, with the given term $-y^2$ one anticipates that $J(y)$ can be negative for some admissible functions.

The Euler equation is

(3.14) $$12\dot{y}^2\ddot{y} = -2y$$

and $y_0(t) = 0$ is a solution having the given end values. No other constant-valued function satisfies this equation; hence, if y is any solution distinct from y_0, then dy does not vanish identically. For any values of t such that $(dy)(t) \neq 0$, (3.14) implies that

$$(3.15) \qquad 12\mathring{y}^3 \frac{d\mathring{y}}{dy} = -2y \qquad or \qquad 12\mathring{y}^3 \, d\mathring{y} = -2y \, dy.$$

The last form is needed to integrate, but before doing so we remark that the operations performed have introduced extraneous solutions $y(t) = \pm a \neq 0$. All other solutions of this equation satisfy (3.14).

From (3.15) we find by integration that

$$3\mathring{y}^4 = a^2 - y^2,$$

hence that

$$(3.16) \qquad \frac{dt}{\sqrt{3}} = \pm \frac{dy}{(a^2 - y^2)^{1/4}}.$$

The last form conveniently rejects the extraneous solutions. It also rejects the solution $y_0(t) = 0$, but other solutions of (3.14) are now expressible in terms of an integral.

Observe that $f_{rr}(t,y,\mathring{y}) = 12\mathring{y}^2$, consequently that y_0 fails to satisfy III' and hence that Theorem 3.1 for the existence of a field is inapplicable. The hypotheses of this theorem are sufficient for the existence of a local field but not necessary as shown by Example 3.3, for which III' also fails.

A numerically oriented reader may think of tabulating values for

$$\sqrt{3} \, t = \pm \int_0^y \frac{ds}{(a^2 - s^2)^{1/4}} - 1$$

for a spread of values of a and y in search of evidence of the existence of a field about y_0. This would furnish presumptive evidence, not a proof, but might suggest the structure of a proof. If there were such a field, the strict convexity of the integrand in $\mathring{y}$ and Theorem 3.4 would assure that $J(y_0)$ is a proper strong local minimum. We shall see that this gambit cannot succeed, but it might do so for some other problem.

Another move is to add the penalty term $k^2\mathring{y}^2$ to the integrand. Although y_0 satisfies necessary conditions I and III' for the integral J_k defined by (3.12), y_0 does not satisfy the necessary condition IV of Jacobi for values of k near zero, and therefore $J_k(y_0)$ is not even a weak local minimum for such values of k. It follows from the relationship (3.12) between J and J_k that $J(y_0)$ is not a weak local minimum. We also can show this by a judicious selection of particular admissible functions.

Clearly $y_\alpha, y_\alpha(t) \equiv \alpha(t^2 - t)$ is admissible for all real values of α and, since $y_\alpha(t)$ and $\mathring{y}_\alpha(t) = \alpha(2t - 1)$ both converge to zero with α, then y_α

is in any preassigned first-order neighborhood of y_0 provided that $|\alpha|$ is sufficiently small. By elementary calculation we find that

$$J(y_\alpha) = \frac{(6\alpha^2 - 1)\alpha^2}{30};$$

consequently $J(y_\alpha) < 0$ for all α such that $6\alpha^2 < 1$. $J(y_0)$ is not even a weak local minimum.

The preceding examples serve to exhibit some of the gymnastics that one must be prepared to employ.

Exercise 3.2

Investigate each of the given integrals for minima and maxima on the class $\mathscr{Y}$ of smooth functions with the given endpoints. Identify the type of minimum or maximum found in each case or explain why there is no minimum or maximum.

1. $\int (3\dot{y} - \dot{y}^3)\, dt$, $(0,0)$ and $(1,1)$.

2. $\int e^{y\dot{y}}\, dt$, $(0,0)$ and $(1,0)$.

3. $\int t^2\dot{y}^2\, dt$, $(0,0)$ and $(1,0)$.

4. $\int y^2\, dt$, $(-1,0)$ and $(1,0)$.

5. $\int (t^2 - y^2 + y\dot{y})\, dt$, $(-1,0)$ and $(1,0)$.

6. $\displaystyle\int \frac{dt}{\dot{y}^2 - 4}$, $(0,0)$ and $(1,1)$.

7. $\displaystyle\int \frac{dt}{\dot{y}^2 - 4}$, $(0,0)$ and $(1,3)$.

8. $\int t\dot{y}(\dot{y} + y)\, dt$, $(0,0)$ and $(1,0)$.

9. $\displaystyle\int \left(y - t\dot{y} - y^2 + ty\dot{y} - \frac{t^2\dot{y}^2}{4} \right)^2 dt$, $(0,0)$ and $(2,0)$.

   (a) Verify that $y_0(t) = 2t - t^2$ satisfies the end conditions and that $J(y_0) = 0$ by direct substitution of y_0 into the integrand. What conclusions follow?

   (b) Find the envelope of the one-parameter family $y = 2\alpha t - \alpha^2 t^2$.

   (c) Explain why the results under (a) and (b) do not contradict the Jacobi necessary condition IV.

10. $J(y) = \int \dot{y}^3\, dt$ and the fixed endpoints are $(0,0)$ and $(2,0)$. Then $y_0(t) = 0$ satisfies I and the end-conditions. Determine by examination of II'_N the class of functions for which, as a consequence of (3.8), $J_k(y) - J_k(y_0) > 0$. Then explain why, for this example, we cannot conclude by letting $k \to 0$ that y_0 furnishes a local minimum for the original integral $J(y)$.

3.8 SUFFICIENT CONDITIONS WHEN THERE IS A CORNER

Necessary conditions I, II, and III as given in Chapter 2 apply to a minimizing function y_0 with one or more corners. We did not consider the Jacobi condition for such a solution and shall not do so in this book, but it is possible to do so and to obtain sufficiency theorems analogous to those of Section 3.6. We point out briefly how the approach of Section 3.4 can be extended.

It is well known, for the ordinary line integral $J(y) = \int P(t,y)\, dt + Q(t,y)\, dy = \int (P + \dot{y}Q)\, dt$, that $J(y)$ is independent of the choice of a PWS path joining (t_0, h) and (t, y) iff the function W defined by the relation

$$W(t,y) \equiv \int_{(t_0,h)}^{(t,y)} P(\tau,\eta)\, d\tau + Q(\tau,\eta)\, d\eta$$

has the total differential

(3.17) $dW = P(t,y)\, dt + Q(t,y)\, dy.$

Suppose given the problem $J(y) = $ minimum on the class $\mathscr{Y}$ of PWS functions with fixed endpoints and that $y_0 \in \mathscr{Y}$, having a corner at t_2 between t_0 and t_1, satisfies necessary conditions I, II, III, and the Erdmann corner conditions. Suppose further that a one-parameter family $\phi(\cdot, \alpha)$ of solutions of (2.18) has been found, which includes y_0 as a member, such that, for each α, $\phi(\cdot, \alpha)$ has exactly one corner satisfying the Erdmann conditions, such that the corners are all in the set of points (t,y) constituting a function g or a vertical segment and such that a suitable subset S of the strip bounded by t_0 and t_1 is simply covered by this family with y_0 interior to S except for its endpoints on the boundary lines of the strip. The circumstances we have described are indicated by Fig. 3.1.

Let $p(t,y)$ or $q(t,y)$, respectively, denote the slope of that function $\phi(\cdot, \alpha)$ through (t,y) when that point is to the left or right of g, and let $r(t,y)$ denote $p(t,y)$ or $q(t,y)$, whichever applies. We can leave r undefined on g, set $r(t,y) = p(t,y)$ on g, or set $r(t,y) = q(t,y)$ on g.

We thus have a field $\mathscr{F} = (S, r)$ about y_0, which does not precisely fit Section 3.2 because of the corners but extends the notion of that section. That such a family $\phi(\cdot, \alpha)$ and field exist can be proved in a theorem like Theorem 3.1 with IV' now being the Jacobi condition that is not discussed in this book.

Consider W defined by the relation

(3.18) $W(t,y) \equiv \int_{(t_0,h)}^{(t,y)} \{ f[\tau,\eta,r(\tau,\eta)] - r(\tau,\eta) f_r[\ ] \}\, d\tau + f_r[\ ]\, d\eta.$

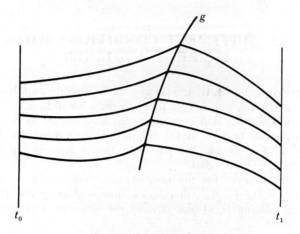

FIGURE 3.1

One finds that relation (3.17) applies to this case and that

$$(3.19) \qquad dW = \{f[t,y,r(t,y)] - r(t,y)f_r[\ ]\} \, dt + f_r[\ ] \, dy.$$

Verify, as a result of our hypothesis that $\phi(\cdot,\alpha)$ satisfies the corner conditions together with the blanket hypothesis on f, that the coefficients of dt and dy are both continuous on g as well as elsewhere in the set S covered by the field.

Integral (3.18) is the Hilbert integral in a slightly different notation than (3.6). The above observations are an outline of a proof that Theorem 3.2 on independence of path of the integral remains valid for such a field as that of Fig. 3.1. Granted this, the proof of (3.8) in Section 3.4 goes over to the present case. Similar remarks apply if y_0 has two or more corners. This section merely opens the subject of conditions on a function y_0 with corners sufficient for $J(y_0)$ to be an extremum.

EXAMPLE 3.6

$J(y) = \int (\dot{y}^2 - 1)^2 \, dt$ with fixed endpoints $(0,0)$ and $(2,0)$.

Discussion

Let y_0 consist of points (t,y) on the broken line from $(0,0)$ to $(1,1)$ to $(2,0)$ and let $\mathscr{F} = (S,p)$ be the field in the large defined by the family y_0 $+\alpha$. The corners all lie on the vertical line $t = 1$. It is suggested to the reader as an exercise that he think through the details associated with (3.18), (3.19), and (3.8) for this example and verify that, as a consequence of (3.8), $J(y_0)$ is a global minimum.

3.9   EXTENSIONS, OTHER METHODS

The fixed-endpoint nonparametric problem in $(n+1)$-space is that of extremizing an integral

$$J(y) = \int_{t_0}^{t_1} f(t,y,\dot{y})\, dt$$

on the class $\mathcal{Y}$ of all PWS vector-valued functions y: $[t_0,t_1] \to R^n$ having fixed endpoints $[t_0,y(t_0)]$, $[t_1,y(t_1)]$. It can be treated along generally similar lines to those presented here for the case $n = 1$. There are necessary conditions I (Euler), II (Weierstrass), III (Legendre–Clebsch), and IV (Jacobi–Mayer) and also sufficiency theorems similar to those of Sections 3.4 and 3.6.

An important difference between the cases $n = 1$ and $n > 1$ occurs in the theory of fields. When $n = 1$, any one-parameter family of solutions of the Euler equation that simply covers a suitable set S to which a particular solution y_0 of that equation is interior generates a field about y_0. Further conditions must be satisfied in the higher-dimensional cases. Such fields, called *Mayer fields*, are treated by Bliss (IX, Chap. 2) for $n = 2$ and (IX, Chap. 4) for a general n.

An introduction to the general Problem of Bolza is given in Chapter 5. Development of the theory for these more complex problems began in the early days of the calculus of variations, at least as far back as Lagrange, but major results and refinements have come in the twentieth century. This area continues to be an active field of research with many unanswered questions. The Problem of Bolza as formulated by Bliss (IX, p. 189) was often mentioned as the most general single integral problem of the calculus of variations, but various more general cases have recently become important because of the wide interest in systems optimization. See, for example, Hestenes (XXI) and (20f).

Sufficiency theorems considered in preceding sections have all been based on fields. A second so-called *expansion method* uses suitable expansions of the integrand such as that of Taylor. This approach has long been effective for weak local extrema. It was first applied to sufficiency for strong local minima by E. Levi in 1911 and has been extended by W. T. Reid (45b,c,d) to Bolza Problems. We shall not include any of this work in the present book but Section 3.12 exhibits a sufficiency theorem for global minima involving expansion. A third type of sufficiency theorem proved by the indirect method of arriving at a contradiction was used by McShane (33k) and by Hestenes (20b,d,e).

The literature on sufficiency has relatively little to say about global minima. A traditional view (X,pp.10–11) has been to the effect that one needs merely to identify all relative minima of $J(y)$ and then to pick out the least of the lot. If we ignore as exceptional the possible case with infinitely many relative minima, there is still a great deal left to the imagination. And yet in many examples a global minimum or maximum is precisely the goal. This is certainly the case in systems optimization. Granted a criterion for optimality, one desires the best of all programs, not merely one that is best in comparison with other nearby programs. Sections 3.11 and 3.12 of this chapter are devoted to some relatively simple sufficiency theorems for global minima.

3.10 CONVEX SETS AND CONVEX POINT-FUNCTIONS

Section 2.3 reminds us that the existence and determination of minima of a function ϕ: $[a,b] \to R$ are related to convexity of ϕ without assuming any knowledge of this topic beyond that provided in elementary calculus. We now examine convexity of subsets of R^n and of ordinary functions in preparation for extension of these ideas to function spaces and to functions J of the calculus of variations. For further information see Hestenes (XXI,pp.45–48) or T. Bonnesen and W. Fenchel, *Theorie der konvexen Körper* (Springer, Berlin, 1934, reprinted by Chelsea, New York, 1948), pages 18–21.

That a subset K of R^n is *convex* means that, for every pair of points x and y of K, the point $z = x + \tau(y - x) = (1 - \tau)x + \tau y$ is in K for every $\tau \in [0,1]$. Such points z constitute the segment in R^n with endpoints x and y. Since $\tau = 0$ and 1 yield points x and y which were chosen in K, the content of the definition is the same if the closed τ-interval is replaced by the open interval (0,1).

Given that K is convex, a function ϕ: $K \to R$ is called *convex on K* if

$$(3.20)\quad \phi(x) + \tau[\phi(y) - \phi(x)] \geqslant \phi[x + \tau(y-x)], \forall x,y \in K, \forall \tau \in [0,1].$$

If $-\phi$ is convex on K, then ϕ is said to be *concave on K*. In contrast with the use of these terms in optics, convexity and concavity of functions are not intrinsic properties of curves and surfaces. For example, $\phi(x) = x^2$ and $\psi(x) = -x^2$ yield the same parabola but differently located in the coordinate plane, and the functions ϕ and ψ are, respectively, convex and concave.

The only convex subsets of the set R of real numbers are intervals of finite or infinite length including R itself, singleton sets, and the empty set. The last two degenerate cases are of little interest here. Most of the figures of elementary plane and solid geometry when regarded as point sets in R^2 and R^3 are seen to contain a segment if they contain its endpoints. One easily visualizes other convex subsets of R^2 or R^3, some bounded and others not. Most sets are not convex, for example, a star-shaped set. The reader should use the definition of a convex set in verifying the following properties of convex subsets of R^n. The intersection of finitely or infinitely many convex subsets of R^n. is a convex subset of R^n, possibly degenerate. Unions and differences of convex sets are in general not convex.

If K is a convex subset of R^n, then any subset H of K obtained by fixing the same coordinates of all points of K is convex. For instance the circular disc $K \equiv \{(x,y) \in R^2: x^2 + y^2 < 25\}$ is clearly convex and the subset $H \equiv \{(x,y) \in R^2: x^2 + y^2 < 25, x = 3\}$ is a segment, hence also a convex subset of R^2. Moreover, such sets as K and H can be regarded as convex subsets of $R^p, p \geqslant 3$.

If K_m and K_n are convex subsets of R^m and R^n, then the cartesian product $K_m \times K_n$ is a convex subset of $R^m \times R^n = R^{m+n}$.

Definition (3.20) of a convex function ϕ says that the segment joining two points $[x,\phi(x)]$ and $[y,\phi(y)]$ of ϕ contains no point $\{x + \tau(y-x), \phi(x) + \tau[\phi(y) - \phi(x)]\}$ that is below the corresponding point $\{x + \tau(y-x), \phi[x + \tau(y-x)]\}$ of the function. We can thus easily distinguish among examples that are convex, concave, or neither, provided that $n = 1$ or 2 and the graph of ϕ is simple enough to visualize. In general, we must depend upon definition (3.20). The following properties of convex functions, which are direct consequences of that definition, are useful.

If ϕ and ψ are both convex on K, then $\phi + \psi$ is convex on K. Thus if $K = R$, $\phi(x) = x^2$, and $\psi(x) = \cosh x$; then $\phi + \psi$ is convex on R. If ϕ and ψ are convex on respective subsets K_m and K_n of R^m and R^n, then $\phi + \psi$ is convex on the convex subset $K_m \times K_n$ of R^{m+n}. For example, if $\phi(x) = x^2, x \in R$ and $\psi(y) = -\cos y, y \in [-\pi/2, \pi/2]$, then the function $f \equiv \phi + \psi$ with values $f(x,y) = x^2 - \cos y$ is convex on the set $\{(x,y) \in R^2: x \in R, y \in [-\pi/2, \pi/2]\}$. Differences $\phi - \psi$ and products $\phi\psi$ of convex functions are generally not convex.

If $\phi: K \to R$ is convex on K, then ϕ restricted to any convex subset H of K is convex on H. If K is a subset of R^n, then ϕ can also be regarded as a convex function from a subset of $R^p, p > n$, to R. For example, if $\phi: R^2 \to R$ has values $\phi(x,y) = x^2 + 4y^2$, then $\phi(x,3)$ gives the values of a convex function from R to R. Also the function $\phi: R \to R$ with values $\phi(x) = x^2$, visualized as a parabola, can also be considered as a convex function with domain R^2 and values $\phi(x,y) = x^2$ that are independent of y and which we visualize as a parabolic cylinder in R^3.

Although definition (3.20) mentions neither the continuity nor differentiability of a convex function ϕ, we shall point out that continuity of ϕ at most points of K together with the existence and finiteness of certain first-order derivatives is a consequence of (3.20).

Given an interior point x of the convex subset K of R^n and a function ϕ that is convex on K, let a and b be distinct points of K such that $x = (a+b)/2$ and set $y = (a-b)/2$. Then $x - y$ and $x + y$ are in K as is $x + hy$

if $h \in (0,1)$ is near 0 or 1. With h so near 0 that $x + hy$ is in K, replace $x, y,$ and τ in (3.20) by $x - y, x + hy,$ and $1/(1 + h)$ and find that

$$(3.21) \qquad \phi(x) - \phi(x - y) \leqslant [\phi(x + hy) - \phi(x)]/h \equiv Q.$$

With y and τ in (3.20) replaced by $x + y$ and h, we obtain the inequality

$$(3.22) \qquad Q \equiv [\phi(x + hy) - \phi(x)]/h \leqslant \phi(x + y) - \phi(x).$$

With points $x, a,$ and b of K fixed, we see from (3.21) and (3.22) that the quotient Q is bounded. If y in (3.22) is replaced by $\theta y, \theta \in (0,1)$ and we divide through by θ, we see that Q is nondecreasing in h and hence that Q has a finite limit as $h \to 0+$. Since points a and b chosen above in K are distinct, $y = (a - b)/2$ is not the zero-vector; hence quotient Q in (3.21) and (3.22) can be written in terms of the unit vector $y/|y|$ as

$$\phi[x + h|y|(y/|y|)] - \phi(x)]/h.$$

One sees with reference to the usual definition of a directional derivative that the finite limit of $Q/|y|$ as $h \to 0+$ is the directional derivative of ϕ at x in the direction of y. We denote this derivative by $\phi'(x;y)$.

In the case $n = 1$, x and y are real numbers and $\phi'(x;y)$ is the right derivative $\phi'^{+}(x)$ or the negative of $\phi'^{-}(x)$ defined in Section 1.7 according as the fixed y is positive or negative. Since these derivatives are finite, ϕ is both right and left continuous at x and hence continuous at x, an arbitrary interior point of K. If the interval K includes an endpoint c, elementary examples show that ϕ need not be continuous at c. Moreover, if ϕ is not continuous at an endpoint c of its domain, then $\phi'^{-}(c) = \infty$ or $\phi'^{+}(c) = -\infty$ according as c is a right or left endpoint of K. If $x, y,$ and z are in K and $x < y < z$, it can be proved that

$$(3.23) \qquad \phi'^{+}(x) \leqslant \phi'^{-}(y) \leqslant \phi'^{+}(y) \leqslant \phi'^{-}(z).$$

It also can be proved that the left and right derivatives are equal and hence that the derivative $\phi'(x)$ exists at all but a countable set of points of K. Granted the nondecreasing character (3.23) of ϕ'^{+} and ϕ'^{-}, these functions are then Riemann integrable over any subinterval of K on which their values are bounded and, if a and x are endpoints of such an interval, it can further be proved that

$$(3.24) \qquad \phi(x) = \int_a^x \phi'^{+}(s) \, ds + \phi(a) = \int_a^x \phi'^{-}(s) \, ds + \phi(a).$$

Conversely, an indefinite integral of any bounded nondecreasing function is convex.

Although $\phi'(x)$ can fail to exist at the points x of a countable set, there are many convex functions $\phi: K \to R$ such that $\phi'(x)$ exists and is continuous at every point of K. Given such a function ϕ, it follows from (3.20) and the Mean Value Theorem of the Differential Calculus that, if x and y are in K and $0 < \tau < 1$, then there exists $\theta \in (0,1)$ such that

$$\phi(y) - \phi(x) \geq \{\phi[x + \tau(y-x)] - \phi(x)\}/\tau = \phi'[x + \theta\tau(y-x)](y-x).$$

Since ϕ' is continuous by hypothesis, we can let $\tau \to 0+$ and find that

$$(3.25) \qquad \phi(y) - \phi(x) - \phi'(x)(y-x) \geq 0, \qquad \forall x,y \in K.$$

It follows, by application of the Mean Value Theorem to $\phi(y) - \phi(x)$, that there exists $\theta_1 \in (0,1)$ such that

$$\{\phi'[x + \theta_1(y-x)] - \phi'(x)\}(y-x) \geq 0.$$

In the event that ϕ'' also exists and is continuous on K, we can apply the Mean Value Theorem to the expression in braces to obtain the existence of $\theta_2 \in (0,1)$ such that

$$(3.26) \qquad \phi''[x + \theta_2\theta_1(y-x)](y-x)^2 \geq 0, \qquad \forall x,y \in K.$$

If $x,y \in K$ and $x \neq y$, it follows that $\phi''[x + \theta_2\theta_1(y-x)] \geq 0$. With x fixed, let $y \to x$ and conclude from the continuity of ϕ'' that

$$(3.27) \qquad \phi''(x) \geq 0, \qquad \forall x \in K.$$

If $n > 1$, we can hold all but one of the coordinates of x fixed and obtain similar properties of all first-order partial derivatives of a convex function $\phi: K \subset R^n \to R$. If this function ϕ has continuous first-order derivatives $\phi_i \equiv \partial\phi/\partial x^i$, $i = 1, \cdots, n$ on K, the extension of (3.25) is the condition that

$$(3.28) \qquad \phi(y) - \phi(x) - \phi_i(x)(y^i - x^i) \geq 0, \qquad \forall x,y \in K,$$

with summation on i. This is derived from (3.20) by using the Mean Value Theorem for a function of n-variables. The similar extension of (3.27) for the case in which all second-order derivatives $\phi_{ij} \equiv \partial^2\phi/\partial x^i \partial x^j$ are continuous on K is obtained as follows.

Corresponding to (3.26) and with summation on two indices i and j we find that

$$\phi_{ij}[x + \theta_2\theta_1(y-x)]\theta_1(y^i - x^i)(y^j - x^j) \geq 0, \qquad \theta_1 \in (0,1) \text{ and } \theta_2 \in (0,1).$$

The positive factor θ_1 can be dropped and it follows that

$$\phi_{ij}[x + \theta_2\theta_1(y-x)]\,u^iu^j \geq 0,$$

where $u \equiv (y-x)/|y-x|$ is the unit vector corresponding to any pair x, y in K with $y \neq x$. With x fixed, let $y \to x$ with the direction of $y-x$ fixed and hence the unit vector u fixed. It follows that $\phi_{ij}(x)u^iu^j \geq 0$. If we set $\alpha \equiv \lambda u$, $\lambda \geq 0$, then

(3.29) $\phi_{ij}(x)\alpha^i\alpha^j \geq 0$, $\forall x \in K$, and $\forall \alpha \in R^n$ such that either $\alpha = 0$ or the vector α has the same direction as $y - x$ for some y in K. If x is an interior point of K, this is no restriction on α.

We have shown under respective hypotheses on derivatives that (3.20) implies (3.28) or (3.29). It can be verified that if first derivatives are continuous, then (3.28) implies (3.20), and if second derivatives are continuous that (3.29) implies (3.20).

Convexity on K of a function ϕ as defined by (3.20) or under further restrictions on ϕ by (3.28) or (3.29) is a *global property of ϕ*, that is, a property involving all points of K and the corresponding values of ϕ.

We shall say that ϕ is *convex at x relative to K* (often abbreviated to simply *convex at x*) if corresponding to the fixed $x \in K$ and to each $y \in K$ is a positive real number $\epsilon_{x,y}$ such that

(3.30) $0 < \tau < \epsilon_{x,y} \Rightarrow$ inequality (3.20).

The function ϕ is convex on K if and only if it is convex at x relative to K at every point x of K. The function $\phi: R \to R$ with values $\phi(x) = x^2$ or 1 according as $x^2 \leq 1$ or $x^2 > 1$ is seen from its graph to be convex at 0 relative to R but not convex on R. The function $\phi: R \to R$ with values $\phi(x) = 2x^2 - x^4$ is neither convex on R nor even convex at 0 relative to R, but the restriction of ϕ to $(-\sqrt{2},\sqrt{2})$ is convex at 0 relative to this open interval and the restriction of ϕ to $(-\sqrt{3}/3,\sqrt{3}/3)$ is convex on that interval.

The preceding concepts and results are useful in the remainder of this chapter and in parts of later chapters. We mention in contrast another definition of convexity at x that will not be used. If $n = 1$, if x, y, and z are in the domain K of ϕ with x fixed, and if $y < x < z$, then the condition that

(3.31) $[\phi(x) - \phi(y)]/(x-y) \leq [\phi(z) - \phi(x)]/(z-x)$ $if\, y < x < z$

between slopes of chords certainly states a property commonly associated with convexity at x. The function $\phi: R \to R$ with values $\phi(x) = x^{2/3}$ has property (3.31) with $x = 0$ in that condition, but this function is not convex at 0 in the sense (3.30) relative to any convex subset K of R.

3.11   CONVEXITY OF INTEGRALS AND GLOBAL MINIMA

We have seen in Theorem 3.9 and elsewhere that a global minimum can be identified if a field (S,p) about y_0 covers an infinite strip of the (t,y) plane and if either the figurative is convex in r for each fixed $(t,y) \in S$ or even if $E[t,y,p(t,y),q] \geqslant 0$ for each (t,y) in S and each real q. Thus far in this book this is the only means at our disposal.

If $J(y_0)$ is a global minimum, then $J(y_0)$ is necessarily a weak local minimum; hence in seeking a global minimum it suffices to examine only those admissible functions y_0 that furnish a minimum of the latter type. Given such a y_0 we can then ask for additional conditions sufficient to guarantee that $J(y_0)$ is actually the desired global minimum.

These ideas can be discussed just as easily for the nonparametric fixed-endpoint problem in a general $(n+1)$-space as for the particular case $n = 1$. Let $\mathscr{Y}$ now denote the class of all vector-valued functions $y:[t_0,t_1] \to R^n$, $y = (y^1, \ldots, y^n)$ that are PWS and have fixed endvalues $y(t_0), y(t_1)$. That y is PWS means by definition that each component y^i is PWS in the sense of Section 1.9.

Given $x,y \in \mathscr{Y}$, define a function $z: [0,1] \times [t_0,t_1] \to R^n$ with values

$$(3.32) \qquad z(\tau,t) \equiv (1-\tau)x(t) + \tau y(t)$$
$$= x(t) + \tau[y(t) - x(t)], \tau \in [0,1], t \in [t_0,t_1].$$

For each fixed $\tau \in [0,1]$, we see that $z(\tau,\cdot) \in \mathscr{Y}$. With $x,y \in \mathscr{Y}$ both fixed, the mapping $\zeta: [0,1] \to \mathscr{Y}$, where $\zeta(\tau) \equiv z(\tau,\cdot)$ is called a *deformation*. The function x is deformed continuously onto y as τ traverses its unit interval $[0,1]$.

The function z with values (3.32) is seen to have these properties:

$$(3.33) \qquad z(0,t) = x(t) \quad and \quad z(1,t) = y(t), \qquad t \in [t_0,t_1].$$

$$(3.34) \qquad |z(\tau,t) - x(t)| = \tau\,|y(t) - x(t)| \to 0 \; with \; \tau,$$
$$|y - x| \; being \; the \; euclidean \; distance,$$

and

$$(3.35) \qquad |z_t(\tau,t) - \dot{x}(t)| = \tau|\dot{y}(t) - \dot{x}(t)| \to 0 \; with \; \tau.$$

It is immediate from (3.32) and the fact that x and y are PWS that

(3.36) *The τ-derivatives z_τ and $z_{\tau t}$ of z and its t-derivative exist and are finite for all $\tau \in [0,1]$ and $t \in [t_0,t_1]$ except for those t, if any, that correspond to corners of x or y.*

We are again interested in minima of a function $J: \mathscr{Y} \to R$, where

$$J(y) = \int_{t_0}^{t_1} f(t,y,\dot{y}) \, dt,$$

but f is now a function from $[t_0,t_1] \times R^{2n}$ to R. We understand that f is subject to a blanket continuity and differentiability hypothesis.

The definitions of strong and weak minima stated in Section 2.4 apply to the present case, in which y has n components, provided that $|x-y|$ and $|\dot{x}-\dot{y}|$ are interpreted as lengths of n-vectors $x-y$ and $\dot{x}-\dot{y}$.

To say that J is *convex* or *strictly convex at* $x \in \mathscr{Y}$ *relative to* $\mathscr{Y}$, respectively, will mean that corresponding to x and to every $y \in \mathscr{Y}$, $y \neq x$, is a positive number $e_{x,y}$ such that if

(3.37) $0 < \tau < e_{x,y}$,

then

(3.38) $\Phi(\tau) \equiv J(x) + \tau[J(y) - J(x)] - J[z(\tau, \cdot)] \geq 0$ (or > 0).

Theorem 3.10

If $J(y_0)$ is a weak local minimum of $J(y)$ on $\mathscr{Y}$ and J is convex at y_0 relative to $\mathscr{Y}$, then $J(y_0)$ is a global minimum.

PROOF

Let y be an arbitrary but fixed function in $\mathscr{Y}$. Since J is convex at y_0 and τ satisfies (3.37), then

(3.39) $\tau[J(y) - J(y_0)] \geq J[z(\tau, \cdot)] - J(y_0)$.

By properties (3.34) and (3.35) of z, there must exist a positive number d, depending on y_0, y, and δ such that, if

$$0 < \tau < d,$$

then $z(\tau, \cdot)$ is in the first-order neighborhood $U_1(\delta,y_0)$ of the definition of a weak local minimum. With τ satisfying these inequalities as well as (3.37), relation (3.39) holds and the right member is nonnegative. Since $\tau > 0$ we see that $J(y) - J(y_0) \geq 0$ and, since y is in $\mathscr{Y}$ but otherwise arbitrary, the proof is complete.

Theorem 3.11

A necessary and sufficient condition for $J(y_0)$ to be a proper global minimum of $J(y)$ on $\mathscr{Y}$ is that (i) $J(y_0)$ be a weak local minimum and (ii) J be strictly convex at y_0 relative to $\mathscr{Y}$.

PROOF

To show that (i) and (ii) imply that $J(y_0)$ is a proper global minimum, follow the preceding proof with a strict inequality in (3.39). For the converse, one finds by differentiation of (3.38) that

$$(3.40) \quad \Phi'(0) = \int_{t_0}^{t_1} [f(t,y,\dot{y}) - f(t,y_0,\dot{y}_0) - f_y(t,y_0,\dot{y}_0) \cdot z_\tau - f_r(t,y_0,\dot{y}_0) \cdot z_{\tau t}] \, dt,$$

where $f_y \cdot z_\tau$ denotes the scalar product of the n-vector whose components are the partials of f with respect to components of y by the n-vector z_τ, with a similar description for the other dot product.

After an integration by parts like that in Section 2.6 of the next to last term, we use the Euler condition (now a system of n Euler equations found in the treatment of problem 12, Exercise 2.3) in form (2.18), which y_0 must satisfy under the present hypothesis that $J(y_0)$ is a weak local minimum. From this property of y_0 and the fact that $z_\tau(0,t_1) = y(t_1) - y_0(t_1) = 0$, relation (3.40) reduces to

$$(3.41) \quad \Phi'(0) = J(y) - J(y_0).$$

The right member is positive if the function y is distinct from y_0, since $J(y_0)$ is a proper global minimum. By inspection of (3.38), $\Phi(0) = 0$ and, by (3.41), $\Phi'(0) > 0$; hence (3.38) must hold with $>$ for sufficiently small positive τ. This completes the proof of conclusion (ii). To verify (i), simply observe that $J(y_0)$ being a proper global minimum implies that $J(y_0)$ is a weak local minimum.

A sufficient condition for convexity or strict convexity of J at $y_0 \in \mathcal{Y}$ is that the integrand $f(t,y,r)$ be convex in (y,r) or strictly convex in (y,r) at $[y_0(t),\dot{y}_0(t)]$ for each $t \in [t_0,t_1]$. In applying (3.20) and subsequent remarks to f, the number $2n$ replaces the previous n and the $2n$ components of the pair (y,r) replace the n components of x.

EXAMPLE 3.7

$\int (\dot{y}^2 + y\dot{y} + y^2) \, dt$ with endpoints $(0,0)$ and $(1,1)$.

Discussion

One finds readily that y_0, where

$$(e^2 - 1)y_0(t) = e^{t+1} - e^{-(t-1)},$$

satisfies conditions I, III′, and IV′, hence that y_0 furnishes a weak local minimum by Theorem 3.8. The quadratic form (3.29) for the present integrand, $f(t,y,r) = r^2 + ry + y^2$, is $Q(\alpha) \equiv 2\alpha_1^2 + \alpha_1\alpha_2 + \alpha_2\alpha_1 + 2\alpha_2^2 = 2(\alpha_1^2 + \alpha_1\alpha_2 + \alpha_2^2)$. We observe that $Q(\alpha)$ is free of t and has a negative discriminant; hence $Q(\alpha) \neq 0$, $\alpha \neq 0$, and, since it is clearly positive for $(\alpha_1,\alpha_2) = (1,0)$, it is positive if $\alpha \neq 0$. It follows that $f(t,y,r)$ is strictly convex in (y,r), hence strictly convex at $(y,r) = [y_0(t),\dot{y}_0(t)]$ in the sense of (3.30). By Theorem 3.11, $J(y_0)$ is a proper global minimum.

We shall avoid higher-dimensional examples. The reader who may wish to consider such cases will find criteria for identifying positive definite and semi-definite quadratic forms in many books.

Integrands that are convex or even convex in (y,r) in the sense of (3.30) are of rare occurrence among all possible integrands $f(t,y,r)$. They often are met in the Hamiltonian integrals considered in Chapter 4, in the accessory minimum problem related to the Jacobi condition, and in certain optimization problems from control theory.

In view of the fact that Theorem 3.11 gives necessary and sufficient conditions, it would be pleasing if we had a less stringent sufficient condition for the convexity of J than convexity of f in (y,r).

3.12   A NAIVE EXPANSION METHOD

Let x, y, p, and q be n-vectors and define

$$(3.42) \quad G(t,x,y,p,q) \equiv f(t,y,q) - f(t,x,p)$$
$$- (y-x) \cdot f_y(t,x,p) - (q-p) \cdot f_r(t,x,p),$$

in which dots again indicate scalar products. The right member is the difference between $f(t,y,q)$ and the first $2n+1$ terms in a Taylor expansion of $f(t,y,q)$ for fixed t about the point $(y,q) = (x,p)$ of $2n$-space.

The problem $J(y) = minimum$ on $\mathscr{Y}$ is again the nonparametric fixed endpoint problem in $(n+1)$-space. The Euler necessary condition I in form (2.18) for this problem is the vector equation

$$(3.43) \qquad f_r[t,y(t),\dot{y}(t)] = \int_{t_0}^{t} f_y[\tau,y(\tau),\dot{y}(\tau)]\,d\tau + c,$$

and the following theorem is somewhat similar to Theorem 3.3.

Theorem 3.12

If $y_0 \in \mathscr{Y}$ satisfies the Euler necessary condition (3.43) and y is an arbitrary admissible function, then

$$(3.44) \qquad J(y) - J(y_0) = \int_{t_0}^{t_1} G[t,y_0(t),y(t),\dot{y}_0(t),\dot{y}(t)]\,dt.$$

PROOF

With function y_0 fixed, define an integral J^* similar to and yet distinct from the Hilbert invariant integral,

$$(3.45) \quad J^*(y) \equiv \int_{t_0}^{t_1} [f(t,y_0,\dot{y}_0) + (y-y_0) \cdot f_y(t,y_0,\dot{y}_0)$$

$$+ (\dot{y}-\dot{y}_0) \cdot f_r(t,y_0,\dot{y}_0)] \, dt.$$

If we integrate the second term by parts and use (3.43) together with the fact that y and y_0 have common end values, we find that

$$(3.46) \qquad\qquad J^*(y) = J(y_0).$$

We are interested in the left member $J(y) - J(y_0)$ of (3.44), which, in view of (3.46), is equal to $J(y) - J^*(y)$, and the latter is precisely the right member of (3.44).

It is convenient to state the next theorem in terms of conditions that we denote by II_R^*, $\text{II}_R^{*\prime}$ and define as follows.

$\text{II}_R^* \quad G[t,y_0(t),y,\dot{y}_0(t),q] \geq 0, \qquad \forall t \in [t_0,t_1], \ \forall (y,q) \in R^2.$

$\text{II}_R^{*\prime} \quad G[t,y_0(t),y,\dot{y}_0(t),q] > 0, \qquad \forall t \in [t_0,t_1], \ \forall (y,q) \neq [y_0(t),\dot{y}_0(t)].$

Theorem 3.13

If $y_0 \in \mathcal{Y}$ satisfies conditions I and $\text{II}_R^*(\text{II}_R^{*\prime})$, then $J(y_0)$ is the global minimum (proper global minimum) of $J(y)$ on $\mathcal{Y}$.

PROOF

This is an intermediate corollary to Theorem 3.12.

With $n = 1$, the gap between hypotheses II_R^* or $\text{II}_R^{*\prime}$ of Theorem 3.13 and necessary condition II given in Chapter 2 is greater than that between sufficient conditions in Section 3.6 and the combined necessary conditions of Chapter 2. A similar remark applies to Theorem 3.13 and the higher-dimensional analogues of Section 3.6.

In the event that $f_y[t,y_0(t),\dot{y}_0(t)]$ vanishes identically, which necessarily is the case if f is free of y, conditions II_R^* and $\text{II}_R^{*\prime}$ reduce to II_R and II_R' stated in Section 3.1.

It follows from problem 6, Exercise 3.3, which follows, that II_R^* is equivalent to the statement that the surface $u = f(t,y,r)$ of parameter t in (y,r,u)-space include, for no fixed t, any point below the plane tangent to that surface at $[x,p,f(t,x,p)]$. That $f(t,y,r)$ be convex in (y,r) at $(x,p) = [y_0(t),\dot{y}_0(t)]$ for each t is a condition that implies II_R^*. Similarly that $f(t,y,r)$ be strictly convex in (y,r) at $(x,p) = [y_0(t),\dot{y}_0(t)]$ implies $\text{II}_R^{*\prime}$.

Theorem 3.13, in contrast with Theorem 3.11, does not include the hypothesis that $J(y_0)$ be a weak local minimum. There is no requirement in the proof of Theorem 3.13 that y_0 be smooth. Although the class of problems for which this theorem is an effective tool is limited, its simplicity recommends it. When it does apply, it is useful. We shall extend this theorem to certain Bolza Problems in Chapter 5 and apply the extension in characterizing the global extremum for a problem in missile trajectory optimization. An extension of Theorem 3.13 to problems with time delays has been considered by D. K. Hughes (21a).

EXAMPLE 3.8

$\int \; (\dot{y}^2 + y\dot{y} + y^2) \; dt$ with endpoints $(0,0)$ and $(1,1)$.

Discussion

A function y_0 satisfying the Euler condition I is given under Example 3.7. Since the integrand $f(t,y,r)$ is strictly convex in (y,r), we know, by Theorem 3.13, that $J(y_0)$ is a proper global minimum.

Examples simple enough to be analyzed readily and completely by one method often yield to other methods. It is not difficult to verify, for this one, that there exists a field in the large about y_0 and to conclude either from Theorem 3.9 or directly from the Fundamental Theorem 3.4 that $J(y_0)$ is a global minimum. This procedure is, however, much longer than that in the preceding paragraph or that under Example 3.7.

EXAMPLE 3.9

$\int \; (\dot{y}^2 - y^2) \; dt$ with endpoints $(\pi/6,1/2)$ and $(\pi/2,1)$.

Discussion

The familiar hyperbolic paraboloid or saddle surface with equation $u = r^2 - y^2$ is neither convex nor concave. Neither the function f with values $f(y,r) = r^2 - y^2$ nor its negative is even convex in the sense of condition II$^*_R$; hence Theorem 3.13 is not sufficiently discriminating to tell us whether the function $y_0, y_0(t) = \sin t$, satisfying the Euler equation and the given end conditions furnishes an extreme value. Another approach that yields a partial analysis of this example is to transform the integral into one for which Theorem 3.12 is effective.

Set $y = (-z)^{1/2}$, $z < 0$. Then $z = -y^2$ and there is a one-one correspondence between positive-valued functions $y: [\pi/6,\pi/2] \to R$ with end values $\frac{1}{2}$ and 1 and negative-valued functions $z: [\pi/6,\pi/2] \to R$ with end values $-\frac{1}{4}$ and -1. Moreover, $\dot{y} = -\dot{z}/2(-z)^{1/2}$; hence y is PWS iff z is PWS. The original integral $J(y)$ transforms into

$$I(z) \equiv \int \; [(-\dot{z}^2/4z) + z] \; dt.$$

The Euler equation for $I(z)$ is $2z\ddot{z} - \dot{z}^2 + 4z^2 = 0$ and the solution $z_0(t) = -\sin^2 t$ corresponds to $y_0(t) = \sin t$ for $J(y)$. The integrand of (3.44) for $I(z)$ is found to be

(3.47) $\qquad G(t,z_0,z,\dot{z}_0,\dot{z}) = -[(\dot{z} - 2z \cot t)^2/4z]$

and, if z is any negative-valued admissible function for $I(z)$ with values $z(t)$ bounded away from 0, then the right member of (3.47) is non-negative. With this restriction on z, we verify from the proof of Theorem 3.12 that (3.44) is applicable and therefore that $I(z_0) < I(z)$ if $z(t)$ is negative definite and hence, by the continuity of z on $[\pi/6, \pi/2]$, bounded away from 0. It follows that $J(y_0) < J(y)$ if y is admissible and positive definite.

This is a more illuminating conclusion than that obtainable by application of Theorem 3.6 or 3.7 to the present example. We would then know only that $J(y_0) < J(y)$ if y is admissible and in some zero-order neighborhood $U_0(\delta, y_0)$ without necessarily knowing the size of the neighborhood. If, however, we follow the hint of problem 2, Exercise 3.1, with possible reference to Theorem 3.9, it can be verified that $J(y_0) \leq J(y)$ for every admissible y and indeed that the strict inequality holds if y is distinct from y_0.

Exercise 3.3

Draw upon all of Chapters 2 and 3 in investigating the existence and nature of minimizing PWS functions for problems 1 through 5. Some of these examples have pathological features.

1. $J(y) = \int (e^{-\dot{y}} - t y \dot{y})\, dt$ with endpoints $(0,0)$ and $(1,0)$.

2. $J(y) = \int (\dot{y}^2 + 2t\dot{y} + y^2)\, dt$ with endpoints $(0,1)$ and $(1,1)$.

3. $J(y) = \int \sqrt{\dot{y}}\, \sqrt{1 + \dot{y}^2}\, dt$ with endpoints $(0,0)$ and $(1,1)$.

4. $J(y) = \int (t\dot{y}^3 - 3y\dot{y}^2)\, dt$. Choose a pair of endpoints.

5. $J(y) = \int (\sqrt{1 + \dot{y}^2} - \sqrt{1 - y^2}\, dt$ with endpoints $(0,0)$ and $(1,0)$.

6. Consider the "surface" in $(2n+1)$-space with equation $u = f(t,y,r)$ with t fixed and (y,r,u) variable. Establish conclusions similar to those of Section 2.10. Show that, if f is convex in (y,r) for each fixed t, then condition II$^*_\text{}$ holds as a consequence of the blanket hypothesis on f and results in Section 3.10.

7. If $\phi:K \to R^n$ and $\psi:K \to R^n$ are convex on the common convex domain K and α and β are positive, show that $\alpha\phi + \beta\psi$ is convex on K.

8. Discuss the conclusions available from Theorem 3.13 and problem 6 above for $J(y) = \int (\dot{y}^2 + y^2)\, dt$.

9. Given that y_0 satisfies the hypothesis of Theorem 3.13. Does it follow that y_0 satisfies all the necessary conditions I, II, III, and IV. Give reasons for an affirmative answer or construct a counter example, whichever is possible.

10. We have remarked under Example 3.9 that $r^2 - y^2$ is not convex in (r,y). Given the endpoints $(0,0)$ and $(\pi/2, 1)$ of problem 2, Exercise

3.1, and the conclusion for that problem, is J locally convex at the admissible y_0 through the given endpoints relative to $\mathscr{Y}$ or not, and why?

11. Verify that a linear integrand $P(t,y) + Q(t,y)\dot{y}$ is both convex and concave in $\dot{y}$. Investigate extrema of an integral with this integrand distinguishing between the cases $P_y(t,y) \equiv Q_t(t,y)$ and $P_y(t,y) \not\equiv Q_t(t,y)$ with reference to Sections 3.7 and 3.12. Finally, discuss the example $\int (t^2 + y^2 + y\dot{y}) \, dt$ with endpoints $(-1,0)$ and $(1,0)$.

12. Investigate the integral $J(y) \equiv \int [a(t)\dot{y}^2 + b(t)\dot{y} + c(t)] \, dt$ by the methods of Section 3.12 given only that the coefficients are continuous and $a(t) > 0$ on $[t_0, t_1]$.

13. Point out precisely what the Weierstrass necessary condition II says- in terms of convexity of f. Do the same for conditions II′, II′$_N$, and II′$_R$.

Chapter 4

VARIATIONS AND
HAMILTON'S
PRINCIPLE

4.1 INTRODUCTION

This chapter arbitrarily groups together two subjects that are not necessarily dependent but which are often so grouped in books on rational mechanics or other topics from physics or engineering. The reader who has encountered this approach to the calculus of variations elsewhere may have been surprised at not finding it earlier in the present book. It has been deferred until this point so that it can be viewed with the perspective provided by Chapters 2 and 3.

Since derivatives and differentials have played an important part in the investigation of extrema of point-functions ϕ, it would be natural to ask whether one can introduce similarly useful concepts for a function J. Although this question does not appear to have received attention by Euler and his predecessors, Lagrange introduced an operator δ analogous to differentiation around 1760. For many years thereafter it was traditional to define this operator more or less along the lines of our next section and to use it extensively. It was called a variation and the collection of techniques associated with it became known as the calculus of variations. That it is by no means essential is illustrated by its complete absence from Chapters 2 and 3 and indeed from all chapters except the present one. We shall point out by examples that it can be misleading.

This chapter represents both an interlude and a change of pace and can be omitted with little effect on the study of the remainder of the book.

The intent of the first part of the chapter (Sections 4.2 through 4.5) is to acquaint the reader with the notation and flavor of the highly formal approach to variational problems that was common in the early literature and is also to be found in some of the books and articles of recent date. If one hopes to make unrestricted use of available material on the calculus of variations and its applications, he must be prepared to meet and follow these methods and yet at the same time to be aware of their limitations.

4.2 THE OPERATOR δ

Recall with reference to Section 2.6 that in deriving the Euler necessary condition we used comparison functions of the form

$$(4.1) \qquad\qquad y = y_0 + \epsilon\eta.$$

The term $\epsilon\eta$, which is a functional increment analogous to the numerical increment Δx in the definition of a derivative, will now be called the *variation of y_0* or simply the variation of y with the subscript suppressed and will be denoted by δy. Thus

$$(4.2) \qquad\qquad \delta y \equiv \epsilon\eta.$$

By differentiation of (4.1), $dy = dy_0 + \epsilon\dot{\eta}\,dt$. The last term is the increment of dy due to the functional increment $\epsilon\eta = \delta y$ added to y. This suggests the introduction of

$$(4.3) \qquad\qquad \delta\,dy \equiv \epsilon\dot{\eta}\,dt,$$

called the *variation of dy*. The right member of (4.3) is clearly the differential of the right member of (4.2); hence we have the theorem based on definitions (4.2) and (4.3) that

$$(4.4) \qquad\qquad \delta\,dy = d\,\delta y,$$

which is described by saying that operators δ and d commute.

We also define the variation of $\dot{y}$ by the statement that

$$\delta\dot{y} \equiv \epsilon\dot{\eta},$$

from which it follows that

$$\delta\dot{y} = \frac{d(\epsilon\eta)}{dt} = \frac{d(\delta y)}{dt}.$$

Using the operator D to denote differentiation with respect to t, we see from the last result that

$$(4.5) \qquad\qquad \delta\, Dy = D\, \delta y,$$

that is, that δ and D also commute.

Next set

$$(4.6) \qquad\qquad \phi\,(\epsilon) \equiv f\,(t, y + \epsilon\eta, \dot{y} + \epsilon\dot{\eta}\,)$$

and restrict attention to those integrands f such that ϕ is equal to the sum of its Maclaurin series provided that $|\epsilon|$ is sufficiently small.

For such ϵ,

$$\phi\,(\epsilon) = \phi\,(0) + \phi'\,(0)\epsilon + \frac{\phi''(0)\,\epsilon^2}{2} + \cdots.$$

Define the kth *variation* $\delta^k f$ *of* f by the statement that

$$(4.8) \qquad\qquad \delta^k f \equiv \phi^{(k)}(0)\epsilon^k$$

and the *total variation* Δf of f as the difference

$$(4.9) \qquad\qquad \Delta f \equiv \phi\,(\epsilon) - \phi\,(0).$$

As a theorem that follows from (4.6) through (4.9),

$$\Delta f = \delta f + \frac{\delta^2 f}{2!} + \frac{\delta^3 f}{3!} + \cdots.$$

For the case $k = 1$, it follows from (4.6) and (4.8) that

$$(4.10) \qquad\qquad \delta f = f_y(t, y, \dot{y})\, \delta y + f_r(t, y, \dot{y})\, \delta\dot{y},$$

a form that is easily remembered by its similarity to the pattern of the total differential of a function of two variables. Similarly,

$$(4.11) \qquad\qquad \delta^2 f = f_{yy}(\delta y)^2 + 2f_{ry}(\delta y)\,(\delta\dot{y}) + f_{rr}(\delta\dot{y})^2,$$

in which arguments $(t, y, \dot{y})$ of f_{yy}, etc., are suppressed.

We have followed Bolza (X, pp. 15–20) and (XI, pp. 20–21) in phrasing these definitions. Practice is not uniform. Akhiezer (I, pp. 92–93) and others include a factor ϵ on the right in (4.10), ϵ^2 in (4.11), etc., but these are differences in detail and need not concern us here even though they must be watched in consulting different sources.

One can define the kth *variation of an integral* J and its *total variation* as follows. Given $J = \int f\, dt$ with limits t_0, t_1 suppressed,

(4.12)
$$\delta^k J \equiv \int \delta^k f \, dt$$

and

(4.13)
$$\Delta J \equiv \int \Delta f \, dt.$$

These definitions are so phrased that operators δ^k and Δ both commute with the integral sign.

An operation is merely another name for a function or mapping. If one pauses to identify the domain and range of function δ, without which the discussion is hazy, he sees that δ is applied to a variety of different kinds of mathematical objects.

We have now constructed a formalism somewhat comparable to that associated with differentiation and with which one becomes proficient after a little practice.

4.3 FORMAL DERIVATION OF THE EULER EQUATION

Starting with an integral

$$J(y) = \int_{t_0}^{t_1} f(t, y, \dot{y}) \, dt,$$

suppose that $J(y)$ is a minimum or maximum and apply δ to each side. We find that

$$\delta J = \int_{t_0}^{t_1} \delta f \, dt = \int_{t_0}^{t_1} (f_y \delta y + f_r \delta \dot{y}) \, dt.$$

By definition (4.2) and the condition $\eta(t_0) = \eta(t_1) = 0$ of Section 2.6, we find that, with y fixed, δy is a mapping from $[t_0, t_1]$ to the reals with values $(\delta y)(t_0) = (\delta y)(t_1) = 0$.

Although we could proceed as in Section 2.6 to a du Bois Reymond integration by parts of the first term of the integrand, we elect to use a Lagrange integration by parts, that is, of the last term. We find that

$$\int_{t_0}^{t_1} f_r \, \delta \dot{y} \, dt = f_r[t, y(t), \dot{y}(t)](\delta y)(t) \Big]_{t_0}^{t_1} - \int_{t_0}^{t_1} \left(\frac{d}{dt} f_r \right) \delta y \, dt.$$

The first term on the right vanishes and hence

(4.14)
$$\delta J = \int_{t_0}^{t_1} \left(f_y - \frac{d}{dt} f_r \right) \delta y \, dt.$$

The right member is an expression for ϵ times the $F'(0)$ of Section 2.6. Consequently, if $J(y)$ is either a minimum of maximum, it is necessary that $\delta J = 0$, a condition analogous to the vanishing of the first derivative $\phi'(a)$ of a point-function $\phi: R \to R$ when $\phi(a)$ is an extremum.

Suppose that for some interior point t of $[t_0,t_1]$,

(4.15) $$f_y[t,y(t);\dot{y}(t)] - \frac{d}{dt}f_r[t,y(t),\dot{y}(t)] \neq 0.$$

The left member, being continuous in t under our blanket hypothesis on f, then differs from zero on some open interval I. Since δy vanishes at t_0 and t_1 but is otherwise arbitrary, we can choose δy so that $(\delta y)(t) \neq 0$ on I and $(\delta y)(t) = 0$ on $[t_0,t_1] - I$. In addition to these properties, we can also require that δy be smooth on $[t_0,t_1]$ if we wish. With such a δy, the right member of (4.14) cannot vanish and we contradict the condition $\delta J = 0$. Hence we must infer that condition (4.15) can hold nowhere on the open interval (t_0,t_1), or equivalently that the Euler equation (2.24) must hold on (t_0,t_1).

We have not stated explicitly what functions y are admitted. If the domain $\mathscr{Y}$ of J consists of those $y:[t_0,t_1] \to R$ that are smooth and have fixed end values, then all steps in the preceding derivation are valid provided that δy is chosen to be smooth on $[t_0,t_1]$. Moreover, the left member of (4.15) is continuous at t_0 and at t_1 and, since (2.24) must hold on (t_0,t_1), (2.24) must also hold at t_0 and t_1, hence on the closed interval $[t_0,t_1]$.

With reference to Theorem 2.3 we see that the Lagrange integration by parts and the continuity argument applied to (4.15) tacitly assume that the extremizing function y is smooth. Under this hypothesis, now made explicit, we can still admit all PWS comparison functions y with the fixed end values, as was done in Chapter 2, and the derivation of this section becomes an alternative proof that the Euler equation in form (2.24) is a necessary condition on a smooth extremizing function.

It is typical of old-style formal calculus of variations that it proceeds often without saying precisely what extremum problem is being considered, and it is then necessary for the reader to supply a problem for which the steps in the derivation are meaningful and valid.

Exercise 4.1

1. Given $J(y) = \int (\dot{y}^2 + 2ty + y^2)\, dt$ with end values of y fixed, derive the Euler equation in form (2.24) by applying δ and tracing through all the steps.

2. Given $J(y) = \int f(t,y,\dot{y})\, dt$ with end values of y fixed, where y now has n components $y^1, \ldots, y^n$, let δy now mean the vector with components $\delta y^i = \epsilon \eta^i$, $i = 1, \ldots, n$. Derive the system of n Euler equations (4.15) using the δ-calculus.

3. Given $J(y) = \int f(t,y,\dot{y},\ddot{y})\, dt$ on the class of functions $y:[t_0,t_1] \to R$ having continuous first and second derivatives on $[t_0,t_1]$ and such that $y(t)$ and $\dot{y}(t)$ have assigned values at t_0 and t_1, define $\delta\ddot{y}$ and derive the Euler necessary condition
$$f_y - \frac{d}{dt}f_{\dot{y}} + \frac{d^2}{dt^2}f_{\ddot{y}} = 0$$

by the δ-process. Also derive this condition by an extension of the methods of Section 2.6.

4.4 THE SECOND VARIATION

As a consequence of (4.13) and other definitions in Section 4.2,

$$\Delta J = \int_{t_0}^{t_1} (f_y\,\delta y + f_r\delta\dot{y})\,dt + \frac{1}{2!}\int_{t_0}^{t_1} [f_{yy}(\delta y)^2 + 2f_{yr}(\delta y)(\delta\dot{y}) + f_{rr}(\delta\dot{y})^2]dt + \cdots.$$

Given that $\delta J = 0$, it is plausible by analogy with the theory of minima of point-functions based on Taylor expansions to anticipate the additional necessary condition that $\delta^2 J \geq 0$. That this is indeed correct can be seen from Section 2.12. We are also inclined to guess that if the pair of conditions $\delta J = 0$ and $\delta^2 J > 0$ both hold for a certain y, then this should be sufficient to guarantee that $J(y)$ is some sort of a minimum.

That this is not a valid conclusion will be pointed out by an example. Before doing so, we call attention to the fact that the variations $\delta y = \epsilon\eta$ that we have used are so-called *weak variations* because of the fact that both $\delta\dot{y} = \epsilon\dot{\eta}$ and $\delta y = \epsilon\eta$ converge to zero with ϵ. Thus, given any positive e, $y + \delta y$ will be in the first-order neighborhood $U_1(e,y)$ of y provided that $|\epsilon|$ is sufficiently small. The most for which we can hope from the combined conditions $\delta J = 0$ and $\delta^2 J > 0$ is then that conceivably they may imply that $J(y)$ is a weak local minimum. The following is a counterexample.

EXAMPLE 4.1

$J(y) \equiv \int (y^2 - \dot{y}^3)\,dt$ *with endpoints* (0,0) *and* (1,0).

Discussion

The Euler equation in form (2.25) is $3\dot{y}\ddot{y} + y = 0$ and, by inspection, the function y_0, $y_0(t) \equiv 0$ is a solution through the given endpoints. We then know that $(\delta J)(y_0) = 0$ without using the δ-technique. We next find that

$$(\delta^2 J)(y_0) = \int_0^1 2(\delta y)^2\,dt.$$

This is clearly positive unless the continuous function $\delta y\colon [0,1] \to R$ is identically zero, and it would be quite difficult to convince someone who has been indoctrinated with pre-Jacobi calculus of variations that $J(y_0)$ is not a weak local minimum.

However, in the light of Chapters 2 and 3, we notice that the figurative $u = y^2 - r^3$ is neither convex nor concave in r. We see from Section 2.10 that $E[t,y_0(t),\dot{y}_0(t),q]$ can be negative for slopes q arbitrarily near $\dot{y}_0(t)$;

hence Theorem 3.3 suggests the possibility of functions y that are near to y_0 even in terms of first-order distance and such that $J(y) < J(y_0)$. The example was picked for these features and also for the fact that $f_{rr}[t, y_0(t), \dot{y}_0(t)] = -6\dot{y}_0(t) = 0$ for all t. The strengthened Legendre condition III' fails with a vengeance; neither the theory of the Jacobi condition, for which III' is a hypothesis, nor any sufficiency theorem in Section 3.6 is applicable. We may think of Section 3.7, but it will be of no help for this example. It is also relevant to verify with reference to Section 4.2 that the total variation ΔJ for this example reduces to two terms, namely, that

$$\Delta J = \tfrac{1}{2}\delta^2 J + \tfrac{1}{6}\delta^3 J = \int_0^1 \left[(\delta y)^2 - (\delta \dot{y})^3 \right] dt.$$

In order that $J(y_0)$ be a weak relative minimum, it is necessary that ΔJ be positive for all δy such that $|(\delta y)(t)|$ and $|(\delta \dot{y})(t)|$ are sufficiently small. The form of the integrand above is such as to make this eventuality appear doubtful.

These observations serve to disturb confidence in the conclusion suggested by the form of $\delta^2 J$, but they are not enough to confirm nor deny it. We turn to special admissible functions $y_\epsilon : [0, 1] \to R$,

$$y_\epsilon(t) \equiv \begin{cases} \epsilon t, & 0 \le t \le \epsilon^2, \\ \epsilon^3(t-1)/(\epsilon^2 - 1), & \epsilon^2 < t \le 1, \end{cases}$$

for which the derivative is

$$\dot{y}_\epsilon(t) \equiv \begin{cases} \epsilon, & 0 \le t < \epsilon^2, \\ \epsilon^3/(\epsilon^2 - 1), & \epsilon^2 < t \le 1. \end{cases}$$

One finds by calculation of the given integral for the function y_ϵ that

$$J(y_\epsilon) = -\epsilon^5 [1 - \epsilon/3 - \epsilon^4/(\epsilon^2 - 1)^2].$$

For $|\epsilon|$ sufficiently small, the bracketed expression remains positive; consequently, $J(y)$ changes sign with ϵ for values of ϵ arbitrarily near zero. We have confirmed the suspicions raised above. Positiveness of the second variation does not imply even a weak local minimum.

A second but less subtle way in which one can be misled by the second variation is illustrated by the next example.

EXAMPLE 4.2.

$J(y) = \int (\dot{y}^2 - y^2) \, dt$ with endpoints $(0,0)$ and $(2,0)$.

Discussion

We find that

$$\delta^2 J = \int_0^2 \left[(\delta \dot{y})^2 - (\delta y)^2 \right] dt.$$

The integrand viewed simply as a real quadratic form in δy and $\delta \dot{y}$ is clearly not of fixed sign for all choices of values for these quantities, and one might jump to the conclusion that $\delta^2 J$ is not of fixed sign and hence that $J(y)$ can have neither a minimum nor a maximum value. However, $\delta \dot{y}$ and δy are not independent, as one sees from the definitions in Section 4.2. We again turn to Chapters 2 and 3. The general solution of (2.25) is $y = a \cos t + b \sin t$ and the unique solution through the given endpoints is y_0, $y_0(t) \equiv 0$. This is found to satisfy the hypotheses of Theorem 3.6 and other sufficiency theorems of Section 3.6 that guarantee a strong local minimum. That $J(y_0)$ is actually a global minimum can be shown by constructing a field in the large.

Exercise 4.2

1. Show for Example 4.2 how to select a one-parameter family of solutions of (2.25) that generates a field on the strip bounded by $t = 0$ and $t = 2$ and discuss the application of Theorem 3.4 to this example.

2. Given $J(y) = \int (y^3 - \dot{y}^4) \, dt$ and the fixed endpoints $(0,0)$ and $(1,0)$, observe that y_0, $y_0(t) = 0$ satisfies the Euler equation and express the general solution of the Euler equation in terms of an integral with a variable upper limit. Examine the expression for $\delta^2 J$ and $\delta^3 J$ and guess whether or not $J(y_0)$ is an extremum. Then try to prove or disprove your conjecture with reference to Chapters 2 and 3.

3. Investigate similarly $J(y) = \int (y^4 + \dot{y}^4) \, dt$ with the endpoints $(0,0)$ and $(1,0)$, this time also examining $\delta^4 J$.

4. With η as a fixed PWS function on $[0,1]$ and $\eta(0) = \eta(1) = 0$, consider the function F defined by (2.19) for the particular integrand $y^2 - \dot{y}^3$ of Example 4.1. Verify that if $y_0(t) = 0$, then $F'(0) = 0$ and $F''(0) > 0$, and state precisely what conclusion concerning the value $J(y_0)$ is implied by (2.6). This may appear to contradict our analysis of Example 4.1. Explain why it does not.

5. Apply the necessary condition of problem 4, Exercise 2.4, to Example 4.1.

4.5   CONCLUDING REMARKS ON THE δ-CALCULUS

The δ-formalism is a technique for generating Euler equations in form (2.24). For problems with variable endpoints (see Exercise 2.6, problem 11, and also Section 5.2) one can also obtain the so-called transversality conditions with the use of δ, but this is not done in the present book. All such results can be obtained without use of the operator δ and, once obtained, can be applied directly to particular examples.

Those who have developed facility with the δ-technique tend to repeat the full ritual exhibited in Section 4.3, including a Lagrange integration by parts with every example. This is not only inefficient repetition that can be omitted by simply writing the Euler equation but is conducive to errors. For complex problems like those of Chapter 5, a great many steps are required to do the job that corresponds to Section 4.3 for the simple problem. It is very easy to overlook a term or make some other elementary blunder in the course of a sequence of equations running through several pages.

The second variation is an inadequate tool for dealing with the important question of sufficiency. This is the fatal weakness of the formal approach to the calculus of variations.

In addition to the weak variations discussed here, there are *strong variations* $\omega(t,\epsilon)$ having derivatives $\omega_t(t,\epsilon)$, which, in contrast with the $\epsilon\dot\eta$ in our definition of $\delta\dot{y}$, need not converge to zero with ϵ. Much attention was given to these matters in the nineteenth century, but such work is not a part of the mainstream. In our derivation of the Weierstrass necessary condition in Section 2.8, the difference $y(t) - y_0(t)$ between the comparison function y of (2.32) and the minimizing function y_0 was a strong variation, but there was no need to use that term. For further information on strong variations, see Bolza (XI, 45–53) or Osgood (XXXI, pp. 357,379).

We also mention that a number of mathematicians, among them V. Volterra and M. Fréchet, have defined derivatives or differentials in the context of general functional analysis. A review of this area up to 1931 is provided by the dissertation of R. G. Sanger (XV, for the years 1931–1932). There is a continuing active interest in the Fréchet differential, which extends the notion of a first variation δJ and which will be found in books on functional analysis.

4.6 INTRODUCTION TO HAMILTON'S PRINCIPLE

The remainder of this chapter, like the first part, is largely independent of the rest of the book. Brief treatment of this subject usually has been included in introductions to the calculus of variations, and we follow that tradition. See, for example, Akhiezer (I, pp. 186–189), Bliss [5(a), pp. 710–714], Bolza (XI, pp. 554–557), Pars (XXXII, pp. 128–136), and Weinstock (XXXVII, pp. 74–92). For a more extensive treatment see Lanczos (XXIII) or Osgood (XXXI, pp. 356–388).

The variational principle usually ascribed to W. R. Hamilton by American and west European authors is called the Hamilton–Ostrogradski Principle by Akhiezer (I, p. 186), with the name of M. V. Ostrogradski, a contemporary of Hamilton, included.

A *particle* or *point-mass* is an idealization consisting of a point with positive mass. For certain purposes of analysis an extended body of mass m can be replaced by a particle of mass m located at the center of mass of the body.

Given a system of n particles with respective positions (x_i,y_i,z_i) at time t in a cartesian coordinate system and with respective fixed masses m_i, $i = 1, \ldots ,n$, the *kinetic energy* T of the system can be defined by the relation

$$(4.16) \qquad T \equiv \tfrac{1}{2}m_i(\dot{x}_i^2 + \dot{y}_i^2 + \dot{z}_i^2), \quad summed\ on\ i\ from\ 1\ to\ n.$$

There may or may not exist a function U called a *potential* from a subset S of R^{3n+1} to R with values $U(t,x_1,y_1,z_1, \ldots ,x_n,y_n,z_n)$ and such that values of its partial derivatives

$$U_{x_i}, U_{y_i}, U_{z_i}, \qquad i = 1, \ldots ,n$$

are the components of force acting on the ith particle in the respective directions of x, y, and z coordinate axes at time t.

If and only if there exists a potential U, the particles are said to be in a *conservative field* and Hamilton's Principle asserts that the motions of all the particles will be such that the $3n$ Euler equations in form (2.25) for the integral

$$(4.17) \qquad J(x_1,y_1,z_1, \ldots ,x_n,y_n,z_n) \equiv \int (T+U)\, dt$$

will hold on every time interval.

The present discussion is in the spirit of classical mechanics, which excludes discontinuities of derivatives. We accordingly ignore the integral form (2.18) of the Euler condition and generally use (2.25) but may occasionally use (2.24).

It is shown at the end of Section 4.8 that Hamilton's Principle is essentially equivalent to a set of differential equations for a dynamical system that are obtained from Newton's law of motion $F = Ma$. The principle is a concise alternative way of stating something that is already a part of newtonian mechanics.

The principle frequently is stated in terms of the operator δ and a function $V \equiv -U$ by the equation

$$(4.18) \qquad \delta \int (T-V)\, dt = 0.$$

There seems to be no general agreement on notation or signs in potential theory. One writer's potential may be our U, another's our V, or the meanings of these symbols may exchange the meanings given above.

If $y(t)$ is the position of a particle moving on the y axis and if the velocity $\dot{y}(t_1) = 0$, this situation is described by saying that the particle is stationary at t_1 or that y is stationary at t_1. By imperfect analogy, an integral J is often said to be *stationary* at y_0 or $J(y_0)$ is called a *stationary value* of the integral if the first variation of J is zero at y_0. This stationarity of a functional has no intuitive interpretation comparably simple to that of an object that is "instantaneously at rest" in the sense that its velocity vanishes at a particular time. That J is stationary at y_0 means that $(\delta J)(y_0) = 0$, that is, that y_0 satisfies the Euler necessary condition for an extremum, no more and no less.

It is frequently, although incorrectly, stated that Hamilton's Principle requires motions that minimize the integral in (4.18), for example, in (X, p. 554) or (XXXVII, p. 74). This is true for sufficiently short time intervals but not in general, as will be seen from Example 4.4. The only part of variational theory required by Hamilton's Principle is the Euler necessary condition in form (2.24) or (2.25) for the Hamiltonion integral (4.17). This determines the motions. There is no need to investigate other necessary conditions nor sufficiency criteria to use Hamilton's Principle.

4.7 EXAMPLES

EXAMPLE 4.3

Investigate the idealized planar motion of a particle of mass m in the vicinity of a flat motionless earth with constant downward gravitational acceleration g.

Discussion

The n of Section 4.6 is unity. Choose the plane of motion as the (x,y) plane with upward-directed y axis. Then $z(t) = 0$ and z is eliminated from the analysis. By (4.16),

$$T = \tfrac{1}{2}m(\dot{x}^2 + \dot{y}^2).$$

In order to seek a potential U, we can ask what work is required to lift a particle of weight mg from any fixed height y_1 to height y. The result, by elementary physics, is

$$U(x,y) = mg(y - y_1).$$

One verifies that $U_x = 0$, $U_y = mg$, and hence that the components of force acting on the particle are 0 and $-mg$ in the respective coordinate directions, as they should be. Observe that this result is independent of the choice of y_1, therefore that we may as well take $y_1 = 0$ for simplicity and use the Hamilton integral

(4.19) $J(x,y) = \int \left[\tfrac{1}{2}m(\dot{x}^2 + \dot{y}^2) - mgy\right] dt.$

The condition for stationarity of J, $\delta J = 0$, is equivalent to the pair of Euler equations

$$f_x - \frac{d}{dt}f_{\dot{x}} = 0, \qquad f_y - \frac{d}{dt}f_{\dot{y}} = 0,$$

which, for integral (4.19), are

$$m\ddot{x} = 0, \qquad m\ddot{y} + mg = 0.$$

These familiar equations are derived in many elementary books on physics or calculus without any need to mention Hamilton. Our object here is simply to see how such a result fits into the present framework.

EXAMPLE 4.4

Apply Hamilton's Principle to the motion of a particle of mass m on a frictionless x axis, the only force being directed toward the origin and with magnitude proportional to the displacement from the origin.

Discussion

Everyone knows that the answer is the differential equation is $\ddot{x} + kx = 0$ for simple harmonic motion, but we wish to see how to get it by the present method. To find the work we must integrate force kx with respect to distance. Thus

$$V = \int kx \, dx = \tfrac{1}{2}kx^2,$$

where again we take the most convenient constant of integration, namely, zero. Clearly $T = \tfrac{1}{2}m\dot{x}^2$ and integral (4.18) is

(4.20) $$J(x) = \int \tfrac{1}{2}(m\dot{x}^2 - kx^2) \, dt,$$

for which the Euler equation is the one stated above.

This concludes the use of Hamilton's Principle in this example, but we now investigate when a smooth function x_0: $[t_0, t_1] \to R$ satisfying condition I for (4.20) will minimize $J(x)$ on the class of PWS functions each of which is coterminal with x_0. To simplify the details, take $m = k = 1$. This is essentially the integral of Example 4.2. The solution $x_0(t) = 0$ on the interval $[0,2]$ that minimizes $J(x)$ according to Example 4.2 has a trivial interpretation. The particle remains at rest. Nevertheless, this solution $x_0(t)$ on various intervals $[0,t_1]$ will serve our purpose.

The figuratives $u = \tfrac{1}{2}(r^2 - x^2)$ constitute a family with parameter x. They are convex parabolas and $f_{rr}(t,x,r) = 1$, a constant; hence condition III_R holds and this implies III' and II'_N. The solution $\Delta(t,0)$ of Jacobi's equation, vanishing at $t = 0$, is

$$\Delta(t,0) = -\sin t;$$

hence if $t_1 < \pi$, x_0 satisfies I, II'_N, III', and IV' and, by Theorem 3.6, $J(x_0)$ is a proper strong local minimum. It can be shown that a field in the large exists and that actually $J(x_0)$ is a global minimum. If, however, $t_1 > \pi$, then IV does not hold. This condition is necessary as a consequence of III' and of Theorem 2.10; therefore, $J(x_0)$ is not a minimum of any type when $t_1 > \pi$.

However, $x_0(t) \equiv 0$ describes the only state of the idealized physical system that can occur with boundary values $x_0(0) = 0$, $x_0(t_1) = 0$ if t_1 is not of the form $n\pi$. Pars discusses the exceptional case (XXXII, pp. 130–133) pointing out, for $n = 1$, that all functions x, $x(t) = b \sin t$ yield the minimum of the integral.

4.8 SIDE-CONDITIONS AND NEW COORDINATES

Frequently the domain of J is the subclass of smooth $3n$-tuples consisting of those that satisfy given relations

$$(4.21) \qquad \phi_\beta(x_1, y_1, z_1, \ldots, x_n, y_n, z_n) = 0, \qquad \beta = 1, \ldots, m < 3n,$$

called *side-conditions* or *constraints* in addition to possible endpoint conditions. If m in (4.21) were more than $3n$ and the constraints were independent, as we have tacitly supposed, then the $3n$ real numbers $x_1, \ldots, z_n$ would be overdetermined; that is, the constraints would be inconsistent. If m were $3n$ and the constraints were consistent, then the positions of the particles would be completely determined by the constraints, and such positions if unique would correspond to both a degenerate global minimum and global maximum of the Hamiltonian integral.

If $m < 3n$ and system (4.21) can be solved for any m of the $3n$ arguments in terms of the others, we can eliminate these from the Hamiltonian integral. It is frequently best tactics to replace the original rectangular coordinates of the n particles by new variables and to express T and V in terms of these variables.

EXAMPLE 4.5

Apply Hamilton's Principle to the ideal double pendulum of Example 1.4 in Section 1.14.

Discussion

It is clear, with reference to Fig. 1.1, that

$$(4.22) \qquad T = \tfrac{1}{2}m_1(\dot{x}_1^2 + \dot{y}_1^2) + \tfrac{1}{2}m_2(\dot{x}_2^2 + \dot{y}_2^2),$$

that

$$(4.23) \qquad V = m_1 g(r_1 - y_1) + m_2 g[(r_1 - y_1) + (r_2 - y_2)],$$

and that there are two side-conditions

(4.24) $x_1^2 + y_1^2 = r_1^2, \qquad (x_2 - x_1)^2 + (y_2 - y_1)^2 = r_2^2.$

Since there are four coordinates x_1, y_1, x_2, y_2 and two conditions (4.24), the mechanism is said to have $4 - 2 = 2$ *degrees of freedom.*

Figure 1.1 suggests the possibility of using θ_1, θ_2 as new variables and we see from the figure that

(4.25)
$$x_1 = r_1 \sin \theta_1, \qquad y_1 = r_1 \cos \theta_1,$$
$$x_2 = r_1 \sin \theta_1 + r_2 \sin \theta_2, \qquad y_2 = r_1 \cos \theta_1 + r_2 \cos \theta_2.$$

After substituting these expressions into (4.22) and (4.23), we find that

$$T = \tfrac{1}{2} m_1 r_1^2 \dot\theta_1^2 + \tfrac{1}{2} m_2 [r_1^2 \dot\theta_1^2 + r_2^2 \dot\theta_2^2 + 2 r_1 r_2 \dot\theta_1 \dot\theta_2 \cos(\theta_2 - \theta_1)]$$
and
$$V = m_1 g r_1 (1 - \cos \theta_1) + m_2 g [r_1 (1 - \cos \theta_1) + r_2 (1 - \cos \theta_2)].$$

Observe that the side-conditions (4.24) are incorporated in these expressions for T and V and that we need make no further explicit use of the side-conditions.

With $f = T - V$, the condition $\delta J = 0$ is equivalent to the pair of Euler equations

$$f_{\theta_i} = \frac{d}{dt} f_{\dot\theta_i}, \qquad i = 1, 2.$$

Exercise 4.3

1. Verify that the differential equations for the double pendulum are

$$(m_1 + m_2) r_1 \ddot\theta_1 + m_2 r_2 \ddot\theta_2 \cos(\theta_2 - \theta_1)$$
$$- m_2 r_2 \dot\theta_2^2 \sin(\theta_2 - \theta_1) + (m_1 + m_2) g \sin \theta_1 = 0,$$
$$r_2 \ddot\theta_2 + r_1 \ddot\theta_1 \cos(\theta_2 - \theta_1) + r_1 \dot\theta_1^2 \sin(\theta_2 - \theta_1) + g \sin \theta_2 = 0.$$

2. Investigate the ideal simple pendulum along the lines of Example 4.5, finding a differential equation for θ, the signed angular displacement of the cord from the vertical, by Hamilton's Principle.

3. Apply Hamilton's Principle to the problem of two bodies of masses m_1, m_2 that are subject to a mutual attraction of magnitude $km_1 m_2 / r^2$ but to no other forces.

We turn now to the relationship between Hamilton's Principle (4.18) and Newton's second law of motion $F = Ma$. Given particles with fixed masses m_i and positions (x_i, y_i, z_i) at time t, $i = 1, \ldots, n$, let X_i, Y_i, Z_i denote the components of force at time t acting upon the particle of mass m_i in the directions of the coordinate axes. Then, by Newton's law,

(4.26) $\qquad m_i \ddot{x}_i = X_i, \qquad m_i \ddot{y}_i = Y_i, \qquad and \qquad m_i \ddot{z}_i = Z_i.$

Suppose that all coordinates (x_i, y_i, z_i), $i = 1, \ldots, n$, are expressible in terms of *generalized coordinates* $q_1, \ldots, q_k$ by equations of the form

$$x_i = x_i(q_1, \ldots, q_k, t),$$
(4.27) $$y_i = y_i(q_1, \ldots, q_k, t),$$
$$z_i = z_i(q_1, \ldots, q_k, t).$$

We suppose further that all first- and second-order partial derivatives of x_i, y_i, and z_i that may be convenient to this discussion exist and are finite and that the new coordinates are independent of each other, so that no smaller number than k new coordinates will serve. The dynamical system is then said to have k *degrees of freedom*. In Example 4.5 there are two degrees of freedom represented by the generalized coordinates θ_1 and θ_2, and equations (4.25) represent a special instance of (4.27) with right members that are free of t and with no equations for z_i stated since it is understood that $z_1 = z_2 = 0$.

It follows from (4.27) that

(4.28) $\qquad \dot{x}_i = (\partial x_i / \partial q_r) \dot{q}_r + \partial x_i / \partial t \qquad$ *with summation on r.*

Since the right members of equations (4.27) do not involve derivatives $\dot{q}_r$, we see from (4.28) that

(4.29) $\qquad \partial \dot{x}_i / \partial \dot{q}_r = \partial x_i / \partial q_r, \qquad i = 1, \ldots, n \qquad and \qquad r = 1, \ldots, k.$

We also have equations like (4.28) and (4.29) with y and z in place of x.

From expression (4.16) for the kinetic energy T we find by differentiation that

$$\partial T / \partial \dot{q}_r = m_i (\dot{x}_i \partial \dot{x}_i / \partial \dot{q}_r + \dot{y}_i \partial \dot{y}_i / \partial \dot{q}_r + \dot{z}_i \partial \dot{z}_i / \partial \dot{q}_r),$$
$$summed\ on\ i\ with\ r = 1, \ldots, k.$$

It follows with the aid of (4.29) and the companion equations for y and z that

$$\partial T / \partial \dot{q}_r = m_i (\dot{x}_i \partial x_i / \partial q_r + \dot{y}_i \partial y_i / \partial q_r + \dot{z}_i \partial z_i / \partial q_r).$$

Taking the derivative of each side with respect to t, we have that

(4.30) $\quad d(\partial T / \partial \dot{q}_r)/dt = m_i (\ddot{x}_i \partial x_i / \partial q_r + \ddot{y}_i \partial y_i / \partial q_r + \ddot{z}_i \partial z_i / \partial q_r)$
$\qquad\qquad\qquad + m_i [\dot{x}_i d(\partial x_i / \partial q_r)/dt + \dot{y}_i d(\partial y_i / \partial q_r)/dt + \dot{z}_i d(\partial z_i / \partial q_r)/dt].$

We find by differentiation of expression (4.16) for T with respect to q_r that

$$(4.31) \qquad \partial T/\partial q_r = m_i\,(\dot{x}_i\,\partial \dot{x}_i/\partial q_r + \dot{y}_i\,\partial \dot{y}_i/\partial q_r + \dot{z}_i\,\partial \dot{z}_i/\partial q_r).$$

It is a further consequence from (4.27) that

$$\partial \dot{x}_i/\partial q_r = d\,(\partial x_i/\partial q_r)/dt$$

with similar equations in y and z; hence from (4.31),

$$(4.32) \qquad \partial T/\partial q_r = m_i\left[\dot{x}_i\frac{d}{dt}(\partial x_i/\partial q_r) + \dot{y}_i\frac{d}{dt}(\partial y_i/\partial q_r) + \dot{z}_i\frac{d}{dt}(\partial z_i/\partial q_r)\right].$$

Suppose finally that there exists a potential $U = U(t,x_1,y_1,z_1,\ldots,x_n,y_n,z_n)$, hence that the right members of equations (4.26) are the partial derivatives of U with respect to x_i, y_i, and z_i. We can express U in terms of $q_1,\ldots,q_k$ by means of (4.27) and

$$(4.33) \qquad \begin{aligned} \partial U/\partial q_r &= (\partial U/\partial x_i)\,(\partial x_i/\partial q_r) + (\partial U/\partial y_i)\,(\partial y_i/\partial q_r) + (\partial U/\partial z_i)\,(\partial z_i/\partial q_r) \\ &= X_i\partial x_i/\partial q_r + Y_i\partial y_i/\partial q_r + Z_i\partial q_r, \end{aligned}$$

while

$$(4.34) \qquad \partial U/\partial \dot{q}_r = 0$$

since the expression for U described above is free of $\dot{q}_r$.

It follows from (4.30) through (4.34) that

$$(4.35) \qquad \frac{d}{dt}(\partial T/\partial \dot{q}_r) - \partial T/\partial q_r = \partial U/\partial q_r = -\frac{d}{dt}(\partial U/\partial \dot{q}_r) + \partial U/\partial q_r,$$
$$r = 1,\ldots,k.$$

These are the Euler equations for the integral (4.17) in terms of the independent coordinates q_r. With $-V$ in place of U, we obtain the Euler equations for the integral (4.18).

The case $k = 3n$ in which

$$x_i = q_{3i-2}, \quad y_i = q_{3i-1}, \quad z_i = q_{3i}, \qquad i = 1,\ldots,n,$$

is included under (4.27) and the preceding discussion. One verifies for this case with the aid of the third member of (4.33) and the fact that

$$T = \tfrac{1}{2}m_i\,(\dot{q}_{3i-2}^2 + \dot{q}_{3i-1}^2 + \dot{q}_{3i}^2)$$

that equations (4.35) reduce to the equations (4.26) expressing Newton's second law. The examples in Section 4.7 involve no constraints and hence fall under this case.

When constraints are present, as in Example 4.5, one hopes to use them so as to express integral (4.17) or (4.18) in terms of a set of independent coordinates before introducing the variation of the integral. This is an essential feature of Euler equations (4.35). Unless the constraints are of simple form, it will not be possible to obtain a system of equations (4.27) with right members that are combinations of elementary functions.

4.9   THE GENERALIZED HAMILTON PRINCIPLE

The extension of (4.10) to the case in which the y of that relation has many components can be applied to a potential V. We now have $3n$ components $x_1, y_1, z_1, \ldots, x_n, y_n, z_n$ and, by the extension of (4.10),

$$(4.36) \qquad \delta V = (\partial V/\partial x_i)\, \delta x_i + (\partial V/\partial y_i)\, \delta y_i + (\partial V/\partial z_i)\, \delta z_i,$$
$$\text{with summation on } i \text{ from 1 to } n.$$

A potential V has the dimensionality of work, hence so also does δV, which can be thought of as work against a force-field that results from giving the respective n particles displacements $\delta x_i, \delta y_i, \delta z_i$ in the respective coordinate directions. However, such displacements are not actually executed by the moving particles. Given their positions $[x_i(t), y_i(t), z_i(t)]$, $i = 1, \ldots, n$ at time t on a possible set of paths, then $[x_i(t) + (\delta x_i)(t), y_i(t) + (\delta y_i)(t), z_i(t) + (\delta z_i)(t)]$ represents positions for this same time t on another set of paths for the n particles. Since the particles will not actually move in the directions corresponding to $\delta x_i, \delta y_i, \delta z_i$, (4.36) is called *virtual work*.

Condition (4.18) and relation (4.36) are restricted to conservative systems, for which the forces are the negative partial derivatives of V. There must be no frictional or other dissipative forces. For a system that includes such forces, let X_i, Y_i, Z_i denote components of the resultant force affecting the particle of mass m_i and set

$$(4.37) \qquad W_\delta \equiv X_i\, \delta x_i + Y_i\, \delta y_i + Z_i\, \delta z_i.$$

This reduces to (4.36) iff the only forces are conservative. We have used the symbol W_δ rather than δW to avoid the question whether the right member is or is not the variation (4.10) of some function.

It can be shown that the particles will move in such a way that

$$(4.38) \qquad \int (\delta T - W_\delta)\, dt = 0.$$

This is *Hamilton's Principle* for a general, not necessarily conservative system. This condition does not say that a functional J must have a minimum or even, as in the special case (4.18), that a functional must be stationary.

EXAMPLE 4.6

A particle moves on an upward directed y axis subject to its weight and to a frictional force of respective magnitudes mg and $k|\dot{y}|$. Apply (4.38) to obtain the equations of motion.

Discussion

Clearly $T = \frac{1}{2}m\dot{y}^2$. The resultant force acting on the particle is $mg + k\dot{y}$; hence

$$W_\delta = (mg + k\dot{y})\,\delta y,$$

and (4.38) is the equation

(4.39) $$\int [m\dot{y}\,\delta\dot{y} - (mg + k\dot{y})\,\delta y] = 0.$$

Since we did not start with an integral $J(y)$, we do not have an integrand f for which to write an Euler equation but must integrate the term in $\delta\dot{y}$ by parts. After doing so and using end-conditions $(\delta y)(t_0) = (\delta y)(t_1) = 0$, as was done in Section 4.3, relation (4.39) reduces to

$$\int -(m\ddot{y} + mg + k\dot{y})\,\delta y = 0,$$

and, as a consequence of the arbitrary nature of δy, we conclude as in the derivation of (2.24) by the procedure of Section 4.3 that

$$m\ddot{y} + k\dot{y} + mg = 0.$$

Given a second-order ordinary differential equation, there is (XI, pp. 37–39) a large class of integrands f for which the given equation is the Euler equation (2.25). One such in the present instance is

$$f(t,y,r) = \exp[-k/m^2 g(mr + ky)].$$

This has no immediate physical interpretation comparable to $T - V$ in (4.18). Conceivably, the general Hamiltonian integral (4.38) could always be replaced by a condition $\delta \int f\,dt = 0$, in form (4.18), for which f has a useful physical interpretation, but whether such is the case is not known to the author.

4.10 APPLICATIONS TO ELECTRIC NETWORKS

An idealized lumped parameter electric network consists of inter-connected *resistors, inductors,* and *capacitors* with respective associated positive constants denoted by R, L, and C, together with energy sources. We consider the case of a voltage source characterized by a real-valued function E defined for nonnegative time t.

Given an electric network and a time t, each capacitor will carry a charge $q(t)$ and each branch will carry a current $i(t)$. The quantity

$$(4.40) \qquad V = \tfrac{1}{2} S_j q_j^2, \qquad summed\ on\ j,$$

where elastance S_j is the reciprocal of the capacitance C_j of the jth capacitor, is analogous to the potential of a mechanical system. Whether a branch contains a capacitor or not, we can always replace the current i by the symbol $\dot{q}$ and the quantity

$$(4.41) \qquad T = \tfrac{1}{2} L_k \dot{q}_k^2, \qquad summed\ on\ k,$$

is analogous to the total kinetic energy of all masses in a mechanism.

Given a conservative network and hence one which consists only of L's and C's with no resistors or voltage sources, Hamilton's Principle applies in form (4.18) with the V and T given above. If resistors or voltage sources are present, we need the general form (4.38). A resistor R and a voltage E contribute respective terms

$$R\dot{q}_k \delta q_k \qquad and \qquad -E_k \delta q_k$$

to the expression for W_δ.

EXAMPLE 4.7

Given a closed loop consisting of L, R, C, and E connected consecutively in series, we have one term (4.41) *and*

$$W_\delta = \delta(\tfrac{1}{2}Sq^2) + (R\dot{q}-E)\,\delta q;$$

therefore (4.38) *has the form*

$$\int \{\delta(\tfrac{1}{2}L\dot{q}^2) - \delta(\tfrac{1}{2}Sq^2) - (R\dot{q}-E)\,\delta q\}\,dt.$$

After carrying through the details, we find the familiar differential equation

$$(4.42) \qquad L\ddot{q} + R\dot{q} + Sq = E.$$

Exercise 4.4

1. Verify (4.42) as a consequence of the vanishing of the integral (4.38).

2. A single loop consists of L and C in series. Use Hamilton's Principle to obtain the equation $LC\ddot{q} + q = 0$.

3. Given the two-loop circuit in Fig. 4.1 with $\dot{q}_1$, $\dot{q}_2$ now being so-called circulating currents so that the downward current in the resistor is $\dot{q}_1 - \dot{q}_2$, verify that (4.38) is of the form

$$\int \{ \tfrac{1}{2}\delta(L_1\dot{q}_1^2 + L_2\dot{q}_2^2) - \tfrac{1}{2}\delta(S_1q_1^2 + S_2q_2^2)$$
$$+ R(\dot{q}_1 - \dot{q}_2)\,\delta(q_1 - q_2) - E\delta q_1\} \, dt = 0;$$

then find two differential equations.

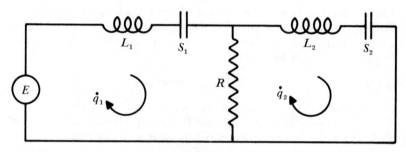

FIGURE 4.1

4.11 CONCLUDING REMARKS

With the exception of Example 4.5 we have followed the common textbook custom of giving simple examples for which the differential equations of the system are more easily obtained in other ways. For such examples from mechanics one can identify the forces and use Newton's second law; for the electrical examples, Kirchhoff's laws suffice.

When we turn to more complex examples, Hamilton's Principle can offer tactical advantages. To obtain the differential equations for problem 1, Exercise 4.3, directly from Fig. 1.1 and Newton's law $F = Ma$, one must include the centrifugal force acting at an arbitrary time t on the lower particle, a step that is somewhat tricky.

One must not, however, expect Hamilton's Principle to be a panacea for attacking particular examples. Its major contribution is to the conceptual and theoretical framework of certain parts of mathematical physics.

The principle can be extended to systems for which the domain of J is a class of functions of more than one variable. It then leads us to Euler equations for multiple integrals, to be mentioned in Chapter 12.

In dealing with the generalized Hamilton Principle, the operator δ is a convenient tool. Although it has been used throughout the chapter to familiarize the reader with it, it could just as well be eliminated from all but the last two sections.

Chapter 5

THE
NONPARAMETRIC
PROBLEM OF BOLZA

5.1 INTRODUCTION

We shall now consider an extension of the ideas of Chapters 2 and 3 to a type of problem published by Bolza (7a) in 1913. The domain $\mathcal{Y}$ of the functional J consists of all continuous PWS vector-valued functions $y = (y^1, \ldots, y^n)$ each of which is from some interval $[t_0, t_1]$ to R^n and each of which satisfies given constraints (side-conditions) $\phi(t, y, \dot{y}) = 0$ together with given end-conditions (initial and terminal conditions). The end-conditions may or may not fix the interval $[t_0, t_1]$, the initial value $y(t_0)$, or the terminal value $y(t_1)$.

Any dynamical system for which the mathematical model is a system of ordinary differential equations is a possible source of Bolza Problems of the type to be discussed. These dynamical equations are supposed to be given together with a suitable function J that is a measure of performance of the system. One then desires a minimum or maximum value of J.

5.2   EXAMPLES

EXAMPLE 5.1

Given an ideal pure inductor with coefficient of inductance L in series with a capacitor of capacitance C and a voltage V, we have in accord with standard electric circuit analysis the system of equations

$$(5.1) \qquad L\dot{I} + Q/C = V \quad \textit{and} \quad \dot{Q} = I,$$

in which Q denotes the charge on the capacitor and I the current. These are the side-conditions or constraints. As end-conditions take

$$(5.2) \qquad t_0 = 0, \quad t_1 = \pi, \quad Q(t_0) = 0, \quad I(t_0) = 0, \quad Q(t_1) = a^2 > 0.$$

Does there exist among all triples (Q, I, V) of continuous PWS functions satisfying conditions (5.1) and (5.2) a triple (Q_0, I_0, V_0) such that the terminal energy $\frac{1}{2}LI_0^2(t_1)$ in the magnetic field of the inductor is a minimum? If such a minimizing triple does exist, what are the functions Q_0, I_0, V_0, and are they unique?

The state of this system at time t is given by the pair of values $Q(t), I(t)$, while we think of $V(t)$ as controllable. The *criterion of optimality* is the function J with values

$$J(Q, I, V) \equiv \frac{1}{2}LI_0^2(t_1).$$

As stated above, the objective is a minimum value for J. In another problem we might desire a maximum of a function.

This problem is possibly of no practical importance but will be used for further illustration under Example 5.8 in Section 5.9.

EXAMPLE 5.2

See Example 1.5, Section 1.14. This example is essentially that of reference (29a). We are interested in the existence and characterization of a triple (x_0, y_0, p_0) among all continuous PWS triples (x, y, p) satisfying the constraints (1.39) and end-conditions (1.40) and such that the time T required for the state $[x(t), y(t)]$ to go from (a, b) to (h, k) have the least possible value. We thus desire the global minimum of

$$(5.3) \qquad J_1(x, y, p) \equiv T,$$

or, alternatively stated after integrating the first equation (1.39), the global maximum of

$$(5.4) \qquad J_2(x, y, p) = \int_{t_0}^{t_1} \left(\frac{\dot{x}}{Ax} \right) p^{-m} \, dt.$$

The problem will be examined further as Example 5.9 in Section 5.9.

5.3   FORMULATION OF THE PROBLEM OF BOLZA

In stating or treating an example generated by a potential application, it is helpful to use established symbols from the field in which the example arises as in Example 5.1. In discussing the theory we use a neutral vector notation.

Let $\mathscr{Y}$ denote the class of all PWS functions y: $[t_0, t_1] \rightarrow R^n$ satisfying side-conditions

$$(5.5) \qquad \phi_\beta(t, y, \dot{y}) = 0, \qquad \beta = 1, \ldots, m, \qquad 0 \leqslant m < n,$$

and end-conditions

$$(5.6) \quad \psi_\mu[t_0, y(t_0), t_1, y(t_1)] = 0, \qquad \mu = 1, \ldots, p, 2 \leqslant p \leqslant 2n + 2.$$

Functions $y \in \mathscr{Y}$ are called *admissible*. They and only they are admitted to competition. Given a functional J,

$$(5.7) \qquad J(y) \equiv g[t_0, y(t_0), t_1, y(t_1)] + \int_{t_0}^{t_1} f(t, y, \dot{y}) \, dt,$$

the Problem of Bolza is by definition that of the existence and characterization of admissible functions y_0 such that $J(y_0) \leqslant J(y)$ or $\geqslant J(y)$ for all $y \in \mathscr{Y}$ or such that $J(y_0)$ is a strong or weak local extremum, terms that will be defined presently in such a way as to include as special cases the meanings given in Section 2.4.

Having exhibited an outline of the problem, we now fill in some essential details. That y is PWS means that every component y^i of y is PWS under the definition of Section 1.9; hence each y^i is continuous, and again we understand continuity to be included under the descriptive term PWS. If at least one component of y has a corner in the sense of Section 1.9, then y is said to have a corner. That a PWS function y satisfies a differential equation (5.5) on an interval means that $\phi_\beta[t, y(t), \dot{y}(t)] = 0$ for all t not corresponding to corners and that, if t corresponds to a corner, the equation holds with $\dot{y}(t)$ interpreted as either $\dot{y}^-(t)$ or $\dot{y}^+(t)$.

Our formulation is essentially but not precisely that of Bliss (IX, pp. 189, 193–194). The domain of f and of each function ϕ_β is understood to be a cartesian product $(a, b) \times R^n \times R^n$, where the interval (a, b) contains all intervals $[t_0, t_1]$ consistent with end-conditions (5.6). The domain of g and of each ψ_μ is an open subset of R^{2n+2} large enough to contain all $(2n+2)$-tuples $[t_0, y(t_0), t_1, y(t_1)]$ that are attainable. That such a set of end values is *attainable* means that there exists a PWS function y satisfying system (5.5) in the sense described above and that the domain $[t_0, t_1]$ of y and the end values $y(t_0), y(t_1)$ satisfy system (5.6). Consequently, if each equation (5.5) is a differential equation as the notation $\phi_\beta(t, y, \dot{y}) = 0$ suggests, if all partial derivatives of f, ϕ_β, g, and ψ_μ that we may wish to introduce exist and are continuous and if each of the systems (5.5) and (5.6) consists of independent equations, then all major results in Bliss (IX, part II) apply to the present problem. We call attention in Exercise 5.3, problem 7, to the case in which constraints (5.5) are free of $\dot{y}$.

We adopt a *blanket hypothesis* to the effect that all partial derivatives of f, ϕ_β, g, and ψ_μ that may be needed at a given stage of the discussion exist and are continuous. We must be prepared to distinguish between subscripts β and μ that are indices and additional subscripts that denote partial differentiation. Notation becomes heavy in places, and although such symbols as y_0 and y_1 are generally reserved for particular admissible functions, they will be used occasionally as abbreviations for $y(t_0)$ and $y(t_1)$ in such expressions as the left member of (5.6) or in the g term of (5.7).

The formulation has not required explicitly that the $m + p$ conditions (5.5) and (5.6) be consistent. If they are not, then the class $\mathscr{Y}$ is empty and there can exist no optimizing admissible function. In practice, if one starts with an idealized physical system and chooses side and end-conditions carefully, they usually turn out to be consistent and also independent. Bliss imposes conditions (IX, p. 193, Sec. 70) on the ranks of certain matrices that ensure independence of the m conditions (5.5) and of the p conditions (5.6). Consistency and independence are less likely to occur if an example is constructed by choosing functions ϕ_β and ψ_μ more or less arbitrarily.

In the event that $f(t, y, r)$ vanishes identically, the Bolza Problem is called a *Problem of Mayer* after Adolph Mayer, who published on such problems in 1878. If $g(t_0, y_0, t_1, y_1)$ is identically zero, the problem becomes a *Problem of Lagrange*. Particular examples of this type were studied by Euler, but Lagrange is given credit because of his systematic investigation and use of what are usually called Lagrange multipliers, even though his proofs were faulty. Only as recently as the 1940s did the theory reach a relatively complete state. Among major contributors have been O. Bolza, G. A. Bliss, C. Carathéodory, L. M. Graves, M. R. Hestenes, E. J. McShane, Marston Morse, and W. T. Reid. For references see Bliss (5a, pp. 743–744), (IX, pp. 287–291) and the bibliography at the end of this book.

5.4   ALTERNATIVE FORMS OF A PROBLEM

Although the Problems of Mayer and Lagrange are special cases of that of Bolza, the latter is equivalent to either a Mayer or a Lagrange Problem if we are willing to add a component to y.

Given the problem of Section 5.3, introduce the additional side-condition $\dot{y}^{n+1} = 0$ and the additional end-condition

$$y^{n+1}(t_0) = g(t_0,y_0,t_1,y_1)/(t_1 - t_0).$$

The original Bolza Problem is then equivalent to the Lagrange Problem

$$J_1(y) \equiv \int_{t_0}^{t_1} [f(t,y,\dot{y}) + y^{n+1}]\, dt = \textit{extremum on } \mathcal{Y}_1,$$

where $\mathcal{Y}_1$ consists of all PWS functions $(y^1, \ldots, y^n, y^{n+1})$ satisfying (5.5) and (5.6), respectively, augmented by the side- and end-conditions given above.

Again given the Bolza Problem we can adjoin the condition

$$\dot{y}^{n+1} - f(t,y,\dot{y}) = 0$$

to (5.5) and $y^{n+1}(t_0) = 0$ to (5.6). As a consequence of these conditions,

$$y^{n+1}(t) = \int_{t_0}^{t} f[\tau,y(\tau),y(\tau)]\, d\tau$$

and the original problem is equivalent to the Mayer Problem

$$J_2(y) = g(t_0,y_0,t_1,y_1) + y_1^{n+1} = \textit{extremum on } \mathcal{Y}_2,$$

where $\mathcal{Y}_2$ consists of all PWS vector-valued functions with $n+1$ components satisfying the augmented systems (5.5) and (5.6).

Sometimes it is possible to choose among alternative forms of a problem without increasing the values of n, m, and p as illustrated by the two functionals in Example 5.2. As another example, suppose given a problem with the side-conditions

(5.8) $m\dot{v} + c\dot{m} + kv|v| + mg = 0 \quad\textit{and}\quad \dot{y} - v = 0,$
          with c, k, and g being positive constants.

If for the moment we regard v and y as known, then (5.8) is a differential equation of the first order in m and, after division by m and integration,

(5.9) $c \ln \dfrac{m(t_0)}{m(t_1)} = \displaystyle\int_{t_0}^{t_1} \left(\dfrac{kv|v|}{m} + \dot{v} + g \right) dt$

$= \displaystyle\int_{t_0}^{t_1} \dfrac{kv|v|}{m} dt + v(t_1) - v(t_0) + (t_1 - t_0)g.$

To minimize the ratio $m(t_0)/m(t_1)$ subject to side-conditions (5.8) and a given set of end-conditions is then equivalent to minimizing any of the three expressions (5.9). The three minimum problems are, respectively, examples of the Problems of Mayer, Lagrange, and Bolza.

Each of the three is a mathematical model for the problem of programming the vertical motion near the earth of a rocket propelled vehicle so as to minimize the ratio of initial to terminal mass. Under the classical formulation of Section 5.3 with PWS vector-valued functions (y,v,m), each of these equivalent minimum problems will, like Example 1.6, fail to have a minimizing triple (y_0,v_0,m_0) except when initial and terminal values of v are chosen in a special manner.

5.5 CONSTRAINED EXTREMA OF POINT-FUNCTIONS

Given a finite set of functions $\phi_\alpha: R^n \to R$, $\alpha = 0, 1, \ldots, m < n$, each having finite first-order partial derivatives, let S denote the subset of R^n consisting of all points $x = (x^1, \ldots, x^n)$ satisfying the system of equations

(5.10) $\phi_\alpha(x) = 0, \qquad \alpha = 1, \ldots, m < n.$

Consider the problem $\phi_0(x) = extremum$ on S. The case $m \geq n$ in which S is generally empty or at most a finite set need not concern us.

Theorem 5.1

Given the functions $\phi_\alpha: R^n \to R$, $\alpha = 0,1, \ldots, m$ described above and given $x_0 \in R^n$ satisfying the system of equations (5.10), then a necessary condition for $\phi_0(x_0)$ to be a relative extremum on S is that there exist constants $\lambda_0, \lambda_1, \ldots, \lambda_m$ not all zero such that the partial derivatives $\partial F/\partial x^i$, $i = 1, \ldots, n$ of $F \equiv \lambda_0\phi_0 + \cdots + \lambda_m\phi_m$ all vanish at x_0.

PROOF

(Bliss). The equation

(5.11) $\phi_0(x) = \phi_0(x_0) + u$

clearly holds for $(x,u) = (x_0,0)$. Consider the following matrix, in which

$\phi_{\alpha i}$ denotes $\partial \phi_\alpha / \partial x^i$, $\alpha = 0,1,\ldots,m$ and $i = 1,\ldots,n > m$:

(5.12)

$$\begin{bmatrix} \phi_{01}(x_0) & \phi_{02}(x_0) & \cdots & \phi_{0n}(x_0) \\ \phi_{11}(x_0) & \phi_{12}(x_0) & \cdots & \phi_{1n}(x_0) \\ \cdot & & & \\ \cdot & & & \\ \cdot & & & \\ \phi_{m1}(x_0) & \phi_{m2}(x_0) & \cdots & \phi_{mn}(x_0) \end{bmatrix}$$

If the determinant of the first $m+1$ columns is not zero, then, by a standard existence theorem for implicit functions (the extension of Theorem 1.1 to $m+1$ variables), the system of $m+1$ equations (5.11) and (5.10) determines a unique $(m+1)$-tuple $(x^1,\ldots,x^{m+1})$ corresponding to each point $(x^{m+2},\ldots,x^n,u)$ near $(x_0^{m+2},\ldots,x_0^n,0)$, for some of which $u > 0$ and for others of which $u < 0$; consequently $\phi_0(x_0)$ can be neither a minimum nor a maximum. This same conclusion is similarly reached if any other combination of $m+1$ columns of (5.12) has a nonvanishing determinant. We therefore infer that, in order for $\phi_0(x_0)$ to be an extremum, it is necessary that the rank of matrix (5.12) be below $m+1$. This is known to be a necessary and sufficient condition for the n-vectors whose components are rows of the matrix to be linearly dependent, which means by definition that there exist constants $\lambda_0, \lambda_1,\ldots,\lambda_m$, not all zero, such that

(5.13) $\lambda_0 \phi_{0i}(x_0) + \lambda_1 \phi_{1i}(x_0) + \cdots + \lambda_m \phi_{mi}(x_0) = 0, \quad i = 1,\ldots,n.$

This is the stated conclusion.

Given $m+1$ functions, then, by choice of labels, any one of them can be the ϕ_0 of the preceding theorem; hence the conclusion is independent of which function ϕ_α is to be maximized or minimized so long as the others equated to zero provide the constraints or side-conditions. This observation is the Reciprocity Theorem for constrained extrema.

If (5.13) holds for a given set of Lagrange multipliers, it also holds for any set obtained by multiplying all those of the first set by the same constant. Of course (5.13) holds trivially if $\lambda_\alpha = 0$, $\alpha = 0,\ldots,m$. If there exists no nontrivial set of multipliers satisfying (5.13) with $\lambda_0 = 0$, then the point $x_0 \in R^n$ is called *normal*. If there is a set of multipliers satisfying (5.13) with $\lambda_0 = 0$ and some $\lambda_\alpha \neq 0$, then x_0 is *abnormal*. In the normal case we can always replace a nontrivial set of multipliers by $1, \lambda_1/\lambda_0,\ldots,\lambda_m/\lambda_0$ or realize the same end by simply setting $\lambda_0 = 1$.

The normal case, as its name suggests, is the one most frequently encountered. It takes a little care to construct an example that involves an abnormal point.

EXAMPLE 5.3.

$x - y = extremum$ *subject to the constraint* $x^2 + y^2 = 0$. Set

$$F(x,y,\lambda_0,\lambda_1) \equiv \lambda_0(x-y) + \lambda_1(x^2 + y^2).$$

Then, by (5.13),

(5.14)
$$\lambda_0 + 2\lambda_1 x = 0,$$
$$-\lambda_0 + 2\lambda_1 y = 0,$$

whence by addition, $2\lambda_1(x+y) = 0$. Now we can choose $\lambda_1 \neq 0$, since otherwise (5.14) requires that $\lambda_0 = \lambda_1 = 0$, contrary to Theorem 5.1. It follows that $x + y = 0$, hence from the constraint and (5.14) that $(x_0,y_0) = (0,0)$ is an abnormal point and that $\lambda_0 = 0$. Observe that the set S determined by the side-condition consists of the single point $(0,0)$.

Lagrange multipliers are treated inadequately in a number of textbooks. Frequently no proof of the existence of multipliers is attempted. The function F with λ_0 arbitrarily chosen as unity is simply presented along with one or two examples that fall under the normal case. Sometimes there is a proof with the hypotheses loaded so as to exclude the abnormal case.

We shall not discuss sufficient conditions for constrained extrema of point-functions.

Exercise 5.1

Find *critical points*, that is, points satisfying necessary condition (5.13) and the given constraint.

1. $x^2 + y^2 = extremum$, $4x^2 + 3y^2 = 12$.
2. $x + 2y - 3z^2 = extremum$ with the constraints $x - y = 2$ and $x + 2y = 4$.
3. $x + 2y = extremum$ with constraining inequalities $x + y - 1 > 0$ and $y - x - 1 < 0$. To bring this under Theorem 5.1, set $x + y - 1 = w^2$ and $y - x - 1 = -z^2$.

5.6   DIFFERENT KINDS OF EXTREMA

We now extend the concepts of Section 2.4 to the Bolza Problem (5.7). That $y_0 \in \mathcal{Y}$ furnishes a global minimum means, as in Chapter 2, that

(5.15)
$$J(y_0) \leq J(y), \qquad \forall \, y \in \mathcal{Y}.$$

We also wish to define local minima.

Let x: $[t_0,t_1] \to R^n$ and y: $[u_0,u_1] \to R^n$ be two vector-valued functions in the class $\mathcal{Y}$ of all such functions that are PWS and satisfy side-conditions (5.5) and end-conditions (5.6). The domains of x and y are in general distinct intervals. Although there conceivably may be occasions in which it becomes desirable to include among admissible functions degenerate cases for which the interval consists of a single point, we confine attention here to those whose domains are intervals of positive length. A suitable modification of what follows would serve to include the degenerate functions.

Let h: $[t_0,t_1] \to [u_0,u_1]$ be the linear function with values

$$(5.16) \qquad h(t) = u_0 + \frac{u_1 - u_0}{t_1 - t_0}(t - t_0).$$

Define a *distance of order zero* $d_0(x,y)$ between x and y by the statement that

$$(5.17) \qquad d_0(x,y) \equiv \sup\{|t - h(t)| : t \in [t_0,t_1]\}$$
$$+ \sup\{|x(t) - y[h(t)]| : t \in [t_0,t_1]\}.$$

The first expression with bars is an ordinary absolute value, the second is euclidean distance between two points in n-space. Both x and the composite function $y \circ h$ are continuous on $[t_0,t_1]$; hence the euclidean distance $|x(t) - y[h(t)]|$ is continuous in t on $[t_0,t_1]$, and there must exist $t^* \in [t_0,t_1]$ at which the supremum is realized. A similar remark applies to the first term on the right in (5.17). Thus each supremum in (5.17) is actually a maximum.

It can be verified that $d_0(x,y)$ has all of the properties required in the definition (Section 1.10) of a metric space. In the special case $[u_0,u_1] = [t_0,t_1]$, the function h reduces to the identity mapping $h(t) = t$ and (5.17) becomes the extension to n-space of the zero-order distance of Section 2.4.

Also define a *first-order distance*

$$(5.18) \qquad d_1(x,y) \equiv d_0(x,y) + \sup\{|\dot{x}(t) - \dot{y}[h(t)]| : t \in [t_0,t_1]^*\},$$

where $[t_0,t_1]^*$ denotes the interval $[t_0,t_1]$ less the finite set of points corresponding to possible corners of x or y, at which values of t either $\dot{x}(t)$, $\dot{y}[h(t)]$ or both fail to exist.

We shall understand neighborhoods of orders zero and one to be defined by the same statements used in Section 2.4 but now for distances (5.17) and (5.18).

Our choice of the linear mapping h in definitions (5.17) and (5.18) is arbitrary. It happens to be the simplest strictly increasing function from an interval of positive length to an interval of positive length and it suffices for our purposes.

Any other distance $d_0^*(x,y)$ so related to $d_0(x,y)$ that every neighborhood $U_0^*(\delta_1,x)$ contains a neighborhood $U_0(\delta_2,x)$ and vice versa would serve as well. A similar remark serves to characterize the class of acceptable first-order distances.

If there is a function $y_0 \in \mathcal{Y}$ and a positive real number δ such that

$$(5.19) \qquad J(y_0) \le J(y), \qquad \forall y \in \mathcal{Y} \cap U_0(\delta,y_0),$$

then $J(y_0)$ is a *strong local minimum*. Similarly, if

$$(5.20) \qquad J(y_0) \le J(y), \qquad \forall y \in \mathcal{Y} \cap U_1(\delta,y_0),$$

then $J(y_0)$ is a *weak local minimum*. These definitions, identical in form with (2.10) and (2.11), extend the content of (2.10) and (2.11) to the present more general problem. Three types of maxima are defined by reversing the inequalities in (5.15), (5.19), and (5.20).

Exercise 5.2

1. $x(t) = t, 0 \le t \le 1$, and $y(u) = u - 1, 1 \le u \le 2$. Find $d_0(x,y)$ and $d_1(x,y)$.
2. Show that the set of all PWS functions y from a compact interval of positive length to R^n together with the distance of order zero defined by (5.17) constitutes a metric space.
3. Denote the second term on the right in (5.18) by $\alpha(x,y)$. Show that $\alpha(x,y)$ lacks exactly one of the properties (1.29) but that $d_1(x,y)$ has all properties (1.29).
4. Let $y_0: [-1,1] \to R$ be the fixed function with values $y_0(t) = t$. Describe the class of all functions $y: [u_0,u_1] \to R$ that are in the neighborhood $U_0(\delta,y_0)$.

5.7 THE MULTIPLIER RULE

We now investigate the analogue of Theorem 5.1 for the Problem of Bolza formulated in Section 5.3. We shall need constant multipliers $\lambda_0, e_1, \ldots, e_p$ together with multipliers $\lambda_1(t), \ldots, \lambda_m(t)$.

Results are stated with the aid of auxiliary functions F and G defined as follows:

$$(5.21) \qquad F(t,y,r,\lambda) \equiv \lambda_0 f(t,y,r) + \lambda_\beta(t)\phi_\beta(t,y,r),$$

$$(5.22) \qquad G(t_0,y_0,t_1,y_1) \equiv \lambda_0 g(t_0,y_0,t_1,y_1) + e_\mu\psi_\mu(t_0,y_0,t_1,y_1).$$

Repeated indices β in (5.21) and μ in (5.22) are understood to indicate summation from 1 to m and 1 to p, respectively. Although the right member of (5.22) involves λ_0, $e_1, \ldots, e_p$, it is convenient to suppress these arguments in the symbol on the left for a value of G.

Theorem 5.2, which follows, is the Euler necessary condition for our Bolza Problem. Theorem 5.3 is the *transversality condition*. The combined results of these theorems are called the *Multiplier Rule* by Bliss (IX, p. 202). At a time when only fixed-endpoint problems had been investigated, the same term referred only to Theorem 5.2.

Theorem 5.2

If x: $[t_0, t_1] \to R^n$ minimizes $J(y)$ on $\mathscr{Y}$, there exist multipliers

(5.23) $$\lambda_0, \lambda_1(t), \ldots, \lambda_m(t)$$

with the following properties:

(i) If $\lambda_0 = 0$, then $\lambda_1(t), \ldots, \lambda_m(t)$ do not vanish simultaneously anywhere on $[t_0, t_1]$.

(ii) $\lambda_1, \ldots, \lambda_m$ are continuous on $[t_0, t_1]$ except possibly at t values where $\dot{x}$ is discontinuous.

(iii) They satisfy with x and a certain constant n-vector c the vector-equation

(5.24) $$F_r[t, x(t), \dot{x}(t), \lambda(t)] = \int_{t_0}^{t} F_y[\tau, x(\tau), \dot{x}(\tau), \lambda(\tau)] \, d\tau + c.$$

Symbols F_r and F_y denote n-vectors whose components are the partial derivatives of F with respect to components of r and of y, respectively. Equation (5.24) is equivalent to a system of n scalar equations.

For any t at which $\dot{x}$ and hence also λ are continuous, we can differentiate (5.24) to obtain the vector equation

(5.25) $$F_y[t, x(t), \dot{x}(t), \lambda(t)] = \frac{d}{dt} F_r[t, x(t), \dot{x}(t), \lambda(t)].$$

In stating the next theorem we use the abbreviation $x_0^i = x^i(t_0)$, $i = 1, \ldots, n$ and similarly for x_1^i, $\dot{x}_0^i$, and $\dot{x}_1^i$. We also set

$$[F]_j \equiv F[t_j, x(t_j), \dot{x}(t_j), \lambda(t_j)], \qquad j = 0, 1$$

and use

$$[F_i]_j, \qquad i = 1, \ldots, n \text{ and } j = 0, 1$$

for the components of F_r evaluated at t_0 and t_1.

Theorem 5.3

If x: $[t_0, t_1] \to R^n$ minimizes $J(y)$ on $\mathscr{Y}$, there exist constant multipliers

(5.26) $$\lambda_0, e_1, \ldots, e_p$$

with the following properties:

(i) $\lambda_0, \lambda_1(t), \ldots, \lambda_n(t)$ *have all properties stated in the preceding theorem.*

(ii) $\lambda_0, e_1, \ldots, e_p$ *are not all zero.*

(iii) *The following $2n + 2$ transversality equations hold*:

$$\partial G/\partial t_0 + \dot{x}_0^i \partial G/\partial x_0^i - [F]_0 = 0, \qquad \text{with summation on } i \text{ from } 1 \text{ to } n,$$

$$\partial G/\partial t_1 + \dot{x}_1^i \partial G/\partial x_1^i + [F]_1 = 0, \qquad \text{with summation on } i \text{ from } 1 \text{ to } n,$$

(5.27)

$$\partial G/\partial x_0^i - [F_i]_0 = 0, \qquad i = 1, \ldots, n,$$

$$\partial G/\partial x_1^i + [F_i]_1 = 0, \qquad i = 1, \ldots, n.$$

System (5.27) is equivalent to the Transversality Condition stated by Bliss[IX, p. 202, (74.9)] in a more compact form. The reader is asked to verify that (5.27) also is equivalent to the system of $(2n + 2)$ equations

$$[F]_0 - \dot{x}_0^i [F_i]_0 = \partial G/\partial t_0, \qquad \text{summed on } i,$$

$$[F]_1 - \dot{x}_1^i [F_i]_1 = -\partial G/\partial t_1, \qquad \text{summed on } i,$$

(5.27*)

$$[F_i]_0 = \partial G/\partial x_0^i, \qquad i = 1, \ldots, n,$$

$$[F_i]_1 = -\partial G/\partial x_1^i, \qquad i = 1, \ldots, n.$$

In applications of the Multiplier Rule to particular examples (see Sections 5.9 and 5.12), it seems to make little difference whether one uses form (5.27) or (5.27*). However the $(n + 1)$-vector with components

$$[F]_j - \dot{x}_j^i [F_i]_j, \qquad \text{summed on } i, \qquad \text{and} \qquad [F_i]_j, \qquad j = 1, 2,$$

known as the *transversality vector*, has a geometrical interpretation that can be helpful. If, for example, we look ahead to Section 5.9, Example 5.4, we see from the classical form (5.34) of the transversality condition for the case $n = 1$ [also given in a slightly different notation by (2.57) in problem 11, Exercise 2.6], that this condition requires the transversality vector to be perpendicular to the graph of Y.

Proof of the Multiplier Rule (Theorems 5.2 and 5.3) is both formidable and tedious, and we shall sketch only a part of it here. After completing this chapter, a reader will have acquired a feeling for what the rule says and does and it will then be easier to endure the details of a complete proof. Those who have a serious interest in variational theory must, sooner or later, study some of the proofs. Two important sources are a long paper (5a) of Bliss and a portion (IX, pp. 201–203) of his Lectures, which supersede the original paper (7a) of Bolza. A recent book (XXI) and paper (20f) of Hestenes give corresponding results for what he calls (20f, p. 24) the *optimal control formulation of the Problem of Bolza.*

The special case under the Bolza Problem of Section 5.3 in which both endpoints are fixed and there are no side-conditions (5.5) is the problem in $(n + 1)$-space to which problem 12, Exercise 2.3, and problem 2, Exercise 4.1, have directed attention. The function F of (5.21) now reduces to the integrand f and the Euler condition is (5.24) with F reduced to f. The ith one of the system of n equations to which the vector equation (5.24) is equivalent can be derived from $J(x^1, \ldots, x^i + \epsilon\eta, \ldots, x^n)$ by following Section 2.6. Observe that a variation $\epsilon\eta$ is added only to the ith argument. With $n = 1$ we are back to the problem of Chapters 2 and 3.

Comparison functions of the simple form $x + \epsilon\eta$ will not in general satisfy nonlinear side-conditions or end-conditions. Granted, however, that $x: [t_0, t_1] \to R^n$ is admissible, Bliss proves (IX, pp. 194–201, 213–215) that, if x is normal under the definition given in Section 5.8, then it can be embedded in a family $y(\cdot, b): [t_0(b), t_1(b)] \to R^n$ of parameter $b = (b^1, \ldots, b^{p+1})$ such that $y(\cdot, b)$ is admissible and such that $y(\cdot, 0) = x$, hence such that the domain $[t_0(0), t_1(0)]$ of $y(\cdot, 0)$ is the interval $[t_0, t_1]$ of x. Our admissible functions are those in the class $\mathfrak{M}$ of Bliss (IX, p. 194). That $y(\cdot, b)$ is admissible in the sense of Section 5.3 means that, for each b, $y(\cdot, b)$ is PWS, that

$$\phi_\beta[t, y(t,b), \mathfrak{y}(t,b)] = 0, \qquad \beta = 1, \ldots, m,$$

and that

$$\psi_\mu\{t_0(b), y[t_0(b), b], t_1(b), y[t_1(b), b]\} = 0, \qquad \mu = 1, \ldots, p.$$

It then follows from definitions (5.21) and (5.22) that

$$(5.28) \qquad F[t, y(t,b), \mathfrak{y}(t,b), \lambda] = \lambda_0 f[t, y(t,b), \mathfrak{y}(t,b)]$$

and that

$$(5.29) \qquad G\{t_0(b), y[t_0(b), b], t_1(b), y[t_1(b), b]\}$$
$$= \lambda_0 g\{t_0(b), y[t_0(b), b], t_1(b), y[t_1(b), b]\}.$$

Consider the real-valued function $\mathscr{J}$ with values

$$(5.30) \qquad \mathscr{J}(b) \equiv G\{ \ \} + \int_{t_0(b)}^{t_1(b)} F[\ ] \, dt,$$

in which arguments of the left members of (5.28) and (5.29) have been omitted. By hypothesis in Theorems 5.2 and 5.3, $J[y(\cdot, 0)]$ is a minimum, that is, at least a weak local minimum; consequently, $\mathscr{J}(0)$ is a minimum of $\mathscr{J}(b)$.

Our blanket hypothesis of Section 5.3 and various omitted details ensure that $\mathscr{J}$ has partial derivatives at $b = 0$ with respect to all components b^σ of b. It is convenient to abbreviate $\partial \mathscr{J}/\partial b^\sigma$ by $\mathscr{J}_\sigma$. Since $\mathscr{J}(0)$ is a minimum, we have the necessary condition

$$(5.31) \qquad \mathscr{J}_\sigma(0) = 0, \qquad \sigma = 1, \ldots, p+1.$$

In writing expressions for the partial derivative $\mathscr{J}_\sigma$ we shall use the further abbreviations:

$$G_j \equiv \partial G/\partial t_j(b)\big|_{b=0}, \qquad\qquad j = 0, 1.$$

$$G_{ij} \equiv \partial G/\partial y^i[t_j(b),b]\big|_{b=0}, \qquad j = 0, 1, i = 1, \ldots, n.$$

$$y^i_\sigma \equiv \partial y^i(t,b)/\partial b^\sigma\big|_{b=0}, \qquad i = 1, \ldots, n, \sigma = 1, \ldots, p+1.$$

$$\dot{y}^i_\sigma \equiv \partial \dot{y}^i(t,b)/\partial b^\sigma\big|_{b=0}, \qquad i = 1, \ldots, n, \sigma = 1, \ldots, p+1.$$

$$t_{j\sigma} \equiv \partial t_j(b)/\partial b^\sigma\big|_{b=0}, \qquad j = 0, 1, \sigma = 1, \ldots, p+1.$$

In various expressions to follow, a repeated index i in a term will imply summation from 1 to n.

By differentiating $\mathscr{J}$ with respect to b^σ and then setting $b = 0$ we find that

$$\mathscr{J}_\sigma(0) = G_0 t_{0\sigma} + G_{i0}\{\dot{y}^i[t_0(0),0]t_{0\sigma} + y^i_\sigma[t_0(0),0]\} + G_1 t_{1\sigma}$$

$$+ G_{i1}\{\dot{y}^i[t_1(0),0]t_{1\sigma} + y^i_\sigma[t_1(0),0]\} + [F]_1 t_{1\sigma} - [F]_0 t_{0\sigma}$$

$$+ \int_{t_0(0)}^{t_1(0)} \{F_{y^i}[t,y(t,0),\dot{y}(t,0),\lambda(t)]y^i_\sigma(t,0) + F_{r^i}[\ ]\dot{y}^i_\sigma(t,0)\}dt.$$

Integration by parts of the first term of the integrand yields the expression

$$y^i_\sigma[t_1(0),0]\int_{t_0(0)}^{t_1(0)} F_{y^i}\,dt - \int_{t_0(0)}^{t_1(0)}\Big(\dot{y}^i_\sigma(t,0)\int_{t_0(0)}^{t} F_{y^i}\,ds\Big)dt.$$

After regrouping of terms we then have that

$$\mathscr{J}_\sigma(0) = \{G_0 + \dot{y}^i_0 G_{i0} - [F]_0\}t_{0\sigma} + \{G_1 + \dot{y}^i_1 G_{i1} + [F]_1\}t_{1\sigma}$$

$$+ G_{i0}y^i_\sigma[t_0(0),0] + G_{i1}y^i_\sigma[t_1(0),0] + y^i_\sigma[t_1(0),0]\int_{t_0(0)}^{t_1(0)}F_{y^i}dt$$

$$+ \int_{t_0(0)}^{t_1(0)}\Big(F_{r^i} - \int_{t_0(0)}^{t} F_{y^i}\,ds\Big)\dot{y}^i_\sigma(t,0)\,dt.$$

The expression for $\mathscr{J}_\sigma(0)$ vanishes as a consequence of (5.31) for $\sigma = 1, \ldots, p+1$.

It can be shown (IX, pp. 199–200) that, given any real number λ_0 and any vector $c = (c^1, \ldots, c^n)$ with real components, there exists a unique set of multipliers $\lambda_1(t), \ldots, \lambda_m(t)$ continuous except possibly at those t corresponding to corners of the minimizing function $y(\cdot, 0) = x$ and such that (5.24) holds.

We then see by setting $t = t_0$ in (5.24) that $c^i = [F_i]_0$, hence that

$$\int_{t_0(0)}^{t_1(0)} F_{y^i}\, dt = [F_i]_1 - [F_i]_0.$$

Returning to the last expression for $\mathscr{J}_\sigma(0)$, replace the expression in parentheses to the right of the last integral sign by c^i. The integral is then seen to have the value $c^i\{y_\sigma^i[t_1(0),0] - y_\sigma^i[t_0(0),0]\}$. With the aid of these results it follows that

$$\mathscr{J}_\sigma(0) = \{\ \}t_{0\sigma} + \{\ \}t_{1\sigma} + \{G_{i0} - [F_i]_0\}y_\sigma^i[t_0(0),0]$$
$$+ \{G_{i1} + [F_i]_1\}\, y_\sigma^i[t_1(0),0],$$

in which the first two terms are abbreviations for those in the preceding expression for $\mathscr{J}_\sigma(0)$. By (5.31), $\mathscr{J}_\sigma(0) = 0$, $\sigma = 1, \ldots, p+1$. From this, with the aid of other omitted details, it can be shown that the multipliers λ_β and e_μ can be so chosen that all expressions in braces in the last equation vanish and that moreover this occurs for a set of constant multipliers (5.26) not all zero and a set of multipliers (5.23) not all zero for any $t \in [t_0(0), t_1(0)]$ and for which the Euler equations (5.24) continue to hold. The equations so obtained are precisely the $(2n+2)$ transversality equations (5.27).

5.8 NORMALITY

Given one set of multipliers (5.23) and (5.26) for which (5.24) and (5.27) hold, another such set of multipliers is obtained by multiplying all the first ones by the same constant. If that constant is zero, we get a trivial set of multipliers.

If there are no nontrivial multipliers (5.23) with $\lambda_0 = 0$, the admissible function $x: [t_0, t_1] \to R^n$ is called *normal*, otherwise *abnormal*. In the normal case, which is the one usually encountered, we can divide all multipliers (5.23), (5.26) by λ_0 or alternatively set $\lambda_0 = 1$ as in Section 5.5.

In the event that the Bolza Problem (5.7) has a smooth normal minimizing function x, then x satisfies (5.25) on its interval. We now have the combined system of

n	Euler equations (5.25),
$2n + 2$	transversality equations (5.27),
m	side-conditions (5.5),
p	end-conditions (5.6),

for the determination of

n	components of $x(t)$,
$2n + 2$	end values $t_0, x^i(t_0), t_1, x^i(t_1)$,
m	multipliers $\lambda_1(t), \ldots, \lambda_m(t)$,
p	multipliers $e_1, \ldots, e_p$.

It is reasonable to anticipate that the $(m + 3n + p + 2)$ equations will determine the like number of enumerated mathematical objects but that the latter may not be unique.

If the minimizing function x has corners, then n constants c^i also enter the discussion. If x is abnormal with or without corners there appear to be various possibilities, which we shall not attempt to classify here.

The problem of solving the system of $(m + 3n + p + 2)$ equations can be formidable even in those exceptional cases where the components of x and λ turn out to be expressible in elementary closed form or in quadratures. In an example to be given in Section 5.12, $m + 3n + p + 2 = 22$. This and our other examples have been selected with care.

An optimization question arising in engineering can easily lead to a Bolza problem with large values of m, n, and p and such other complicactions as corners. The mass of detail required by an application of the Multiplier Rule easily can excede the ability and patience of an analyst. Moreover, the Euler equations (5.24) or (5.25) and transversality equations (5.27) will usually involve nonlinearities that exclude the possibility of an elementary solution in closed form, and one must accept numerical methods and approximations. These complexities have long stood in the way of realization of the full potentiality of variational theory for optimization problems with applications.

5.9   APPLICATION OF THE MULTIPLIER RULE TO EXAMPLES

EXAMPLE 5.4

$J(y) = \int f(t, y, \dot{y}) \, dt$ with $n = 1$, $m = 0$, and $p = 3$. *The three end-conditions are*

$$t_0 = a, \quad y(t_0) = b, \quad and \quad y(t_1) = Y(t_1)$$

where $Y: R \rightarrow R$ *is a given smooth function.*

This is the classic nonparametric problem in the plane with variable right endpoint. (See problem 11, Exercise 2.6.)

Discussion

With reference to Theorems 5.2 and 5.3 take

$$F(t,y,r,\lambda) = \lambda_0 f(t,y,r)$$

and

$$G(t_0,y_0,t_1,y_1) = e_1(t_0-a) + e_2(y_0-b) + e_3[y_1-Y(t_1)].$$

There is one Euler equation,

$$(5.32) \qquad \lambda_0 f_r = \lambda_0 \int_{t_0}^{t} f_y \, d\tau + c,$$

and there are four transversality equations,

$$(5.33) \qquad \begin{aligned} e_1 + e_2 \dot{y}_0 - \lambda_0 f(t_0,y_0,\dot{y}_0) &= 0, \\ -e_3 \dot{Y}_1 + e_3 \dot{y}_1 + \lambda_0 f(t_1,y_1,\dot{y}_1) &= 0, \\ e_2 - \lambda_0 f_r(t_0,y_0,\dot{y}_0) &= 0, \\ e_3 + \lambda_0 f_r(t_1,y_1,\dot{y}_1) &= 0. \end{aligned}$$

Observe first that if $\lambda_0 = 0$, then, by (5.33), $e_1 = e_2 = e_3 = 0$. But Theorem 5.3 assures that values for these multipliers exist that are not all zero; hence, if there is an extremizing function y, it must be normal. We accordingly set $\lambda_0 = 1$. Next eliminate e_3 between (5.33_2) and (5.33_4). This yields the prototype transversality condition (2.57), which in the present notation is

$$(5.34) \qquad f(t_1,y_1,\dot{y}_1) + (\dot{Y}_1 - \dot{y}_1) f_r(t_1,y_1,\dot{y}_1) = 0.$$

If one thinks of the sketch that goes with this example it will be clear why the name of the condition was chosen. It is a condition on the minimizing function where it cuts across the fixed function Y.

EXAMPLE 5.5

$J(y) = \int_{t_0}^{t_1} f(t,y,\dot{y}) \, dt$, with $n = 1$, *a side condition of the form*

$$(5.35) \qquad \int_{t_0}^{t_1} h(t,y,\dot{y}) \, dt = k = \text{const.},$$

with fixed endpoints $t_0 = a_0$, $y(t_0) = b_0$, $t_1 = a_1$, $y(t_1) = b_1$.

Discussion

If $h(t,y,\dot{y}) = (1 + \dot{y}^2)^{1/2}$, consider maximizing the area of the planar region bounded by y, the t axis, and ordinates at $t = t_0$ and t_1. Condition (5.35) now requires that the perimeter of this region be a constant for all $y \in \mathscr{Y}$. This is the original *isoperimetric problem*. It is now customary to call the more general problem above an isoperimetric problem and to speak of (5.35) as an *isoperimetric side-condition*. See Akhiezer (I, pp. 113–117) for the case of several such side-conditions.

The present problem can be restated as a Bolza Problem by setting

$$(5.36) \qquad z(t) \equiv \int_{t_0}^{t} h[\tau,y(\tau),\dot{y}(\tau)]\,d\tau.$$

Then

$$(5.37) \qquad \dot{z}(t) = h[t,y(t),\dot{y}(t)].$$

Equations (5.36) and (5.35) provide end-conditions $z(t_0) = 0$, $z(t_1) = k$. We now have a Bolza Problem with $n = 2$, $m = 1$, $p = 6$.

EXAMPLE 5.6

$J(y,z) = \displaystyle\int_{t_0}^{t_1} \dot{y}^2\,dt$ *subject to one side-condition* $\dot{y} - \dot{z}^2 = 0$ *and five end-conditions.*

$$t_0 = 0, \quad y(t_0) = 0, \quad t_1 = 1, \quad y(t_1) = 1, \quad z(t_1) = 2.$$

Discussion

We illustrate the use of a tactic that is occasionally effective. Consider the new problem obtained by deleting the side-condition and the end-condition on z. By any of several approaches from Chapter 3, we find that there is a unique function $y_0, y_0(t) = t$, furnishing a global minimum for this problem. If we now determine $z_0(t) = t+1$ or $-t+3$ from the two conditions involving z, the vector (y_0, z_0) is seen to furnish a global minimum for the original problem. The effect of the side-condition is to require that an admissible function y have a nonnegative derivative $\dot{y}$. In applying methods of Chapters 2 and 3 we admitted all functions in the class $\mathscr{Y}$ of those chapters. Since the minimizing function y_0 for the modified problem happens to be in the proper subclass $\mathscr{Y}_1$, consisting of those $y \in \mathscr{Y}$ with nonnegative derivatives, then y_0 and the corresponding z_0 automatically constitute a minimizing pair for the original problem.

Be reminded that the Multiplier Rule is only a necessary condition and that we have not investigated sufficient conditions for the Bolza Problem. We are able to give a complete analysis of this example only because of its special features.

EXAMPLE 5.7

[*McShane* (33e, pp. 818–819)]. $J(y,z) = \int_{t_0}^{t_1} (\dot{y}^2 + \dot{z}^2) \, dt$ *subject to the side-condition*

(5.38) $\dot{y} + \dot{z}^3 = 0,$

and the fixed endpoint conditions

$$t_0 = 0, \qquad t_1 = 1,$$

(5.39) $y(t_0) = 0, \qquad y(t_1) = 0,$

$$z(t_0) = 0, \qquad z(t_1) = 0.$$

Discussion

The Euler equations are

$$2\lambda_0 \dot{y} + \lambda_1(t) = c_1,$$

(5.40)

$$2\lambda_0 \dot{z} + 3\lambda_1(t) \dot{z}^2 = c_2.$$

The integrand being a sum of squares, it is clear that inf $J(y,z) \geq 0$ on the class of admissible functions, which is the class of PWS pairs (y,z) satisfying (5.38) and (5.39). By inspection, $y_0(t) = 0$ and $z_0(t) = 0$ satisfy (5.38) and (5.39). This pair also satisfies (5.40) with any value whatever for λ_0 and with $\lambda_1(t) = c_1 = $ const. We can, in particular, choose $\lambda_0 = 0$; hence the pair (y_0, z_0) is abnormal. This might be a source of trouble but, in the present simple instance, we can observe that $J(y_0, z_0) = 0$, which is the infimum mentioned above, and therefore (y_0, z_0) furnishes the global minimum.

EXAMPLE 5.8

Apply the Multiplier Rule to Example 5.1 *of Section* 5.2.

Discussion

We can ignore the factor $\frac{1}{2}$ in seeking a minimum. To simplify the algebra, consider the case $L = C = 1$. Then

$$F(\ ) = \lambda_1(t)(\dot{I} + Q - V) + \lambda_2(t)(\dot{Q} - I)$$

and

$$G(\ ) = \lambda_0 I_1^2 + e_1 t_0 + e_2(t_1 - \pi) + e_3 Q_0 + e_4 I_0 + e_5(Q_1 - a^2).$$

With reference to the statement about notation at the beginning of Section 5.3, we remark that it is essential for each person to adopt a systematic procedure for translating back and forth between such notation as (Q,I,V) and the (x^1,x^2,x^3) of the Multiplier Rule. The details are so complex that it is extremely easy to make costly mistakes.

The three Euler equations are

(5.41)

$$\lambda_2(t) = \int_0^t \lambda_1(\tau)\, d\tau + c_1,$$

$$\lambda_1(t) = -\int_0^t \lambda_2(\tau)\, d\tau + c_2,$$

$$0 = -\int_0^t \lambda_1(\tau)\, d\tau + c_3.$$

Theorem 5.2 ensures only that λ_1 and λ_2 are piecewise continuous, but an integral is continuous with respect to a variable upper limit; hence, by the first two equations (5.41), λ_1 and λ_2 are actually continuous in this instance. We can, therefore, differentiate the third equation finding that $\lambda_1(t) = 0$ and then differentiate the second to find that $\lambda_2(t) = 0$. Theorem 5.2 tells us that $\lambda_0, \lambda_1, \lambda_2$ need not all vanish; therefore we can choose $\lambda_0 = 1$ for simplicity. This is all the information obtainable from the Euler equations for this example.

The $2n + 2 = 8$ transversality equations are

$$e_1 + \dot{Q}_0 e_3 + \dot{I}_0 e_4 = 0,$$

$$e_2 + \dot{Q}_1 e_5 + 2\dot{I}_1 I_1 = 0,$$

$$e_3 = 0, \quad e_4 = 0, \quad 0 = 0, \quad e_5 = 0, \quad 2I_1 = 0, \quad 0 = 0.$$

These require that $e_1 = e_2 = e_3 = e_4 = e_5 = 0$. The one piece of useful information comes from the next to last transversality condition, $2I_1 = 0$, which means that $2I(t_1) = 0$, with $t_1 = \pi$.

This will obviously be the desired solution if it is possible to realize the value $I(\pi) = 0$ by means of a triple (Q,I,V) satisfying the given side- and end-conditions. To investigate this, we find by variation of parameters that the solution of equations (5.1) satisfying the given initial conditions $Q(0) = I(0) = 0$ is

(5.42)

$$Q = \sin t \int_0^t V(\tau) \cos \tau\, d\tau - \cos t \int_0^t V(\tau) \sin d\tau,$$

$$I = \cos t \int_0^t V(\tau) \cos \tau\, d\tau + \sin t \int_0^t V(\tau) \sin \tau\, d\tau.$$

In order that $Q(\pi) = a^2$ and $I(\pi) = 0$, it must happen that

$$\int_0^\pi V(\tau) \sin \tau\, d\tau = a^2, \qquad \int_0^\pi V(\tau) \cos \tau\, d\tau = 0.$$

By inspection, $V(t) \equiv a^2/2$ satisfies these conditions. Finally, from (5.42),

$$Q(t) = (a^2/2)(I - \cos t) \quad and \quad I(t) = (a^2/2)\sin t$$

complete the optimal triple.

We have not developed any general sufficient conditions for the Bolza Problem. That we are able to exhibit the above triple (Q,I,V) and to establish that it furnishes a global minimum $I^2(\pi) = 0$ is because of favorable features built into this particular example. One should always look for such features.

EXAMPLE 5.9

Investigate Example 1.5, Section 1.14.

Discussion

A frontal attack on the Mayer problem $J(x,y,p) = T = minimum$ subject to conditions (1.39) and (1.40) yields Euler and transversality equations that appear to have no elementary solution.

Lefkowitz and Eckman do not proceed in this manner in (29a). They introduce new variables

(5.43) $$v = \ln(a/x) \quad and \quad u = y/x.$$

With reference to Section 1.14, we find that

$$\dot{v} = -\dot{x}/x = Ap^m,$$

hence that

$$dt = \alpha(1-\mu)^{1-\beta}\,dv,$$

with new symbols defined as follows:

$$\beta \equiv \frac{n}{n-m}, \quad \mu = 1 - \frac{Bp^n}{Ap^m}, \quad and \quad \alpha^n \equiv \frac{B^{\beta m}}{A^{\beta n}}.$$

It follows that the total time T for the process is

$$T = \int_0^V (1-\mu)^{1-\beta}\,dv, \quad V \equiv \ln(a/b).$$

The original reaction equations (1.39) imply that

(5.44) $$u' = \mu u + 1, \quad where\ u'\ means\ \frac{du}{dv},$$

and the problem of minimal time T now becomes the Lagrange Problem

$$\int_{v_0}^{v_1}(1-\mu)^{1-\beta}\,dv = minimum$$

subject to one side-condition (5.44) and four end-conditions

(5.45)
$$v_0 = 0, \qquad v_1 = V = \ln(a/b),$$
$$u(v_0) = b/a, \qquad u(v_1) = k/h,$$

with no end-condition on μ.

To apply Theorem 5.2 set

$$F(\ ) = \lambda_0(1-\mu)^{1-\beta} + \lambda_1(v)(u' - \mu u - 1).$$

Since μ' does not appear, the Euler equation for μ reduces to

(5.46)
$$\lambda_0(1-\beta)(1-\mu)^{-\beta} - \lambda_1(v)u = 0,$$

while that for u is

(5.47)
$$\lambda_1(v) = -\int_0^v \lambda_1(s)\mu(s)\,ds + c.$$

If $\lambda_1(v)$ were to vanish anywhere on $[0,V]$, then, from (5.46), $\lambda_0 = 0$. We see from conclusion (i) of Theorem 5.2 that if there is a minimizing pair, it is normal and we set $\lambda_0 = 1$. By (5.46),

$$\lambda_1(v) = (1-\beta)(1-\mu)^{-\beta}/u(v).$$

From (5.47), $\lambda_1'(v) = -\lambda_1(v)\mu(v)$. It follows that

(5.48)
$$\beta u \mu' = 1 - \mu, \qquad \mu' = \frac{d\mu}{dv}$$

and from this and (5.44) that

$$\beta u \frac{d\mu}{du} = \frac{1-\mu}{\mu u + 1}$$

or

$$\frac{du}{d\mu} + \frac{\beta}{\mu - 1}u = \frac{\beta\mu}{1-\mu}u^2.$$

This Bernoulli equation can be solved for $u(\mu)$ in terms of a quadrature, but it is messy. After this result is substituted into (5.48) we can express $v(\mu)$ in terms of two successive quadratures. This classical ritual is not really worth pursuing. One can approximate a solution for $u(v)$ and $\mu(v)$ directly from the pair of equations (5.44) and (5.48) using numerical methods.

We shall leave open the question whether this pair minimizes the integral. It has been shown only that this pair satisfies a necessary condition for a minimum.

Exercise 5.3

1. Find a smooth function y satisfying all conditions of Theorems 5.2 and 5.3 for the integral $\int \dot{y}^2 \, dt$, $n = 1$, with no side-condition and with end-conditions $t_0 = 0$, $y(t_0) = 1$, and $y(t_1) = t_1^2$.

2. Investigate the isoperimetric problem $\int y \, dt = $ minimum with the side-condition $\int (1 + \dot{y}^2)^{1/2} \, dt = 3$ and the fixed endpoints $(0,0)$, $(2,0)$. Point out why if we replace 3 by a constant $k \geq \pi$, the problem can have no solution under the formulation of this chapter.

3. Treat Example 5.6 by first using the side-condition to replace the given integral by $\int \dot{z}^4 \, dt$.

4. Change Example 5.6 by adjoining the additional end-condition $z(t_0) = 1$. Then obtain all the information provided by the Multiplier Rule.

5. Given the problem $\int f(t,y,\dot{y}) \, dt = $ extremum, with $y = (y^1, \ldots, y^n)$, with q isoperimetric side-conditions $\int h_\beta(t,y,\dot{y}) \, dt = k_\beta, \beta = 1, \ldots, q$, and fixed endpoints, prove that if there is a solution y_0, then the multipliers $\lambda_\beta(t)$ of Theorem 5.2 will be of the form $\lambda_\beta(t) = $ const.

6. Theorems 5.2 and 5.3 have both been stated for the case of a minimum. Consider the application of these theorems to the problem $-J(y) = $ minimum, which is equivalent to $J(y) = $ maximum. What changes in the wording of these theorems are needed to obtain companion theorems for maxima?

7. Given $J(y) = \int [\dot{y} - (2t-1)]^2 \, dt$ with fixed endpoints $(0,0)$ and $(1,0)$, we find, by Chapter 3, that $y_0(t) = t^2 - t$ furnishes the global minimum. Suppose, however, that we introduce the constraint $y \geq 0$. To fit the pattern of Section 5.3, use a device of F. A. Valentine (XV for years 1933–1937), replacing the inequality by an equality $y = z^2$. We now have a Bolza Problem with $n = 2$, $m = 1$, $p = 4$ and with a side-condition that is free of derivatives. Granted that the Multiplier Rule applies to this case [for proof, see Bliss (5a, pp. 703–705)] extract all the information that it provides.

8. Given $J(y) = \int_{t_0}^{t_1} (\dot{y}^2 + y^2) \, dt$ with no side-conditions and the end-conditions $t_0 = 0$, $y(t_0) = h_0$, $t_1 = T = $ const., find a function y_0 satisfying the Multiplier Rule.

9. With reference to Section 2.9 and Euler condition (5.24), formulate and prove an Erdmann corner condition for the Bolza Problem.

10. With reference to Sections 2.8 and 2.11 and if necessary to Bliss (IX) or some other book, formulate an E-function for the Bolza Problem in terms of the auxiliary function F and derive necessary conditions of Weierstrass and Legendre for the Bolza Problem.

5.10 FURTHER NECESSARY CONDITIONS, SUFFICIENT CONDITIONS FOR LOCAL EXTREMA

Necessary conditions extending those of Weierstrass, Legendre, and Jacobi from the simple problem of Chapter 2 to that of this chapter are discussed by Bliss in (5a) and (IX, Chap. VIII). For additional information on the Jacobi condition, see Reid (45a); on the Weierstrass and Legendre–Clebsch conditions, see Graves (18c) and McShane (33e). The theorems of McShane in contrast with earlier proofs are not restricted to normal minimizing functions.

Sufficiency theorems for both weak and strong local minima, the statements of which resemble those in our Section 3.6 for the simple problem, are to be found in Bliss (IX, Chap. IX) and (5a, Chap. IV). Also see Hestenes (20a,c), McShane (33k), Morse (38a,b,c), and Reid (45b,c).

All these developments require considerable ingenuity as well as patient detail. We shall not pursue them here.

In the remainder of this chapter we turn to the easier task of extending a sufficiency theorem in Section 3.12 for global extrema to a special class of Bolza Problems. For a similar extension of Section 3.11, see (12h, pp. 104–105).

5.11 SUFFICIENT CONDITIONS FOR GLOBAL EXTREMA

Define, using the vector notation employed in (5.24) together with the function F of (5.21),

$$(5.49) \qquad G(t,x,y,p,q,\lambda) \equiv F(t,y,q,\lambda) - F(t,x,p,\lambda) \\ - (y-x) \cdot F_y(t,x,p,\lambda) - (q-p) \cdot F_r(t,x,p,\lambda).$$

This function G of six arguments generalizing that of (3.42) is not to be confused with the function (5.22), also called G.

The theorems of this section are phrased in terms of the Lagrange Problem. In view of Section 5.4 we can anticipate that they are adaptable to certain other Bolza Problems. They can sometimes be applied to a Bolza Problem with the g term present by way of suboptimization, a device exhibited in the next section of this chapter. We call attention to the fact that although multipliers, the Euler equations (5.24), and the transversality equations (5.27) are used in proving the two theorems that follow, these ingredients enter the discussion as hypotheses. Hence the proofs and conclusions are independent of the fact that the Multiplier Rule is a necessary condition on a minimizing function y and are therefore unaffected by the fact that we have not given complete proofs of Theorems 5.2 and 5.3.

Theorem 5.4

Given a Lagrange Problem for which t_0 and t_1 are fixed and given $y_0 \in \mathcal{Y}$ that satisfies the Euler equation (5.24) with a set of multipliers $\lambda_0 = 1, \lambda_1(t), \ldots,$ $\lambda_m(t)$, then

$$(5.50) \qquad J(y) - J(y_0) = \int_{t_0}^{t_1} G[t, y_0(t), y(t), \dot{y}_0(t), \dot{y}(t), \lambda(t)] \, dt$$
$$+ [y(t_1) - y_0(t_1)] \cdot F_r[t_1, y_0(t_1), \dot{y}_0(t_1), \lambda(t_1)]$$
$$- [y(t_0) - y_0(t_0)] \cdot F_r[t_0, y_0(t_0), \dot{y}_0(t_0), \lambda(t_0)].$$

PROOF

Define an auxiliary integral J^*,

$$(5.51) \qquad J^*(y) \equiv \int_{t_0}^{t_1} [F(t, y_0, \dot{y}_0, \lambda) + (y - y_0) \cdot F_y(\ ) + (\dot{y} - \dot{y}_0) \cdot F_r(\ )] \, dt.$$

After a du Bois Reymond integration by parts (that is, of the middle term), we use the consequence of Euler equations (5.24) that

$$c_i = F_{ri}[t_0, y_0(t_0), \dot{y}_0(t_0), \lambda(t_0)]$$

to find that $J^*(y) - J^*(y_0)$ equals the expression after the integral in (5.50). Now $y_0 \in \mathcal{Y}$ must satisfy side-conditions (5.5) and $\lambda_0 = 1$ by hypothesis. It follows that $J^*(y_0) = J(y_0)$, whence

$$J(y) - J(y_0) = J(y) - J^*(y_0) = [J(y) - J^*(y)] + [J^*(y) - J^*(y_0)].$$

This is the stated result (5.50).

Denote by II_R^* and $\text{II}_R^{*'}$, respectively, the conditions

$$(5.52) \qquad G[t, y_0(t), y, \dot{y}_0(t), q, \lambda(t)] \geq 0, \qquad \textit{for all}$$
$$t \in [t_0, t_1] \textit{ and all } (y, q) \in R^2.$$

$$(5.53) \qquad G[t, y_0(t), y, \dot{y}_0(t), q, \lambda(t)] > 0, \qquad \textit{for all}$$
$$t \in [t_0, t_1] \textit{ and all } (y, q) \neq [y_0(t), \dot{y}_0(t)].$$

At a possible corner we interpret these conditions [in accord with convention (2.17)] to mean that the inequalities hold with $\dot{y}_0(t)$ as either the left or right derivative.

Theorem 5.5

Suppose given a Lagrange Problem with end-conditions (5.6) *that fix t_0 and t_1 and that, for $i = 1, \ldots, n$ and $j = 0, 1$, either fix $y^i(t_j)$ or are independent of $y^i(t_j)$. If, moreover, y_0 satisfies the Multiplier Rule with $\lambda_0 = 1$ and if II_R^* or $II_R^{*\prime}$ holds, then $J(y) - J(y_0) \geq 0$ or > 0, respectively, for all $y \in \mathscr{Y}$ or for all $y \in \mathscr{Y}$ distinct from y_0.*

PROOF

The last two Transversality equations (5.27) applied to a Lagrange Problem [g of (5.7) identically zero] imply that

$$(5.54) \qquad e_\mu \frac{\partial \psi_\mu}{\partial [y_0^i(t_j)]} = \pm F_{r\,i}[t_j, y_0(t_j), \dot{y}_0(t_j), \lambda(t_j)], + \text{ if } j = 0, - \text{ if } j = 1,$$

$$i = 1, \ldots, n, j = 0, 1; \text{ summation on } \mu \text{ from } 1 \text{ to } p.$$

If all end-conditions (5.6) are free of a particular end-value $y^i(t_j)$, the left member of the corresponding equation (5.54) vanishes and the right member of that equation then points out a vanishing term in one of the dot products of (5.50). If the end-conditions either explicitly or implicitly fix the value $y^i(t_j)$ for all $y \in \mathscr{Y}$, then $y^i(t_j) - y_0^i(t_j) = 0$ and again a term is eliminated from (5.50). Under the hypotheses of the theorem, all terms on the right in (5.50) except the integral are eliminated in one or the other of these two ways; hence the alternative conclusions of Theorem 5.5 follow from (5.50).

Sufficiency Theorem 5.5 has quite restrictive hypotheses and hence applies to a restricted class of problems. However, in contrast with much of the sufficiency theory for local minima mentioned in Section 5.10, it has the advantage of not requiring a strengthened Legendre condition or any mention of a Jacobi condition, a field, normality, or nonsingularity (IX, p. 204, [74.12]). Moreover, the function y_0 in Theorem 5.5 need not be smooth.

A complete analysis of a particular example including a precise characterization of all admissible functions y_0 that furnish a global minimum is difficult. Textbook examples are often chosen with integrands f that are sums of squares and with linear side and end-conditions. Theorem 5.5 is effective with such examples and others that have sufficient convexity and linearity, among which are some of the problems termed linear by A. Miele (36b). An example of this type treated in Section 5.12 will also serve to point out how Theorem 5.5 can be used to reduce a given Bolza Problem for which the time interval is not fixed to an ordinary minimum problem.

When Theorem 5.5 does apply, it is recommended not only by its simplicity but because it identifies the real desideratum, namely, a global extremum. In systems optimization one wants the best of all programs or designs, not merely one that is best in comparison with others that are nearby.

One would like to have theorems that extend the content of our Theorem 3.9 to various classes of Bolza Problems even though the applicability of such a theorem to an example would be limited by the difficulty of constructing a field in the large. The author in collaboration with W. R. Haseltine has used this approach (12g) with an example of long standing.

In view of the likely increased use of numerical methods as time goes on, we suggest the desirability of having sufficiency criteria, possibly in terms of probabilistic statements, applicable to the end results of a numerical program.

There has been a tendency to place confidence in the output of a direct numerical procedure intended to approximate an optimizing function for a variational problem if the process appears to converge, if one or more other procedures in some sense independent of the first appear to converge to the same result, and if, when the algorithm is applied to an elementary example with known results in closed form, it converges to these results. In courtroom language this is certainly presumptive evidence in favor of the procedure under test. However, contemplate the conceivable unhappy possibilities that might occur unless one has criteria that exclude them. Suppose, for example, that a global minimum is desired but that it does not exist or that it does exist and yet the process happens to converge to a local minimum distinct from the global minimum, to a local maximum, or to a stationary value of $J(y)$ that is neither a minimum nor a maximum of any type.

EXAMPLE 5.10

Investigate the effectiveness of Theorem 5.5 for the fixed-endpoint isoperimetric problem already replaced by a Bolza Problem under Example 5.5.

Discussion

The single side-condition (5.37) rewritten in the form (5.5) is

$$\phi_1(t,y,z,\dot{y},\dot{z}) \equiv \dot{z} - h(t,y,\dot{y}) = 0$$

and the six end-conditions expressed in the form (5.6) are

$$\begin{aligned}
\psi_1[\ ] &\equiv t_0 - a_0 = 0, & \psi_4[\ ] &\equiv t_1 - a_1 = 0, \\
\psi_2[\ ] &\equiv y(t_0) - b_0 = 0, & \psi_5[\ ] &\equiv y(t_1) - b_1 = 0, \\
\psi_3[\ ] &\equiv z(t_0) = 0, & \psi_6[\ ] &\equiv z(t_1) - k = 0.
\end{aligned}$$

The auxiliary functions (5.21) and (5.22), the second of which is now written G_1 to avoid confusion with (5.49), are

$$F(\ ) \equiv \lambda_0 f(t,y,\dot{y}) + \lambda_1(t) [\dot{z} - h(t,y,\dot{y})]$$

and

$$G_1(\ ) \equiv e_1(t_0 - a_0) + e_2(y_0 - b_0)$$
$$+ e_3 z_0 + e_4(t_1 - a_1) + e_5(y_1 - b_1) + e_6(z_1 - k).$$

One finds with reference to (5.24) that the Euler equations are

$$\lambda_0 f_r(t,y,\dot{y}) - \lambda_1(t) h_r(t,y,\dot{y}) = \int_{a_0}^{t} [\lambda_0 f_y(\ ) - \lambda_1(\tau) h_y(\ )] \, d\tau + c^1$$

and

$$\lambda_1(t) = c^2.$$

By following (5.27) we obtain the transversality equations

$$e_1 + e_2 \dot{y}_0 + e_3 \dot{z}_0 - \lambda_0 f(t_0, y_0, \dot{y}_0) = 0,$$
$$e_4 + e_5 \dot{y}_1 + e_6 \dot{z}_1 + \lambda_0 f(t_1, y_1, \dot{y}_1) = 0,$$
$$e_2 - [\lambda_0 f_r(t_0, y_0, \dot{y}_0) - \lambda_1(t_0) h_r(t_0, y_0, \dot{y}_0)] = 0,$$
$$e_3 - \lambda_1(t_0) = 0,$$
$$e_5 + [\lambda_0 f_r(t_1, y_1, \dot{y}_1) - \lambda_1(t_1) h_r(t_1, y_1, \dot{y}_1)] = 0,$$
$$e_6 + \lambda_1(t_1) = 0.$$

The second Euler equation requires that $\lambda_1(t)$ be constant. Theorem 5.5 will then identify a global minimum according as there does or does not exist a pair (y,z) admissible in the sense of Section 5.3 and satisfying these Euler and transversality equations with some set of multipliers $\lambda_0 \neq 0, \lambda_1, e_1, \dots, e_6$ and hence with a set such that $\lambda_0 = 1$ as required by Theorem 5.5 and moreover such that the function (5.49), now called G_2, has the convexity property (5.52) or the strict convexity (5.53). Alternatively, Theorem 5.5 will establish that $J(y)$ is a global maximum if $-G_2$ has property (5.52) or (5.53).

Consider the special case of the original isoperimetric problem, in which $f(t,y,r) = y$ and $h(t,y,r) = (1+r^2)^{1/2}$. Let the fixed endpoints be $(-1,0)$ and $(1,0)$ and assign the value $k = 3$ to the side-integral (5.35). We look first at the Euler equation for y in the differentiated form (5.25) and suppose, moreover, that the second derivative $\ddot{y}(t)$ exists on $[-1, 1]$. The equation can then be expressed in the form

$$-\lambda_1 \ddot{y} / (1 + \dot{y}^2)^{3/2} = \lambda_0.$$

Since $\lambda_1 = 0$ implies that $\lambda_0 = 0$ and we wish to use Theorem 5.5, which requires that $\lambda_0 = 1$, we consider only the case $\lambda_1 \neq 0$ and set $\lambda_0 = 1$. The Euler equation then says that the signed curvature of y is a constant $-1/\lambda_1$. If we take a circular arc of the required length 3 joining $(-1,0)$ and $(1,0)$, then λ_1 must be positive or negative according as y is above or below the t axis, respectively. The reader is asked to verify that Theorem 5.5 applies as stated to show the y with negative values furnishes a proper global minimum and that the unstated companion theorem for maxima applies to the y with positive values.

Observe that in formulating this particular isoperimetric problem we chose the fixed length 3 to be below the length π of a semicircle of diameter 2. With π in place of 3, one suspects that the two semicircles furnish extreme values but y is no longer PWS because of the fact that the derivatives $\dot{y}(-1)$ and $\dot{y}(1)$ are not finite. The classical theory with PWS admissible functions no longer suffices. If the fixed length exceeded π we could guess that circular arcs greater than semicircles are needed, but no function y as defined in Section 1.2 corresponds to such an arc because of the fact that it is cut by certain vertical lines in two points. There does not exist in the class $\mathscr{Y}$ of all PWS functions $y: [-1,1] \to R$ of fixed length k a particular function y_0 that either minimizes or maximizes the integral $\int_{-1}^{1} y \, dt$ unless k is on the half-open interval $[2,\pi)$.

The question of the existence and characterization of a curve of length $k \geq \pi$ joining $(-1,0)$ and $(1,0)$ and which together with the segment joining these points bounds a subset of the plane of maximal or minimal signed area remains. To treat it by variational methods it should be formulated as a parametric problem of the calculus of variations, a topic introduced in Chapter 6.

Exercise 5.4

1. Investigate the problem $J(y,u) = \int_{t_0}^{t_1} (y^2 + u^2) \, dt =$ global minimum subject to the side-condition $\dot{y} + y + u = 0$ and the end-conditions $t_0 = 0, t_1 = 1$, and $y(t_0) = 1$.
2. Problem 1 modified by adding a fourth end-condition $y(t_1) = 0$.

5.12   ANALYSIS OF A PROBLEM FROM ROCKET PROPULSION

Consider the particle idealization of a rocket-propelled vehicle moving near a fixed flat earth on an upward-directed y axis subject to thrust $-c\dot{m}$ and weight mg but to no other force. Under these simplifying hypotheses, the equation for the trajectory is $m\ddot{y} + c\dot{m} + mg = 0$.

To transform this into a linear equation, set

(5.55) $$l = c \ln m.$$

The derivative $\dot{l} = c\dot{m}/m$ is then, except for sign, the thrust per unit mass.

As a further ingredient in the formulation, we suppose given that

(5.56) $\qquad\qquad -\alpha \leqslant \dot{l} \leqslant 0, \qquad \alpha > 0.$

If the vehicle is to start from rest at the origin with an assigned initial mass M_0 and if the burnout mass is M_1, we wish to determine a function m with these end values and such that the summit altitude is the global maximum.

Our complete mathematical model is the Lagrange Problem

(5.57) $\qquad J(y,v,l,z) = \int_{t_0}^{t_1} \dot{y}\, dt = global\ maximum\ on\ \mathscr{Y},$

where $\mathscr{Y}$ is now the class of all PWS quadruples (y,v,l,z) satisfying side-conditions

(5.58) $$\begin{aligned} \dot{v} + \dot{l} + g &= 0, \\ \dot{y} - v &= 0, \\ \alpha(\sin \dot{z} - 1) - 2\dot{l} &= 0, \end{aligned}$$

and end-conditions

(5.59) $\quad t_0 = 0, \quad y(t_0) = 0, \quad v(t_0) = 0, \quad l(t_0) = L_0, \quad l(t_1) = L_1,$

in which L_0 and L_1 are related to the end values of m by (5.55). We understand that $L_0 > L_1 > 0$.

The third constraint (5.58) is a device mentioned to the author by W. T. Reid for obtaining a single equality equivalent to (5.56) with the aid of an auxiliary variable z. Other devices for replacing (5.56) by equality constraints are in Valentine's dissertation (XV for years 1933–1937) and Miele (36b). It is essential that we use $\dot{z}$ and not z in this constraint. We wish to permit discontinuities of $\dot{l}$. If we had used z, the continuous fourth component of a PWS vector function, then the constraint would restrict $\dot{l}$ to be continuous and the Lagrange Problem would turn out to have no solution. One may not know in advance that this will happen but is playing safe in using $\dot{z}$.

Another ever-present danger is that an essential side- or end-condition has been overlooked. Such a flaw in the formulation may reveal itself through some absurd conclusion after much labor has been squandered on the wrong problem, or worse, it may not be detected at all. The question arises whether an end-condition $v(t_1) = 0$ should have been included in (5.59). The answer, in this case, is that a maximizing quadruple for (5.57) will automatically have this property but such answers are clearer a postiori than at the outset.

This problem involves two state variables y, v and two control variables l, z. It does not meet the restriction of Pontryagin et al. in (XXXIII) to control problems with integrand and side-conditions that are free of derivatives of the control variables or with end-conditions that are free of the control variables.

We wish to use Theorem 5.5, which requires that both t_0 and t_1 be fixed; therefore we study first, not the given problem, but an *auxiliary problem* $J(v,y,l,z) = $ *global maximum* on $\mathscr{Y}_T$ where $\mathscr{Y}_T$ is the subclass of $\mathscr{Y}$ consisting of all quadruples in $\mathscr{Y}$ satisfying the additional end-condition

$$(5.60) \qquad\qquad t_1 = T, \qquad T > 0.$$

Such procedure is called *suboptimization* since we shall be optimizing on the proper subclass $\mathscr{Y}_T$ of the original $\mathscr{Y}$.

To abbreviate the presentation, various details will be suppressed. It turns out that the multiplier λ_0 can have the value unity. We give it that value at the start. Then functions F and G defined by (5.21) and (5.49) are

$$(5.61) \qquad F(\ ) = \dot{y} + \lambda_1(t)(\dot{v} + \dot{l} + g) + \lambda_2(t)(\dot{y} - v) \\ + \lambda_3(t)[\alpha(\sin \dot{z} - 1) - 2\dot{l}],$$

$$(5.62) \qquad G(\ ) = \alpha\lambda_3(t)[\sin \dot{z} - \sin z_0 - (\dot{z} - \dot{z}_0)\cos \dot{z}_0].$$

The Euler equations are found to be

$$1 + \lambda_2(t) = c_1,$$

$$(5.63) \qquad \lambda_1(t) = \int_{t_0}^{t} -\lambda_2(\tau)\, d\tau + c_2,$$

$$\lambda_1(t) - 2\lambda_3(t) = c_3,$$

$$\lambda_3(t)\cos \dot{z}(t) = c_4.$$

Only four of the $2n + 2 = 10$ transversality equations yield useful information. These are

$$\lambda_1(t_1) = 0,$$
$$1 + \lambda_2(t_1) = 0,$$
$$\lambda_3(t_0)\cos \dot{z}(t_0) = 0,$$
$$(5.64) \qquad \lambda_3(t_1)\cos \dot{z}(t_1) = 0.$$

The other seven serve only to determine values of the multipliers e_μ.

Experience with optimization theory for rocket trajectories leads one to expect that the best burning program will consist of maximal thrust from $t = 0$ to burnout time

$$(5.65) \qquad\qquad t_b = (L_0 - L_1)/\alpha,$$

followed by coasting. In terms of constraint (5.58₃) this would mean that

(5.66)
$$\sin \overset{\bullet\bullet}{z}_0 = \begin{cases} -1, & 0 \le t < t_b, \\ 1, & t_b < t \le T. \end{cases}$$

We proceed on the basis of this educated guess. To shorten the exposition, we omit sifting all the quadruples (y,v,l,z) and multipliers $\lambda_1(t)$, $\lambda_2(t)$, $\lambda_3(t)$ that satisfy the combined system of 16 equations (5.58), (5.59), (5.60), (5.63), (5.64), which for an unfamiliar problem would generally be essential to a definitive analysis.

Granted (5.66), we must still make use of the 16 equations to determine the particular quadruple (y_0,v_0,l_0,z_0) and the associated multipliers, which are given below. Alternative expressions (5.67) arranged in two columns apply on the respective intervals $0 < t < t_b$ and $t_b < t < T$. Equations (5.68) apply for $0 \le t \le T$.

(5.67)
$$
\begin{aligned}
y_0(t) &= (\alpha - g)t^2/2, & -gt^2/2 + \alpha t_b(2t - t_b), \\
v_0(t) &= (\alpha - g)t, & -gt + 2\alpha t_b, \\
l_0(t) &= L_0 - \alpha t, & L_1 \text{ or } L_0 - \alpha t_b, \\
z_0(t) &= 3\pi t/2. & (3\pi t_b/2) + \pi(t - t_b)/2.
\end{aligned}
$$

(5.68) $\lambda_1(t) = t - T,$ $\lambda_2(t) = -1,$ $2\lambda_3(t) = t - t_b.$

We find using (5.62), (5.67), and (5.68) that

(5.69)
$$2G(\ ) = \begin{cases} \alpha(t - t_b)(\sin \overset{\bullet\bullet}{z} + 1), & 0 \le t < t_b, \\ \alpha(t - t_b)(\sin \overset{\bullet\bullet}{z} - 1), & t_b < t \le T. \end{cases}$$

It is clear from the form of (5.69) that condition II$_E^*$ holds for $-G$, hence that Theorem 5.5 applies to $-J$, and therefore that $J(v_0,y_0,l_0,z_0)$ is the global maximum on $\mathscr{Y}_T$ of (5.57). We have a definitive conclusion for the auxiliary problem.

To complete the original maximum problem on $\mathscr{Y}$, consider the whole class of auxiliary problems of parameter T corresponding to different classes $\mathscr{Y}_T$, $T > t_b$ of admissible quadruples. One verifies from (5.67) that the maximum of $y_0(T)$ necessarily occurs for $T > t_b$. By the second expression for $y_0(t)$, we discover that the terminal velocity $v_0(T)$ is necessarily zero. It follows that

$$\max_T y_0(T) = (2\alpha - g)\alpha t_b^2/g$$

is the greatest of all summit heights for the original problem.

Figure 5.1 shows three forces, drag, weight, and thrust, acting on the particle idealization of a rocket-propelled vehicle moving in a vertical plane near a flat stationary earth. Notation in the figure shows each force as the product of a scalar times a unit vector with the latter represented by its components in the directions of the respective x and y axes. Symbols k, g, c are positive constants and m is the total mass at time t.

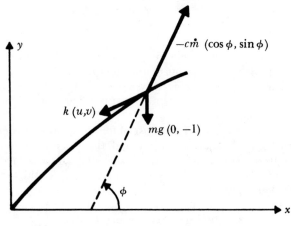

FIGURE 5.1

The differential equations for the motion are

(5.70)
$$\begin{aligned} m\dot{u} + c\dot{m}\cos\phi + ku &= 0, \\ m\dot{v} + c\dot{m}\sin\phi + kv + mg &= 0, \\ \dot{x} - u &= 0, \\ \dot{y} - v &= 0. \end{aligned}$$

Exercise 5.5

1. With m a fixed smooth nonincreasing function whose value goes from an assigned initial value $m(0)$ to a burnout value $m(t_b)$, let t_1 denote a positive time such that $y(t_1) = 0$ and define the range $R(x,y,u,v,\phi)$ as the largest $x(t_1)$. If the idealized vehicle starts from rest at the origin, investigate the problem $R(x,y,u,v,\phi) = $ maximum. For a more general version, see (12i).

5.13   CONCLUDING REMARKS

The Bolza Problem treated in this chapter, which essentially follows the formulation of Bliss (IX, pp. 189, 193–194), includes many nonparametric variational problems that originally arose as separate problems and are so dealt with in various books and papers. There are a number of other Bolza Problems that include or overlap the coverage of this one. See, for example, Hestenes (20b,f), (XXI), Pontryagin et al. (XXXIII), and Reid (45d, f).

The remainder of the book can be used with a degree of flexibility.

Although the order in which the chapters appear is recommended to those who intend to cover them all, they can be read in almost any order if one is willing to do a little cross-referencing or to accept an occasional assertion based on something that he has skipped. Chapter 6 provides some background for Chapter 7 but only Sections 6.3 and 6.4 are essential as a prerequisite. The content of Chapter 8 is essential for Chapter 9, but the reader who is already familiar with the Lebesgue integral can omit Chapter 8 or scan it rapidly for vocabulary, notation, and coverage. Chapters 10 through 12 have little dependence upon each other but draw ideas from certain of the earlier chapters. Since the original printing, sufficient conditions for global extrema of the type treated in Sections 3.12 and 5.11 have been extended and refined in the author's papers 12(j),(k), listed in the Supplementary Bibliography, page 340.

Chapter 6

PARAMETRIC PROBLEMS

6.1 INTRODUCTION

The independent variable has been denoted by t in preceding chapters in deference to the fact that optimization problems suggested by dynamical systems involve functions $y = (y^1, \ldots, y^m)$ with values that depend upon the time. The traditional symbol for a nonparametric integral is

$$(6.1) \qquad \int f(x,y,y') \, dx$$

with x and y', respectively, in place of t and $\dot{y}$.

Given the integral (6.1), we can again introduce the symbol t by thinking of x and y as differentiable functions from a common interval $[t_0, t_1]$ to the reals. Then t is called a parameter and we have the familiar results that

$$y'(x) = \dot{y}(t)/\dot{x}(t) \qquad and \qquad dx = \dot{x}(t) \, dt.$$

If these are substituted into (6.1) in a purely formal manner, we obtain an integral,

$$(6.2) \qquad \int F(x,y,\dot{x},\dot{y}) \, dt \qquad where \qquad F(x,y,\dot{x},\dot{y}) \equiv f(x,y,\dot{y}/\dot{x})\dot{x}.$$

Observe that x has one component but that y in general has $m \geq 1$ components. If $m > 1$ it is usually convenient to express integral (6.2) in the more abbreviated notation

$$(6.3) \qquad \int F(y,\dot{y}) \, dt, \qquad y = y^1, \ldots, y^n, n = m+1,$$

with $y^1, y^2, \ldots, y^n$ respectively, replacing $x, y^1, \ldots, y^m$.

Either (6.2) or (6.3) is called a parametric integral, a term to be defined more precisely in Section 6.6. Familiar examples in the case $n = 2$ are the length integral

(6.4) $$\int [\dot{x}^2 + \dot{y}^2]^{1/2}\, dt$$

and the ordinary curvilinear integral

(6.5) $$\int [P(x,y)\dot{x} + Q(x,y)\dot{y}]\, dt$$

that appears in vector analysis, mathematical physics, and the theory of integration of functions of a complex variable.

The subject of this chapter could be developed in the pattern of Chapters 2 and 3 by proving necessary conditions of Euler, Weierstrass, Legendre, and Jacobi and then sufficiency theorems for local extrema. See for example, Akhiezer (I, pp. 37–45, 63–64), Bliss (IX, Chap. V), Bolza (X, Chap. IV), (XI, Chap. 5), and Pars (XXXII, Chap. IX).

Some of these results will be given but our emphasis is on the nature of curves and of parametric problems together with a fruitful interplay between certain of the latter and corresponding nonparametric problems. The formal relationship between integrals (6.1) and (6.2) must not be mistaken to suggest that a parametric variational problem based on (6.2) is merely a restatement in parametric form of the nonparametric problem of extremizing (6.1). That they are distinct problems will be pointed out.

6.2   WHAT IS A CURVE?

We wish to define the term curve in a manner that is both intuitively acceptable and suitable for the theory of curvilinear integrals. The word is often used loosely, sometimes for the set $\{(t,y) \in R^{n+1}:\ y = y(t),\ t \in [a,b]\}$, which is the function $y\colon [a,b] \to R^n$ under the definition of Section 1.2, and again for the projection of this set into the y-space, that is, for the set $\{y \in R^n:\ y = y(t),\ t \in [a,b]\}$ of image points y.

A number of books and articles in dealing with nonparametric integrals (6.1) speak not of a function y but of the curve y or the curve $y = y(x)$, a practice that is somewhat misleading. In analytic geometry, differential geometry, mathematical physics, or wherever one meets parametric equations $y = y(t)$, the tendency is to identify the word curve with the set of image points $y(t)$ in R^n unless there is an explicit warning against it.

The intuitive notion actually needed is that of the path traced out by a moving point $y(t)$ in R^n as t traverses its interval $[a,b]$ from a to b.

The point $y(t)$ may move through the set of image points in such a way that there is a one-one correspondence between image points and values of t. Again the path may intersect itself in certain points so as to form loops or may retrace parts of itself through reversals in direction.

Consider, for the case $n = 2$ and using (x,y) for a point in the plane, such examples as

(6.6) $x(t) = \cos t, \qquad y(t) = \sin t, \qquad 0 \le t \le \pi,$

(6.7) $x(t) = \cos t, \qquad y(t) = |\sin t|, \qquad 0 \le t \le 2\pi,$

(6.8) $x(t) = -\cos t, \qquad y(t) = \sin t, \qquad 0 \le t \le \pi.$

These all yield the same point set or *graph* in the (x,y) plane, namely, the upper half of the unit circle with center at the origin, but the three parametrizations represent three different paths. The point $[x(t),y(t)]$ traces the graph from right to left under (6.6) but from left to right under (6.8). It traces the graph twice, once in each direction, under (6.7). The reader is asked to try out (6.6) through (6.8) in the length integral (6.4) and such other elementary examples as $\int (x\dot{y} - y\dot{x})\, dt$ or $\int (x\dot{y} + y\dot{x})\, dt$.

Given one continuous parameterization x: $[a,b] \to R^n$ there are always infinitely many others y: $[c,d] \to R^n$ such that the sets $\{x(t) \in R^n: t \in [a,b]\}$ and $\{y(u) \in R^n: u \in [c,d]\}$ are identical and such that the point $y(u)$ moves through this set in the same manner as $x(t)$ when the respective parameters progress through their intervals.

6.3 FRÉCHET DISTANCE BETWEEN MAPPINGS

Given compact intervals $[a,b]$ and $[c,d]$ of positive length, a function h: $[a,b] \to [c,d]$ is called a *homeomorphism* if it is one-one and if h and its inverse h^{-1} are continuous on their respective intervals. Such a homeomorphism is either *sense-preserving* $[h(a) = c, h(b) = d]$ or *sense-reversing* $[h(a) = d, h(b) = c]$. Since the value of an integral (6.3) depends in general upon the direction in which $x(t)$ moves through the image set, we are guided in restricting attention to the sense-preserving homeomorphism, henceforth abbreviated by SPH. Because of this restriction, the curves we shall define are said to be *oriented*.

Fréchet distance between functions x: $[a,b] \to R^n$ and y: $[c,d] \to R^n$, denoted here by $\rho(x,y)$, is defined as follows:

(6.9) $$\rho(x,y) \equiv \inf_h \sup_t |x(t) - y[h(t)]|,$$

in which one first takes the supremum with respect to $t \in [a,b]$, with h

fixed, of the euclidean distance $|x(t) - y[h(t)]|$ defined by (1.30) and then the infimum of all such suprema with respect to h over the class of all SPH's.

Intuition suggests that the two functions

(6.10) $$x: [0,1] \to R^2, \qquad x^1(t) = x^2(t) = 2t$$

and

(6.11) $$y: [0,2] \to R^2, \qquad y^1(u) = y^2(u) = u,$$

ought to represent the same oriented (or directed) curve C, whatever the precise definition of such a curve may be. Each of these functions maps its interval onto the line segment in the plane with endpoints $(0,0)$ and $(2,2)$. Moreover, $x(t)$ and $y(u)$ both move from $(0,0)$ to $(2,2)$ with no reversals of direction. Clearly the function $h: [0,1] \to [0,2]$, $u = h(t) = 2t$, is an SPH and

$$|x(t) - y[h(t)]| = [(2t - 2t)^2 + (2t - 2t)^2]^{1/2} = 0.$$

Consequently the supremum in (6.9) is 0 for the particular h being used. No such supremum can be negative and hence $\rho(x,y) = 0$.

A similar result is obtained any time that there exists an SPH such that $y[h(t)] = x(t)$, for all $t \in [a,b]$. That $\rho(x,y)$ can vanish when there is no such h is shown by the next example.

EXAMPLE 6.1

(6.12) $x: [0,3] \to R^2, \qquad x^1(t) = x^2(t) \equiv \begin{cases} t, & 0 \le t < 1, \\ 1, & 1 \le t \le 2. \\ t-1, & 2 < t \le 3. \end{cases}$

As the second mapping y, take (6.11). One verifies easily that (6.12) yields the same set of image points as do (6.10) and (6.11) and that under mapping (6.12) the point $x(t)$ again moves from $(0,0)$ to $(2,2)$ with no changes of direction. However, point $x(t)$ remains at $(1,1)$ while t is on the interval $[1,2]$. There is no SPH such that $y[h(t)] \equiv x(t)$. Given $e > 0$, consider the SPH, $h_e: [0,3] \to [0,2]$,

(6.13) $h_e(t) \equiv \begin{cases} (1-e)t, & 0 \le t \le 1, \\ (1-e) + 2e(t-1), & 1 < t \le 2, \\ (1+e) + (1-e)(t-2), & 2 < t \le 3. \end{cases}$

This example has been constructed by choosing $x^1 = x^2 = \phi$, where $\phi: [0,3] \to [0,2]$ is not an SPH. The relationship of h_e to ϕ is indicated by Fig. 6.1 The reader should verify that

$$\sup_t |x(t) - y[h_e(t)]| = e\sqrt{2}.$$

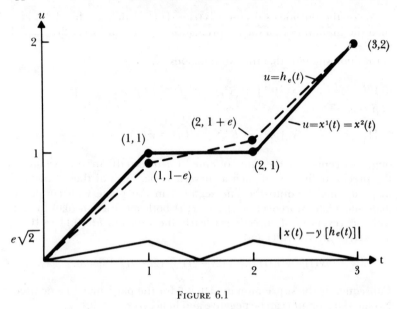

FIGURE 6.1

Since $e > 0$ but otherwise arbitrary, the infimum of all values of $e\sqrt{2}$ is 0; hence $\rho(x,y) = 0$.

Definition (6.9) is so designed that the condition $\rho(x,y) = 0$ makes precise the relation between mappings x and y described in the last paragraph of Section 6.2. Two mappings $x\colon [a,b] \to R^n$ and $y\colon [c,d] \to R^n$ are called *Fréchet-equivalent* iff $\rho(x,y) = 0$. We define a *continuous oriented Fréchet curve in E_n*, henceforth to be called simply a *curve in E_n* or a *curve*, to be any class consisting of all continuous mappings $y\colon [c,d] \to R^n$ each of which is equivalent to a particular such mapping x. Each mapping in such a class is a *parameterization* or *representation* of a curve C; that is, $y\colon [c,d] \to R^n$ is a representative element of the infinite class that constitutes a curve. We speak of a curve in E_n, the euclidean n-space defined in Section 1.10, because of the fact that euclidean distance rather than some other distance between points $x(t)$ and $y[h(t)]$ of R^n has been used in definition (6.9). The term graph of a mapping has been used in Section 6.2. It can be verified that two Fréchet-equivalent mappings have the same graph. Consequently we define the *graph of a curve C* as the graph of an arbitrary representation of C.

This definition of a curve may seem queer until one lives with it for awhile, but it is exactly what is needed in a theory of curvilinear integrals. The use of equivalence classes is, moreover, a time-honored tactic in mathematics. As one of many other instances, a Cantor real number is an equivalence class of Cauchy sequences of rational numbers.

One sees immediately from definition (6.9) that $\rho(x,y) \geq 0$, that

$\rho(x,x) = 0$, and that $\rho(x,y) = \rho(y,x)$. The theorem that follows establishes the triangle inequality. Thus Fréchet distance ρ has all the properties (1.29) except that $\rho(x,y) = 0$ does not imply that y is the same mapping as x. Such a distance ρ is called a *pseudo-metric*.

Theorem 6.1

Fréchet distance ρ between continuous mappings has the triangle property

$$(6.14) \qquad\qquad \rho(x,y) + \rho(y,z) \geq \rho(x,z).$$

PROOF

Let $x: [a,b] \to R^n$, $y: [c,d] \to R^n$, and $z: [e,f] \to R^n$ be continuous mappings of nondegenerate compact intervals into R^n. Given $\epsilon > 0$, there exist by definition (6.9), SPH's

$$h: [a,b] \to [c,d] \qquad and \qquad k: [c,d] \to [e,f]$$

such that

$$(6.15) \qquad\qquad \sup_t |x(t) - y[h(t)]| < \rho(x,y) + \epsilon/3$$

and

$$(6.16) \qquad\qquad \sup_u |y(u) - z[k(u)]| < \rho(y,z) + \epsilon/3.$$

The composite function $k \circ h$ with values $v = (k \circ h)(t) = k[h(t)]$ is necessarily an SPH from $[a,b]$ onto $[e,f]$. From the definition of supremum, there exists $t_\epsilon \in [a,b]$ such that

$$(6.17) \quad |x(t_\epsilon) - (z \circ k \circ h)(t_\epsilon)| > \sup_t |x(t) - (z \circ k \circ h)(t)| - \epsilon/3.$$

Set $u_\epsilon \equiv h(t_\epsilon)$ and $v_\epsilon \equiv k(u_\epsilon)$. Then from (6.15) and (6.16),

$$|x(t_\epsilon) - y(u_\epsilon)| < \rho(x,y) + \epsilon/3$$

and

$$|y(u_\epsilon) - z(v_\epsilon)| < \rho(y,z) + \epsilon/3.$$

From these two relations together with (6.17) and the triangle inequality for the euclidean distance (1.30),

$$\rho(x,y) + \rho(y,z) > \sup_t |x(t) - (z \circ k \circ h)(t)| - \epsilon$$

and the first term on the right dominates $\rho(x,z)$ by definition (6.9). Since ϵ is positive but otherwise arbitrary, the stated conclusion (6.14) follows.

Our restriction to continuous mappings x: $[a,b] \to R^n$ and hence to continuous curves C has been somewhat arbitrary. It is a convenience because there is an extensive theory of continuous curves from which to draw and, moreover, it is the usual practice. However, see R. E. Hughs (22a) for a treatment of discontinuous curves. Our proof of Theorem 6.1 makes no use of continuity. The various steps and hence the conclusion are valid if x,y, and z are bounded but otherwise arbitrary functions from intervals to R^n.

6.4   FRÉCHET DISTANCE BETWEEN CURVES

Let $\mathscr{C}$ denote the class of continuous curves C in E_n. Each such curve is a class $\{x\}$ consisting of all continuous functions x each of which is from some interval of positive finite length to R^n and each pair x_1 and x_2 of which are at Fréchet distance $\rho\,(x_1,x_2) = 0$ from each other.

Given $C_1 = \{x\}$ and $C_2 = \{y\}$ in the class $\mathscr{C}$, we define a distance

(6.18) $d(C_1,C_2) \equiv \rho\,(x,y)$, $x \in C_1$ and $y \in C_2$.

If ξ and η are also respective representations of C_1 and C_2, that is, $\xi \in \{x\}$ and $\eta \in \{y\}$, it then follows from the triangle inequality for ρ that

$$\rho\,(x,y) \le \rho\,(x,\xi) + \rho\,(\xi,\eta) + \rho\,(\eta,y) = \rho\,(\xi,\eta)$$

and

$$\rho\,(\xi,\eta) \le \rho\,(\xi,x) + \rho\,(x,y) + \rho\,(y,\eta) = \rho\,(x,y).$$

Consequently, $\rho\,(\xi,\eta) = \rho\,(x,y)$ and $d(C_1,C_2)$ is seen to depend only on C_1 and C_2, as the symbol (6.18) already indicates.

Exercise 6.1

1. With d defined by (6.18), point out that $(\mathscr{C},d)$ is a metric space.
2. Given x: $[0,1] \to R^2$, $x^1(t) \equiv x^2(t) \equiv t$, and y: $[-1,1] \to R^2$, $y^1(u) \equiv u$, $y^2(u) = u^2 - 1$, find the value of $\rho\,(x,y)$.
3. Given x: $[-1,1] \to R^2$, $x^1(t) \equiv 2$, $x^2(t) \equiv 0$ and y: $[0,3] \to R^2$, $y^1(u) = u$, $y^2(u) = u^2/9$, find $\rho\,(x,y)$.
4. Given x: $[0,\pi] \to R^2$, $x(t) \equiv (\cos t,\, \sin t)$ and y: $[0,\pi] \to R^2$, $y(t) \equiv (\cos^3 t,\, \sin^3 t)$, show that $\rho\,(x,y) = \frac{1}{2}$.
5. Let C be a given Fréchet curve in E_n, let x: $[a,b] \to R^n$ be any one of its representations, and let $[c,d]$ be any closed interval of positive finite length. Show that C has at least one representation y with the given interval $[c,d]$ as its domain.

6. Starting with the cartesian plane R^2 and the non-euclidean distance $d(x,y) \equiv |x^1 - y^1| + |x^2 - y^2|$ between points x and y of R^2, replace (6.9) by the definition

$$\rho(x,y) \equiv \inf_h \sup_t d\{x(t), y[h(t)]\}.$$

What difference if any is there between a Fréchet curve based on this definition of ρ and that based on definition (6.9)?

7. Let (S,d) be an arbitrary metric space, consider continuous mappings $x: [a,b] \to (S,d)$, and define $\rho(x,y) \equiv \inf \sup d\{x(t), y[h(t)]\}$. Discuss the extension of Sections 6.3 and 6.4 to this case.

6.5 PIECEWISE SMOOTH CURVES

A curve C will be called PWS if C has at least one representation x: $[a,b] \to R^n$ that is PWS on $[a,b]$. This means that each component of x is PWS on $[a,b]$ in the sense of Section 1.9. If $h: [c,d] \to [a,b]$ is any PWS SPH, then the composite function $x \circ h: [c,d] \to R^n$ is PWS on $[c,d]$ and $\rho(x, x \circ h) = 0$, hence a PWS curve C actually has infinitely many PWS representations.

In the remainder of this chapter any curve C or representation x that is mentioned will be understood to be PWS unless there is explicit statement to the contrary. This is a sufficient condition for an integral (6.3) with a continuous integrand F to be meaningful as a Riemann integral. We shall relax such restrictions in subsequent chapters.

A PWS curve C also has infinitely many representations that are not PWS. It is not difficult to construct an SPH between two fixed intervals that has infinitely many corners and to see that there are indeed infinitely many different ones having this property and hence that are not PWS. The composite function $x \circ h$ fails to be PWS for each such h. Such representations must be avoided in the present chapter.

We shall want to use partial derivatives of our integrands, but a function F with values $F(x,y,p,q)$ like that in (6.2) generally is not differentiable with respect to p or q at a point of its domain of the form $(x,y,0,0)$. We therefore shall restrict ourselves part of the time to the use of what are called regular representations. By definition, a PWS representation $x: [a,b] \to R^n$ is called *regular* if the derivative vector $\dot{x}(t)$ vanishes nowhere on $[a,b]$ with this understood to mean that, if $x(t)$ is a corner, then neither the left nor the right derivative at t is zero. Every PWS curve C has infinitely many regular and also infinitely many nonregular (singular) representations.

Exercise 6.2

1. Establish that $x(t) \equiv (t^3, t^3)$, $-1 \leq t \leq 1$ and $y(u) \equiv (u, u)$, $-1 \leq u \leq 1$ represent the same smooth curve C and that one of them is regular and the other singular, that is, nonregular.

2. Given that $x: [a,b] \to R^n$ is PWS and regular on $[a,b]$, define

$$s(t) \equiv \int_a^t |\dot{x}(\tau)| \, d\tau.$$

Show that the function $s: [a,b] \to R$ is an SPH and the composite function $x \circ s^{-1}$, where s^{-1} is the inverse of s, is a regular representation equivalent to the given representation.

6.6 PARAMETRIC INTEGRALS AND PROBLEMS

Parametric integrals are encountered under two kinds of circumstances. Questions from geometry involving the notions curve or path of a moving point are most appropriately stated and treated in terms of parametric equations. If, for example, we wish to study the length of plane curves, the nonparametric length integral $\int [1 + (y')^2]^{1/2} \, dx$ is an inadequate tool, since it automatically restricts attention to paths that are the graphs of real-valued functions. Since such paths intersect a line $x = $ const. in at most one point and since paths in general can intersect such lines in many points, one needs the parametric integral (6.4). Certain considerations in mathematical physics are also essentially geometric, hence are properly formulated in terms of the parametric integral (6.5) or its counterpart for three-space. Such situations occur, for example, in the study of electric and magnetic fields. On the other hand, there are questions of which many examples have been given in preceding chapters in which the concern is with real- or vector-valued functions y and not with curves. The initial and seemingly natural formulation of these questions in mathematical language is nonparametric, and yet it is sometimes an aid to the analysis if one shifts to parametric form in the manner indicated by (6.2).

Usually the parametric integrand $F(x,y,p,q)$ corresponding via (6.2) to a well-behaved nonparametric integrand $f(x,y,r)$ is discontinuous at $p = 0$. For instance, if $f(x,y,r) = r^2$, then $F(x,y,p,q) = q^2/p$.

We turn now to a formulation of parametric variational problems so phrased as to include both those of the more traditional geometric type and those generated by nonparametric problems.

Shifting to the notation of (6.3), let A be a nonempty open subset of R^n with points denoted by such symbols as x, y, and z. Let B be a nonempty subset of R^n such that if $r \in B$ and $r \neq \theta \equiv (0,0,\ldots,0)$, the origin of R^n, then the half-line or ray determined by θ and r (possibly including θ and possibly not) is contained in B. We require further of B that it be an open subset of R^n or that it be the union of such a set with the singleton set $\{\theta\}$ consisting of the point θ.

Consider a function $F: A \times B \to R$ subject to a *blanket hypothesis* similar to that in Section 2.2. By this we mean that F is required to be continuous on $A \times B$ together with any partial derivatives of F that may be mentioned in a particular theorem or discussion. Sometimes an expression $F(y,r)$ for values of F is given without explicit description of the sets A and B. When this occurs we shall understand that A and B are the maximal sets with the stated properties such that F has the continuity and differentiability described above.

Suppose given some theorem involving a function F but no derivatives of F. The blanket hypothesis then requires only that F be continuous on $A \times B$. If $n = 2$ and $F(x,y,p,q) = (p^2 + q^2)^{1/2}$, the maximal sets mentioned above are $A = R^2$ and $B = R^2$. For the example $F(x,y,p,q) = q^2/p$, these sets are $A = R^2$ and $B = \{(p,q) \in R^2 : p \neq 0\}$. Some other theorem might require first-order partial derivatives of F. The origin $(0,0)$ would then have to be excluded from the set B for the first example above.

The flexible meaning given to the symbol B and to the blanket hypothesis on F, which has seemed convenient to the author, has two objectives. Many assertions involving derivatives of F are meaningful if and only if those derivatives exist and are continuous and we prefer to abbreviate various statements including steps in proofs by not having to repeat this fact. Second, some statements about F are free of derivatives while others may involve derivatives of the first order, etc.; hence some parts of the theory apply to a larger class of functions F than do others. In comparing results from different books, one needs to check hypotheses of one author against those of another since the practice is not uniform. For example, Hestenes (XXI, p. 79) restricts his discussion at the outset to parametric integrands with continuous second-order derivatives while Bliss (IX, p. 105) requires continuous fourth-order derivatives. Similar remarks apply to the set that we have denoted by B. Some authors explicitly exclude the case in which the origin $r = \theta$ is a point of B as we have not.

Consider an integral (6.3) with integrand F subject to the blanket hypothesis. Let C be any Fréchet curve having a PWS representation $y: [t_0,t_1] \to R^n$ such that

$$(6.19) \qquad [y(t),\dot{y}(t)] \in A \times B, \qquad \forall\, t \in [t_0,t_1].$$

Under the blanket hypothesis, F is continuous, and by (6.19) the composite function $F[y,\dot{y}]$ is meaningful and Riemann integrable over $[t_0,t_1]$.

An integral (6.3) is called a *parametric integral* if its value is the same for all PWS representations having property (6.19) and representing the same Fréchet curve C. The function F is then a *parametric integrand*. Since the value of a parametric integral depends only on the choice of a curve C and not upon the particular choice of a representation with property (6.19), its value will generally be denoted by $J(C)$.

Let $\mathscr{C}$ denote the class of all PWS curves C having PWS representations that satisfy (6.19) together with given end-conditions. The question of existence and characterization of a curve $C_0 \in \mathscr{C}$ such that $J(C_0)$ is a global extremum is then a *parametric problem* of the calculus of variations. We shall define local extrema presently. The curves $C \in \mathscr{C}$ are called *admissible curves* and the representations with property (6.19) of admissible curves are *admissible representations*.

One may wonder at this juncture how parametric integrands and integrals are to be identified. The answer is found in Section 6.7.

6.7 HOMOGENEITY OF PARAMETRIC INTEGRANDS

The theorem of this section is free of derivatives of F; hence, under the blanket hypothesis, F is understood to be continuous on $A \times B$ even though this is not stated in the theorem.

Theorem 6.2

Given an integrand $F: A \times B \to R$, the integral (6.3) has the same value for all PWS representations with property (6.19) of an arbitrary curve C having such a representation if and only if

$$(6.20) \quad F(x,kr) = kF(x,r), \qquad \forall (x,r) \in A \times B \text{ and } \forall k \geq 0 \text{ or } > 0,$$
$$\text{respectively, according as } \theta \in B \text{ or } \theta \notin B.$$

PROOF

Let $y: [t_0,t_1] \to R^n$ be a PWS representation with property (6.19). Let $h: [u_0,u_1] \to [t_0,t_1]$ be an SPH such that $\dot{h}(u)$ exists and is finite and positive on $[t_0,t_1]$. Set

$$(6.21) \quad z(u) \equiv y[h(u)].$$

Then

$$(6.22) \quad \dot{z}(u) = \dot{y}[h(u)]\dot{h}(u).$$

with the equation understood in the sense of (2.17) if $z(u)$ should be a corner point. It is immediate from definition (6.21) of z that it is Fréchet equivalent to y. Since y satisfies (6.19) by hypothesis, it follows from (6.22) and the positiveness of $\dot{h}$ that $\dot{z}(u) \in B$, hence that $[z(u),\dot{z}(u)] \in A \times B$.

Given that integral (6.3) has the same value for every PWS representation y satisfying (6.19) of an otherwise arbitrary curve C, this must be true in particular for every curve represented by a restriction of the given function y to a subinterval $[t_0,t]$ of $[t_0,t_1]$. It follows that the equation

$$\int_{t_0}^{h(u)} F[y(t),\dot{y}(t)]\, dt \equiv \int_{u_0}^{u} F[z(\alpha),\dot{z}(\alpha)]\, d\alpha$$

is an identity in u. We differentiate with respect to u using Theorem 1.3 and the continutiy of F on $A \times B$, finding that

$$(6.23) \qquad F\{y[h(u)],\dot{y}[h(u)]\}\dot{h}(u) = F[z(u),\dot{z}(u)]$$

is an identity in u in the sense (2.17).

In view of the arbitrary nature of the mappings y and h, we can suppose them so chosen that $z(u_1) = y[h(u_1)] = y(t_1)$ is an arbitrary point of A, that $\dot{h}(u_1)$ is an arbitrary positive real number k, and that $\dot{y}(t_1)$ is an arbitrarily selected point r in B, hence that $\dot{z}(u_1) = \dot{y}(t_1)\dot{h}(u_1)$ given by (6.22) is the point kr of B. If we substitute u_1 and $t_1 = h(u_1)$ into (6.23), we obtain the desired conclusion (6.20) with k restricted to be positive.

If the origin $\theta \in B$, then F is continuous at points (x,θ), $x \in A$, under the blanket hypothesis. Let k tend to 0 through positive values in (6.20) and verify both that $F(x,\theta) = 0$ and that (6.20) holds with $k = 0$.

Suppose conversely that (6.20) holds as stated in the two respective cases. Let $y\colon [t_0,t_1] \to R^n$ and $z\colon [u_0,u_1] \to R^n$ be two PWS representations with property (6.19) of the same curve C. Example 6.1 has shown that there may be no SPH that transforms z into y.

With reference to problem 2, Exercise 6.2, we can define

$$(6.24) \qquad s_1(t) \equiv \int_{t_0}^{t} |\dot{y}(\tau)|\, d\tau \qquad and \qquad s_2(u) \equiv \int_{u_0}^{u} |\dot{z}(v)|\, dv.$$

If y and z should happen to be regular, then s_1 and s_2 would both be strictly increasing, and proof that integral (6.3) has the same value for both y and z would be easier. Since y and z are not in general regular, we use a result to be proved in Section 7.6 that every curve C of positive finite length L has a particular representation $X\colon [0,L] \to R^n$ in terms of distance s along the curve as parameter. This representation is PWS when C is PWS.

Granted this representation of the curve C already represented by both y and z, we see with the aid of (6.24) that

$$y(t) = X[s_1(t)] \qquad and \qquad \dot{y}(t) = \dot{X}[s_1(t)]\dot{s}_1(t) = \dot{X}[s_1(t)]|\dot{y}(t)|$$

and similarly that

$$z(u) = X[s_2(u)] \qquad and \qquad \dot{z}(u) = \dot{X}[s_2(u)]\dot{s}_2(u) = \dot{X}[s_2(u)]|\dot{z}(u)|.$$

Now (6.20) holds and y is a representation with property (6.19) by hypothesis. From this and the preceding relations,

$$F[y(t),\dot{y}(t)]\,dt = F\{X[s_1(t)],\dot{X}[s_1(t)]\}\dot{s}_1(t)\,dt = F\{\ \}\,ds_1$$

whether the origin θ of R^n is or is not a point of B. Similarly,

$$F[z(u),\dot{z}(u)]\,du = F\{X[s_2(u)],\dot{X}[s_2(u)]\}\dot{s}_2(u)\,dt = F\{\ \}\,ds_2.$$

In the two expressions at the extreme right, symbols s_1 and s_2 can just as well both be called s. It follows that

$$(6.25) \quad \int_{t_0}^{t_1} F[y(t),\dot{y}(t)]\,dt = \int_0^L F[X(s),\dot{X}(s)]\,ds = \int_{u_0}^{u_1} F[z(u),\dot{z}(u)]\,du.$$

This completes the proof.

One distinguishes between integrals of the form (6.3) that are parametric and those that are not on the basis of Theorem 6.2, that is, on the basis of whether F is or is not homogeneous of the first degree in r as stated in (6.20).

We remark that the restriction in this chapter to PWS curves C, although a customary ingredient in the classical theory of parametric integrals, will be relaxed in Chapter 7. Such a move has certain advantages and is a characteristic of modern calculus of variations.

Exercise 6.3

1. Verify that (6.4) and (6.5) are parametric integrals with the aid of Theorem 6.2.
2. Given $\int F(t,y,\dot{y})\,dt$ with F depending upon t, suppose that this is a parametric integral and obtain a contradiction.

6.8 CONSEQUENCES OF THE HOMOGENEITY OF F

Given that

$$(6.26) \qquad F(y,kr) \equiv kF(y,r), \qquad (y,r) \in A \times B, k > 0,$$

we can differentiate with respect to k under the blanket hypothesis on F and find that

$$(6.27) \qquad F_r(y,kr) \cdot r \equiv F(y,r),$$

in which the dot denotes a scalar product. The restrictions on (y,r) and k stated in (6.26) are to be understood through this section. With $k = 1$, the last identity reduces to a special case of Euler's theorem on homogeneous functions, namely, that

$$(6.28) \qquad F_r(y,r) \cdot r \equiv F(y,r).$$

After writing the left member as a sum, we differentiate with respect to r^j and find that

$$(6.29) \qquad F_{r^i r^j}(y,r) r^i \equiv 0, \qquad summed\ on\ i\ with\ j = 1, \ldots, n.$$

The n equations (6.29) hold for all pairs $(y,r) \in A \times B$ and hence for vectors r distinct from the zero n-vector. Consequently, by a standard theorem on systems of linear homogeneous equations, the determinant

$$(6.30) \qquad |F_{r^i r^j}(y,r)| \equiv 0.$$

For the case $n = 2$ and in the notation of (6.2), identity (6.30) says that

$$F_{\dot{x}\dot{x}}(x,y,\dot{x},\dot{y}) F_{\dot{y}\dot{y}}(\ ) = F_{\dot{x}\dot{y}}^2(\ ).$$

System (6.29) for this case can be written in the form

$$F_{\dot{x}\dot{x}}(x,y,\dot{x},\dot{y})\dot{x} + F_{\dot{x}\dot{y}}(\ )\dot{y} \equiv 0,$$

$$F_{\dot{y}\dot{x}}(\qquad)\dot{x} + F_{\dot{y}\dot{y}}(\ )\dot{y} \equiv 0.$$

After multiplying these respective equations by $\dot{x}$ and $\dot{y}$ and using the fact that $F_{\dot{x}\dot{y}} = F_{\dot{y}\dot{x}}$, a consequence of the blanket hypothesis on F, we find that

$$(6.31) \qquad \dot{x}^2 F_{\dot{x}\dot{x}} = -\dot{x}\dot{y} F_{\dot{x}\dot{y}} = \dot{y}^2 F_{\dot{y}\dot{y}}.$$

One can define a function $F_1: A \times B \to R$,

$$(6.32) \qquad F_1(x,y,\dot{x},\dot{y}) = \begin{cases} F_{\dot{x}\dot{x}}(\ )/\dot{y}^2 & if\ \dot{y} \neq 0, \\ -F_{\dot{x}\dot{y}}(\ )/\dot{x}\dot{y} & if\ \dot{x}\dot{y} \neq 0, \\ F_{\dot{y}\dot{y}}(\ )/\dot{x}^2 & if\ \dot{x} \neq 0. \end{cases}$$

Function F_1 plays a role in the theory of parametric variational problems in the plane [Bolza (X,p.121), (XI,p.196), and other books] analogous to that of f_{rr} for the corresponding nonparametric problem. For the extension of F_1 to the cases $n > 2$, see Carathéodory (XII,p.216).

Differentiating F with respect to a component of r lowers the degree of homogeneity in r by one. It is thus clear from (6.32) that F_1 is positive homogeneous in r of degree -3.

The Weierstrass E-function for a parametric integrand $F(y,r)$ can be defined by the statement that

$$(6.33) \qquad E(y,p,q) \equiv F(y,q) - F(y,p) - (q-p) \cdot F_r(y,p).$$

As a consequence of identity (6.28) we also have the alternative forms

$$E(y,p,q) = F(y,q) - q \cdot F_r(y,p) = q \cdot [F_r(y,q) - F_r(y,p)].$$

For each fixed $y \in A$, the equation

$$(6.34) \qquad\qquad u = F(y,r)$$

determines a "surface" in R^{n+1} called a *figurative*. As a consequence of the positive homogeneity (6.26) of F in r, if (r,u) is a point on the figurative, then so also is (kr,ku) for every positive k. Consequently, the figurative is a ruled surface made up of rays issuing from the origin (r,u) $= (\theta,0)$ of R^{n+1}. In the event that the set B belonging with F includes the origin, $r = \theta$, of R^n, then the origin of R^{n+1} is understood to be included as a point of each of these rays, and the figurative is a cone with the origin as a vertex. There may be other vertices, as the following examples show.

EXAMPLE 6.2

$n = 2$, $F(x,y,p,q) \equiv (p^2 + q^2)^{1/2}$.
The figurative is the upper half of a right circular cone including the vertex $(r,u) = (\theta,0)$.

EXAMPLE 6.3

$n = 2$, $F(x,y,p,q) \equiv 3xp + 2yq$.
Since x and y are present, we have a family of figuratives with x and y as parameters. Each figurative, corresponding to a fixed pair (x,y), is a plane, hence a cone with the origin as a vertex. Every point of a plane is a vertex.

EXAMPLE 6.4

$n = 2$, $F(x,y,p,q) \equiv |q-p|$.
The figurative consists of two half-planes forming a dihedral angle with the line $q - p = 0$, $u = 0$, as edge. Every point of this line, including the origin, is a vertex.

Exercise 6.4

1. With $n = 2$, describe the figuratives $u = F(x,y,p,q)$ in the cases $F(x,y,p,q) \equiv |p|$, $(p^2 - q^2)^{1/2}$, $(p^4 + q^4)^{1/4}$.

2. Given that $n = 2$ and $F(x,y,p,q) \equiv q^2/p$ with B given as the half-plane $p > 0$, describe the figurative.

3. Show for the case $n = 2$ that the difference between the ordinate $u = F(y,r)$ to a figurative and the ordinate to a tangent plane to this figurative is expressible in terms of the E-function. Then investigate the extension of this result to a general n.

6.9 THE CLASSICAL FIXED-ENDPOINT PARAMETRIC PROBLEM

Given a parametric integrand $F : A \times B \to R$, let $\mathscr{C}$ denote the class of all PWS Fréchet curves C each of which has at least one PWS representation $y : [t_0, t_1] \to R^n$ with property (6.19) and with fixed initial and terminal points. Such curves C and such representations y are called *admissible*.

Consider the problem

$$(6.35) \qquad J(C) \equiv \int F(y,\dot{y}) \, dt = minimum \ on \ \mathscr{C}.$$

If C_0 is an admissible curve such that

$$(6.36) \qquad J(C_0) \leqslant J(C), \qquad \forall \, C \in \mathscr{C},$$

then $J(C_0)$ is a *global minimum*, called *proper* if the strict inequality holds.

Fréchet distance (6.18) between curves C_1 and C_2 is independent of derivatives of the respective representations. That the distance $d(C_1,C_2)$ be small implies that given any representation $x : [a,b] \to R^n$ of C_1, there must be, in accord with problem 5, Exercise 6.1, a representation $y : [a,b] \to R^n$ of C_2 with the same domain as x and such that the euclidean distance $|x(t) - y(t)|$ be small for all $t \in [a,b]$. If $d(C_1,C_2) = r$, then the graph of C_2 is a subset of the union of all closed balls of radius r with centers in the graph of C_1 and similarly with the roles of C_1 and C_2 exchanged. The distance $|\dot{x}(t) - \dot{y}(t)|$ between derivatives can be large when $d(C_1,C_2)$ is small. Therefore, Fréchet distance d is a *distance of order zero* denoted in the remainder of this section by d_0 and, if C_0 is a curve in $\mathscr{C}$ such that

$$(6.37) \qquad J(C_0) \leqslant J(C), \qquad \forall \, C \in \mathscr{C} \cap U_0(\delta,C_0),$$

where $U_0(\delta,C_0) \equiv \{C \in \mathscr{C} : d_0(C,C_0) < \delta\}$, then $J(C_0)$ is a *strong local minimum*, again called *proper* if the strict inequality holds. That it is possible to define several first-order distances and neighborhoods that lead to a definition of a weak local minimum has been shown by McShane (33f). However, in the interest of brevity it seems best not to follow this approach. We shall say that $J(C_0)$ is a *weak local minimum* if, given a particular PWS representation y_0: $[a,b] \rightarrow R^n$ of C_0, then

(6.38) $J(C_0) \leqslant J(C)$ *for all $C \in \mathscr{C}$ having representations y: $[a,b] \rightarrow R^n$ with the same domain as y_0 and such that, for some positive δ, $|y(t) - y_0(t)| < \delta$ for all $t \in [a,b]$ and $|\mathring{y}(t) - \mathring{y}_0(t)| < \delta$ for all $t \in [a,b]$ in the sense of convention (2.17).*

6.10 THE CLASSICAL PARAMETRIC PROBLEM OF BOLZA

The parametric problem analogous to the nonparametric problem formulated in Section 5.3 can be stated briefly as follows.

In addition to an integrand F: $A \times B \rightarrow R$ with the nonnegative homogeneity (6.20), we use real-valued functions ϕ_β all having the same domain as F and the same homogeneity property.

Let $\mathscr{C}$ be the class of all PWS curves C having PWS representations y: $[t_0, t_1] \rightarrow R^n$ that satisfy condition (6.19) together with side-conditions

(6.39) $\phi_\beta(y,\mathring{y}) = 0, \qquad \beta = 1, \ldots, m < n - 1,$

and end-conditions

(6.40) $\psi_\mu[y(t_0), y(t_1)] = 0, \qquad \mu = 1, \ldots, p \leqslant 2n.$

Given

$$J(C) \equiv g[y(t_0), y(t_1)] + \int_{t_0}^{t_1} F(y,\mathring{y}) \, dt,$$

the basic problem is that of the existence and characterization of a curve $C_0 \in \mathscr{C}$ that furnishes a global extremum for $J(C)$ on $\mathscr{C}$. Again, we may also be interested in strong or weak local extrema.

In order that this be a curve-problem, not a function-problem, and hence be properly called parametric, observe that in addition to the homogeneity of F and ϕ_β, already stated, all the functions F, g, ϕ_β, and ψ_μ must be free of the parameter t.

For treatment of this problem see M. F. Smiley (XV, years 1933–1937) and various papers in his bibliography.

6.11 THE EULER NECESSARY CONDITION

Suppose given a fixed-endpoint parametric problem as formulated in Section 6.9 and recall that the class $\mathscr{C}$ of admissible curves consists of those with PWS representations y: $[t_0, t_1] \rightarrow R^n$ having property (6.19) and fixed end values.

Theorem 6.3

If $J(C_0)$ is at least a weak local minimum for the integral (6.35) on $\mathscr{C}$, then to each regular admissible representation y_0: $[t_0, t_1] \rightarrow R^n$ of C_0 corresponds a constant vector c such that the vector-equation

$$(6.42) \qquad F_r[y_0(t), \mathring{y}_0(t)] = \int_{t_0}^{t} F_y[y_0(\tau), \mathring{y}_0(\tau)] \, d\tau + c$$

holds in the sense (2.17) on the domain $[t_0, t_1]$ of y_0.

PROOF

The standard proof, which follows that of Theorem 2.2, Section 2.6, need not be given in detail. Let η: $[t_0, t_1] \rightarrow R$ have the properties stated in Section 2.6 and let y be the representation whose jth component is $y_0^j + \epsilon \eta$ but whose other components coincide with those of y_0. The reader should verify that, if $|\epsilon|$ is sufficiently small, such a function y will satisfy the requirements of definition (6.38). We need the openness of the sets A and B, stated in Section 6.6, in order that the representations used in this proof be admissible at least when ϵ is near zero. By following Section 2.6 we obtain the equality of jth components of the respective members of Euler condition (6.42).

As an immediate corollary, we use the continuity in t of the integral in (6.42) and obtain the Weierstrass–Erdmann corner condition

$$(6.43) \qquad F_r[y_0(t), \mathring{y}_0^-(t)] = F_r[y_0(t), \mathring{y}_0^+(t)].$$

Continuing in the pattern of Section 2.6, differentiate (6.42) at any $t \in [t_0, t_1]$ not corresponding to a corner and find that

$$(6.44) \qquad F_y[y_0(t), \mathring{y}_0(t)] = \frac{d}{dt} F_r[y_0(t), \mathring{y}_0(t)].$$

The right member can be expanded, provided that the various derivatives all exist to yield the system of scalar equations

$$(6.45) \quad F_{y^i} = F_{r^i y^j} \mathring{y}_0^j(t) + F_{r^i r^j} \mathring{\mathring{y}}_0^j(t), \quad \text{summed on } j \text{ with } i = 1, \ldots, n.$$

Forms (6.42), (6.44), and (6.45) correspond to the respective forms (2.18), (2.24), and (2.25) of the Euler condition for the problem of Chapter 2.

EXAMPLE 6.5

$n = 2$ and $F(y, r)$ is free of y.

Discussion

One often starts with (6.45), even though cases analogous to those illustrated by examples in Section 2.7 may occur and we would have to turn to (6.44) or (6.42), which hold with fewer restrictions on y_0.

For the present example and with (x,y) in place of $y = (y^1, y^2)$, the two equations (6.45) are

$$F_{\dot{x}\dot{x}}[\dot{x}_0(t), \dot{y}_0(t)]\ddot{x}_0(t) + F_{\dot{x}\dot{y}}[\quad]\ddot{y}_0(t) = 0,$$

$$F_{\dot{x}\dot{y}}[\quad]\ddot{x}_0(t) + F_{\dot{y}\dot{y}}[\quad]\ddot{y}_0(t) = 0.$$

By inspection, these equations hold if $\ddot{x}_0(t) = \ddot{y}_0(t) = 0$ and hence if

$$x_0(t) = a_1 t + b_1 \quad \text{and} \quad y_0(t) = a_2 t + b_2.$$

Since the vector-valued function (x_0, y_0): $[t_0, t_1] \to R^2$ is a representation of a possible extremizing curve C_0, the parameter interval can be chosen at pleasure as any interval of positive finite length—say $[0,1]$ or any other. After making a choice, the given fixed endpoints will determine the coefficients a_1, b_1, a_2, b_2.

The curve C_0 having a regular PWS representation of the above form has infinitely many other such representations, among which are those obtainable from this one by setting $t = \psi(u)$, where ψ: $[u_0, u_1] \to [t_0, t_1]$ is any SPH that is PWS and such that its derivative $\dot{\psi}(u)$ is always positive.

The following theorem gives an alternative form due to Weierstrass of the Euler condition (6.45) in the special case $n = 2$.

Theorem 6.4

If $n = 2$, a regular representation (x,y): $[t_0, t_1] \to R^2$ satisfies the Euler equations (6.45) iff

$$(6.46) \qquad F_{x\dot{y}}(x,y,\dot{x},\dot{y}) - F_{y\dot{x}}(x,y,\dot{x},\dot{y}) + (\dot{x}\ddot{y} - \dot{y}\ddot{x})F_1(x,y,\dot{x},\dot{y}) = 0.$$

PROOF

System (6.45) for the case $n = 2$ and in the (x,y) notation is the pair of equations

$$(6.47) \qquad \begin{aligned} F_x &= \dot{x}F_{\dot{x}x} + \dot{y}F_{\dot{x}y} + \ddot{x}F_{\dot{x}\dot{x}} + \ddot{y}F_{\dot{x}\dot{y}}, \\ F_y &= \dot{x}F_{\dot{y}x} + \dot{y}F_{\dot{y}y} + \ddot{x}F_{\dot{y}\dot{x}} + \ddot{y}F_{\dot{y}\dot{y}}. \end{aligned}$$

By Euler's theorem on homogeneous functions applied to F_x [the relation analogous to (6.28) obtained similarly by starting with the identity $F_x(x, kr) = kF_x(x, r)$], we obtain the identity

$$F_x = \dot{x}F_{x\dot{x}} + \dot{y}F_{x\dot{y}}.$$

Using this and definition (6.32) of F_1, we can express the first equation (6.47) in the form

$$(6.48) \qquad \dot{y}[F_{x\dot{y}} - F_{\dot{x}y} + (\dot{x}\ddot{y} - \dot{y}\ddot{x})F_1] = 0.$$

Similarly, from the second equation (6.47),

$$(6.49) \qquad \dot{x}[F_{\dot{x}y} - F_{x\dot{y}} + (\dot{y}\ddot{x} - \dot{x}\ddot{y})F_1] = 0.$$

Since (x,y) is regular by hypothesis, $\dot{x}(t)$ and $\dot{y}(t)$ cannot vanish simultaneously; consequently (6.48) and (6.49) imply (6.46). Clearly (6.46) implies (6.48) and (6.49), and these two are equivalent to system (6.47).

6.12 NECESSARY CONDITIONS OF WEIERSTRASS AND LEGENDRE

Let $\mathscr{C}$ again be the class of admissible curves for the fixed-endpoint problem of Section 6.9 and recall that admissible representations satisfy (6.19).

Theorem 6.5

If $J(C_0)$ is a strong local or a global minimum for the integral (6.35) on $\mathscr{C}$ and $y_0 : [t_0, t_1] \to R^n$ is a regular admissible representation of C_0, then

$$(6.50) \qquad E[y_0(t), \dot{y}_0(t), q] \geq 0, \qquad \forall\, t \in [t_0, t_1] \text{ and } \forall\, q \in B$$

with the symbol $\dot{y}_0(t)$ understood in the sense of convention (2.17).

PROOF

We follow the proof of Theorem 2.5. The reader can construct his own figure similar to Fig. 2.1 without the axes. The graph of y_0 that one draws now represents a projection into the plane of a graph from R^n and it can have loops and multiple points, which are excluded under the restrictions of Chapter 2.

Given $\tau \in [t_0, t_1)$ and not corresponding to a corner of y_0, select a number $a \in (\tau, t_1)$ that is so near to τ that no parameter value t in $[\tau, a)$ corresponds to a corner of y_0. Let Y denote the vector-valued function with values

$$Y(t) \equiv y_0(\tau) + q(t - \tau),$$

in which q denotes an arbitrary point other than θ of the set B that appears in the formulations of this chapter and is also unequal to $\dot{y}_0(\tau)$.

Given $u \in [\tau,a)$ and that u is so near to τ that $Y(t)$ remains in the set A if $t \in [\tau,u]$, define

$$(6.51) \qquad y(t) \equiv \begin{cases} y_0(t), & t \in [t_0,\tau] \cup [a,t_1], \\ Y(t), & t \in [\tau,u], \\ \phi(t,u), & t \in [u,a], \end{cases}$$

with

$$\phi(t,u) \equiv y_0(t) + \frac{Y(u) - y_0(u)}{a-u}(a-t).$$

The function $y: [t_0,t_1] \to R^n$ with values (6.51) is admissible and coincides with y_0 except generally on the interval (τ,a). When $u = \tau$, $y(t)$ reduces to $y_0(t)$ on the entire interval $[t_0,t_1]$.

Define $\Phi(u) \equiv J(y) - J(y_0)$. The function Φ with these values is differentiable and, as a consequence of the hypothesis on y_0, we have the necessary condition

$$\Phi'(\tau) \geqslant 0.$$

Since $y = y_0$ on $[t_0,\tau] \cup [a,t_1]$ we see that

$$\Phi(u) = \int_\tau^u F(Y,\dot{Y})\,dt + \int_u^a F(\phi,\phi_t)\,dt - \int_\tau^a F(y_0,\dot{y}_0)\,dt.$$

We obtain an expression for $\Phi'(u)$ by means of Theorem 1.3 after observing that the last term is free of u. The result is that

$$\Phi'(u) = F[Y(u),\dot{Y}(u)] - F[\phi(u,u),\phi_t(u,u)]$$
$$+ \int_u^a \{F_y[\phi(t,u),\phi_t(t,u)] \cdot \phi_u(t,u) + F_r[\ \ ] \cdot \phi_{tu}(t,u)\}\,dt.$$

After integrating the first term by parts, we use the fact that $\phi_u(a,u) = 0$, set $u = \tau$, and see that $\phi(t,\tau) = y_0(t)$ must satisfy the Euler equation (6.44) since y_0 has no corners for t in the interval $[\tau,a]$. We then see, with reference to form (6.33) of the E-function, that our necessary condition is the inequality

$$\Phi'(\tau) = E[y_0(\tau),\dot{y}_0(\tau),q] \geqslant 0, \qquad \forall\, q \in B,$$

subject to the restriction at the beginning of the proof that $\tau \neq t_1$ and that $y_0(\tau)$ is not a corner.

To remove these restrictions, use the blanket hypothesis on F as stated in Section 6.6 to obtain the continuity of E. Then let $\tau \to t_1$ from below and find that the above inequality holds with $\tau = t_1$. If $t_2 \in (t_0,t_1)$ corresponds to a corner, let $\tau \to t_2$ from below and also from above. The inequality holds at t_2 in the sense of convention (2.17), that is, with either $\dot{y}_0^-$ or $\dot{y}_0^+$ in place of $\dot{y}_0$. The inequality thus holds for all $\tau \in [t_0,t_1]$.

Finally replace τ by t so as to express the conclusion in the stated form (6.50).

If in the definition of Y we restrict the vector q by the condition that $q - \overset{\bullet}{y}_0(t)$ be of suitably small norm so that y is in the first-order neighborhood of y_0 that appears in our definition of a weak local minimum, then the preceding proof yields the following theorem as a corollary.

Theorem 6.6

If $J(C_0)$ is a weak local minimum for the integral (6.35) and y_0: $[a,b] \to R^n$ is a regular admissible representation of C_0, then there exists a positive real number δ such that

$$(6.52) \qquad E[y_0(t), \overset{\bullet}{y}_0(t), q] \geq 0, \qquad \forall \ t \in [t_0, t_1] \text{ and } \forall \ q \in B$$

$$\text{such that } |q - \overset{\bullet}{y}_0(t)| < \delta.$$

Companion theorems for maxima to Theorems 6.5 and 6.6, obtained by applying these theorems to $-J$, yield (6.50) and (6.52) with reversed inequalities.

Weierstrass necessary conditions for nonparametric variational problems in $(n+1)$-space are available as byproducts from the preceding discussion. We can replace integral (6.35) by (6.1) and hence replace $F(y,r)$ throughout the proof of Theorem 6.5 by $f(x,y,r)$. Since we have not used the homogeneity of F in that proof and the only differentiations are with respect to components of y and r, every step can be rewritten in terms of f. For the case $n = 1$, we recover Theorem 2.5. Comparison of the proof of that theorem in Section 2.8 with the proof of Theorem 6.5 will accentuate and clarify these remarks.

Theorem 6.7

If $J(C_0)$ is at least a weak local minimum of the integral (6.35) on $\mathscr{C}$ and y_0: $[t_0, t_1] \to R^n$ is a regular admissible representation of C_0, then the quadratic form

$$(6.53) \qquad F_{r^i r^j}[y_0(t), \overset{\bullet}{y}_0(t)] v^i v^j \geq 0, \qquad \forall \ t \in [t_0, t_1],$$

$\forall \ n$-vector v not a multiple $\lambda \overset{\bullet}{y}_0(t)$ of $\overset{\bullet}{y}_0(t)$ by a scalar.

PROOF

Under our blanket hypothesis on F, $F(y,q)$ has for a fixed y, the following Taylor expansion with remainder:

$$(6.54) \qquad F(y,q) = F(y,\overset{\bullet}{y}) + (q^i - \overset{\bullet}{y}^i)F_{r^i}(y,\overset{\bullet}{y})$$
$$+ \tfrac{1}{2}(q^i - \overset{\bullet}{y}^i)(q^j - \overset{\bullet}{y}^j)F_{r^i r^j}[y, \overset{\bullet}{y} + \theta(q - \overset{\bullet}{y})], \qquad \theta \in (0,1),$$
$$\text{with summation on } i \text{ and } j \text{ from } 1 \text{ to } n.$$

It follows from definition (6.33) of the E-function that

(6.55) $$E(y,\dot{y},q) = last\ term\ in\ (6.54).$$

To complete the proof, ignore the factor $\frac{1}{2}$, use Theorem 6.6, multiply $q-\dot{y}$ by an arbitrary real number α, and denote the resulting vector by v. Conclusion (6.53) is correct without the stated restriction on v. Observe that, if identity (6.29) is multiplied by r^j and summed on j, the result is identically zero, hence that if $v = \lambda y_0(t)$, then (6.53) holds trivially in the form $0 = 0$.

In the notation of Bliss, the respective necessary conditions (6.42), (6.50), and (6.53) of Euler, Weierstrass, and Legendre are again designated by the Roman numerals I, II, and III. Of course (6.52) is also a Weierstrass necessary condition, but when this term is mentioned without a qualification it is customarily understood to mean form (6.50) for the strong minimum. There is also a Jacobi necessary condition IV for the problem of Section 6.9, but we shall not discuss it. See Bliss (IX, pp. 116–124).

Analogues of these conditions for more general parametric problems including the parametric Bolza Problem of Section 6.10 are also to be found in the literature.

Sufficiency theorems for local extrema of a variety of parametric problems stated in terms of strengthened forms of the four necessary conditions are also available. The statements of such theorems resemble those given in Section 3.6 for the nonparametric problem in the plane. Again see Bliss (IX, pp. 124–132), Pars (XXXII, Chap. IX) for the case $n = 2$, and other standard reference material.

Exercise 6.5

1. With $n = 2$ and using (x,y) for a point in the plane, show that $x_0(t) = y_0(t) = t$, $0 \le t \le 1$, is a representation satisfying the Euler, Weierstrass, and Legendre necessary conditions for the problem, $\int (\dot{x}^2 + \dot{y}^2)^{1/2}\,dt =$ minimum with the fixed endpoints $(0,0)$ and $(1,1)$.

2. Recall that a binary quadratic form $au^2 + 2buv + cv^2$ is of fixed sign iff its discriminant $4(b^2 - ac) \le 0$. Show as a corollary to Theorem 6.7 that, in the case $n = 2$, $F_1[x_0(t),y_0(t),\dot{x}_0(t),\dot{y}_0(t)] \ge 0$ is necessary for a minimum and that ≤ 0 is necessary for a maximum.

3. Investigate the conclusions obtainable from an application of (6.46) to Example 6.5.

4. Discuss for a general n the various conclusions obtained from Sections 6.9, 6.11, and 6.12 for a fixed-endpoint problem based on an integrand F that is free of y.

5. Construct, with possible reference to Chapter 2 for a hint, an example of a parametric problem that has an extremizing curve C_0 with one or more corners.

6. Given the Multiplier Rule as a necessary condition for the nonparametric Bolza Problem, investigate what this implies for an extremizing representation y_0 for the parametric Bolza Problem of Section 6.10.

6.13 RELATED PARAMETRIC AND NONPARAMETRIC PROBLEMS OF LIKE DIMENSIONALITY

Parametric problems arise from the two sources described in Section 6.6 except for artificial examples obtained by constructing continuous homogeneous integrands F at pleasure.

A parametric variational problem of the traditional geometric type accepts both positive and negative values $\dot{x}(t)$ of the derivative of the first coordinate of the moving point (x,y). If, however, we start with a nonparametric integral (6.1), with x understood to run over an interval $[a,b]$, and then pass to the associated parametric integral (6.2), we must restrict x to be increasing in t or else lose an essential feature of the nonparametric problem.

There is a formal one-one correspondence between integrals (6.1), (6.2). Replace symbols $\dot{x}$, $\dot{y}$, and dt of the parametric integral by 1, y' and dx, respectively, to obtain its nonparametric counterpart. There is not, however, a one-one correspondence between the class of all PWS Fréchet curves with fixed endpoints and the class $\mathcal{Y}$ of all PWS functions y with these same endpoints as remarked in Section 6.6. If $\mathcal{C}_1$ denotes the proper subclass of $\mathcal{C}$ consisting of those curves $C \in \mathcal{C}$ each of which has a representation (x,y): $[t_0,t_1] \to R^{m+1}$ such that x is strictly increasing on $[t_0,t_1]$, then there is a one-one correspondence between $\mathcal{C}_1$ and $\mathcal{Y}$.

Parametric and nonparametric fixed-endpoint problems

(6.56) $J_P(C) = minimum\ on\ \mathcal{C}$ and $J_{NP}(y) = minimum\ on\ \mathcal{Y}$

with the integrals formally related as in (6.2) are in general quite distinct problems.

EXAMPLES 6.6

Consider possible minima of the following pairs of integrals (6.56):

(i) $\int x^2|\dot{y}|dt$ and $\int x^2|y'|dx$ *with endpoints* $(-1,-1)$ *and* $(1,1)$,

(ii) $\int (\dot{y}/\dot{x})^2[(\dot{y}/\dot{x}) - 1]^2\dot{x}\ dt$ *and* $\int (y')^2(y' - 1)^2\ dx$ *with endpoints* $(1,1)$ *and* $(2,0)$,

(iii) $\int [\dot{y}^2(\dot{y}-\dot{x})^2]^{1/4} dt$ and $\int [(y')^2(y'-1)^2]^{1/4} dx$ with endpoints (1,1) and (2,0),

(iv) $\int [\dot{x}^2+\dot{y}^2]^{1/2} dt$ and $\int [1+(y')^2]^{1/2} dx$ with endpoints (0,0) and (1,1).

Discussion

The infimum of both integrals (i) with nonnegative integrands is zero. This is realized for the parametric integral $J_P(C)$ by the curve C_0 whose graph is the piecewise linear path from $(-1,-1)$ to $(0,-1)$ to $(0,1)$ to $(1,1)$. The piecewise linear function y_ϵ joining $(-1,-1)$ to $(0-\epsilon,-1)$ to $(0+\epsilon, 1)$ to $(1, 1)$, $\epsilon > 0$, yields the value $J_{NP}(y_\epsilon) = 2\epsilon^2/3$, but there is no function $y_0 \in \mathscr{Y}$ that yields the infimum 0 of $J_{NP}(y)$.

For (ii) let C_ϵ be the admissible curve whose graph is the broken line from $(1,1)$ to $(1-\epsilon,0)$ to $(2,0)$ with $\epsilon > 0$. By elementary integration

$$J_P(C_\epsilon) = -(1/\epsilon)[(1/\epsilon)-1]^2,$$

the infimum of which is $-\infty$. The maximal sets A and B for this parametric integrand are $A = R^2$ and $B = R^2$ less all points with abscissa $p = 0$. Thus the class of admissible representations $(x,y): [t_0,t_1] \to R^2$ excludes by way of condition (6.19) those such that $\dot{x}(t)$ vanishes anywhere on its parameter interval. The value of the parametric integral (ii) for any choice of an admissible representation is a real number; hence the infimum of $J_P(C)$ is realizable by no choice of an admissible representation of an admissible curve. One finds however with reference to Example 2.1, Section 2.7, and to Theorem 3.9 and from the observation that the family of linear functions of slope -1 provides a field in the large that the function y_0; $y_0(x) = 2-x$ furnishes a global minimum for the nonparametric integral.

The two integrands (iii) are nonnegative; hence their respective infima on $\mathscr{C}$ and $\mathscr{Y}$ are nonnegative. The parametric integrand clearly vanishes for any PWS representation of the curve C_0 having as its graph the broken line from $(1,1)$ to $(0,0)$ to $(2,0)$; hence the global minimum of $J_P(C)$ is $J_P(C_0) = 0$. The global minimum of the nonparametric integral also exists, as one finds by the same procedure outlined in the last paragraph. It is $J_{NP}(y_0) = \sqrt{2}$, where y is again the linear function with values $y_0(x) = 2-x$.

The pair of length integrals (iv), in contrast with (i), (ii), and (iii), both have the same global minimum. The admissible representation $x_0(t) = y_0(t) = t \in [0,1]$ satisfies the Euler condition for the parametric integral by Example 6.5. Since we have not discussed sufficient conditions for parametric integrals, let us accept for the moment that the Fréchet curve C_0 having this representation is the curve of least length. Clearly $J_P(C_0) = \sqrt{2}$.

It happens that the minimizing curve C_0 in the class $\mathscr{C}$ of all admissible PWS curves is actually in the subclass that we have called $\mathscr{C}_1$. Granted this result, we have as an immediate corollary that the function $y_0, y_0(x) \equiv x$ corresponding to the graph of C_0 minimizes the corresponding nonparametric integral. That $J_{NP}(y_0) = \sqrt{2}$ is the global minimum of $J(y)$ on $\mathscr{Y}$ can also be established by methods of Chapter 3.

Exercise 6.6

1. Show for the nonparametric integral (ii) under Examples 6.6 that

$$E[x,y,p(x,y),y'] = (y'+1)[(y')^2 - 4y' + 8].$$

   Justify the assertions that have been made concerning this integral by direct use of the E-function and also by considering the figurative. Explain why the minimum value of the integral is proper.

2. Justify the conclusions that have been stated for the nonparametric integral of Example 6.6(iii) by studying the figurative.

6.14 AN ADDENDUM TO THE EULER CONDITION FOR A NONPARAMETRIC INTEGRAL

The fixed-endpoint nonparametric problem of minimizing integral (6.1), as a very special case of the Bolza Problem of Chapter 5, has the Euler necessary condition (5.24).

Consider the related parametric extremum problem on the class $\mathscr{C}_1$ described in Section 6.13. Every curve $C \in \mathscr{C}_1$ has a representation (x,y) such that $x(t) \equiv t$, and this representation is clearly admissible. There is a one-one correspondence between admissible PWS functions $y: [x_0, x_1] \to R^m$ for the nonparametric problem and admissible PWS representations

$$(6.57) \qquad x(t) \equiv t, \qquad y[x(t)] \equiv y(t)$$

for the parametric problem. Moreover, the parametric problem on $\mathscr{C}_1$ is equivalent to the nonparametric problem in the sense that a representation (6.57) furnishes a given type of extremum for the former iff the corresponding function y furnishes that type of minimum for the latter.

It follows that the Euler equation (6.42) is a necessary condition on such a pair (6.57). Starting with the identity

$$F(x,y,\dot{x},\dot{y}) \equiv f(x,y,r)\dot{x}, \qquad \textit{where } r \textit{ is the m-vector } \dot{y}/\dot{x},$$

we see that

(6.58) $F_{\dot{x}}(x,y,\dot{x},\dot{y}) = f(x,y,r) - r \cdot f_r(x,y,r)$ and $F_x(\ ) = f_x(\ )$

and also that

(6.59) $F_{\dot{y}}(x,y,\dot{x},\dot{y}) = f_r(x,y,r)$ and $F_y(\ ) = f_y(\ ).$

The first of the $m+1$ scalar equations equivalent to the vector equation (6.42), written with the aid of (6.58) in the notation of the nonparametric problem, is of the form

(6.60)
$$f[x,y_0(x),\dot{y}_0(x)] - \dot{y}_0(x) \cdot f_r[x,y_0(x),\dot{y}_0(x)] = \int_{x_0}^{x} f_x[\xi,y_0(\xi),\dot{y}_0(\xi)]\, d\xi + c.$$

The remaining m equations obtained similarly from (6.59), in contrast with (6.60), provide nothing new. They are equivalent to equation (5.24) with the present m playing the role of n in Section 5.7.

The integral in (6.60) is continuous in its variable upper limit x. If $x = x_0$ corresponds to a corner it follows that values of the left member with $\dot{y}_0^-(x_0)$ or $\dot{y}_0^+(x_0)$ in place of $\dot{y}_0(x_0)$ must be equal. This is the extension of (2.38) to a general n. The extension of (2.37) is similarly obtained with the aid of (6.59) and (6.42).

6.15 RELATED PARAMETRIC AND NONPARAMETRIC PROBLEMS OF DIFFERENT DIMENSIONALITY

Given a parametric integrand $F(y,\dot{y})$, where y is a vector-valued function with n components, we can regard it as a nonparametric integrand in the $(n+1)$-space of points (t,y) even though F is free of t. Various theorems, methods, and results for nonparametric problems can then be exploited.

Consider, for example, the parametric length integral (6.4) for plane curves and the variational problem with fixed endpoints $(0,0)$ and $(1,1)$ that occurs in the fourth of Examples 6.6. With $[t_0,t_1]$ as an arbitrary but fixed compact interval of positive length, think of the nonparametric problem in three-space:

(6.61) $J(x,y) \equiv \int [\dot{x}^2 + \dot{y}^2]^{1/2}\, dt = minimum$

with the fixed endpoints $(t_0,0,0)$ and $(t_1,1,1)$. Theorem 5.2 for nonparametric problems applies to yield a pair of Euler equations and we are led to the particular solution

(6.62) $x_0(t) = y_0(t) = (t - t_0)/(t_1 - t_0),$

satisfying the given end-conditions.

Because of the convexity of the integrand in $\dot{x}$ and $\dot{y}$, Theorem 3.13 is applicable and we find that $J(x_0, y_0)$ is the global minimum of the nonparametric integral $J(x, y)$. If we now revert to the context of planar PWS Fréchet curves, we know that integral (6.61) is independent of the choice of an admissible representation as a consequence of Theorem 6.2. It follows that the curve C_0, of which (6.62) is one representation, furnishes the least length $\sqrt{2}$ among all PWS curves joining $(0,0)$ and $(1,1)$ in the plane. The gap in our discussion of Example 6.6(iv) is now filled.

This overly simple example serves to suggest that similar ideas also apply to more complex problems.

6.16 CONCLUDING REMARKS

Parametric problems were introduced by Weierstrass about 1872. The literature then tended to separate into two parts, one on parametric problems, the other on nonparametric problems. Methods and conclusions were similar and yet different in important details.

Optimization problems from differential geometry such as the quest for paths of least length or surfaces of least area are appropriately formulated as parametric problems. If, for example, one desires the shortest path joining fixed points in the plane, he is not content to know that a curve C_0 is of smaller length than any other curves C in the class $\mathscr{C}_1$ of Section 6.13. He wants to establish that C_0 has smaller length than any other curve in the larger class $\mathscr{C}$ of Sections 6.9 and 6.13 or better still that it has smaller length than any other continuous Fréchet curve having the given endpoints. This last objective cannot be achieved under the classical restriction of PWS curves, but it is covered by Theorem 7.19.

Variational problems concerning the design or control of dynamical systems may involve t explicitly and usually lack the homogeneity property (6.20) and hence are generally nonparametric. Although this book is oriented toward applications, we cannot afford to ignore parametric problems. McShane achieved (33b,c) advances in existence theory for nonparametric problems with the aid of associated parametric problems and obtained (33k) sufficient conditions for a nonparametric Bolza Problem with the aid of a parametric problem. In Section 10.2 we replace a novel nonparametric Mayer Problem by a classical parametric Mayer Problem. It seems likely that further opportunities can be found for exploiting known methods or results for one type of problem as tools in the analysis of one of the other type.

Chapter 7

DIRECT METHODS

7.1 INTRODUCTION

The work in earlier chapters has depended heavily upon the use of Euler equations and upon the existence and properties of families of solutions. Such traditional methods in the calculus of variations are called indirect. In contrast, procedures that avoid the intervention of differential equations, hence of dependence upon the theory of such equations, and that emphasize convergence properties of functionals and of classes of admissible curves and functions are called direct.

The term *direct methods* is applied to the approach to existence theory initiated by Hilbert (X,pp.245–263; XI,pp.428–436) and developed by Tonelli (XXXV, XXXVI), McShane, and others. Various computationally oriented procedures intended to converge to an optimizing function are also described as direct.

This chapter is concerned with the theory of length, the sequential compactness of important classes of curves and functions, the semi-continuity of functionals, and the existence of global extrema. Such topics mark a shift in content and viewpoint to ideas that have come to the fore in the twentieth century. We exhibit few optimizing functions for particular problems. Our major objective is to investigate the structure of problems and moreover to do so under materially weaker restrictions on integrands and on admissible functions.

7.2 GLOBAL EXTREMA OF REAL-VALUED FUNCTIONS

Recall the classic theorem that, if the function ϕ: $[a,b] \to R$ is continuous on the compact interval $[a,b]$, then there exists $x_* \in [a,b]$ such that $\phi(x_*) \leqslant \phi(x)$, $\forall x \in [a,b]$ and also $x^* \in [a,b]$ such that $\phi(x^*) \geqslant \phi(x)$, $\forall x \in [a,b]$. The usual proofs with minor changes in wording extend to the case of the global minimum and maximum of a continuous function ϕ: $K \subset R^n \to R$, where the domain K is any nonempty compact subset of R^n.

We would like to have theorems of this kind for *curve-functions* J: $\mathscr{C} \to R$, but unfortunately the only such functions that turn out to be continuous in a sense to be defined presently are those of the form

$$J(C) = \int P[x(t)] \cdot \dot{x}(t) \, dt,$$

with integrands that are linear in the components of $\dot{x}(t)$.

However, many interesting functionals J are found to be lower or upper semi-continuous, and these properties with other suitable hypotheses suffice for theorems on the existence of the global minimum and global maximum, respectively.

Let K now denote a set of elements of any kind for which there is a suitable notion of convergence to a limit. Among other things K can be a subset of R^n, or a metric space (S,d) whose elements are functions y or curves C. Suppose further that K is *sequentially compact*, by which we mean that every sequence in K has at least one subsequence converging to an element of K. It is well known and not difficult to prove that every bounded and closed subset of R^p is sequentially compact. We prove that certain important classes of functions y and of curves C are sequentially compact in Sections 7.7 through 7.9.

Theorem 7.1

Given a nonempty sequentially compact set K and a function J: $K \to R$ that is bounded below and lower semi-continuous on K, there exists $x_* \in K$ such that

$$(7.1) \qquad\qquad J(x_*) \leqslant J(x), \qquad \forall x \in K.$$

PROOF

Set $\gamma \equiv \inf\{J(x): x \in K\}$. By hypothesis J has a lower bound, hence $\gamma > -\infty$. As a consequence of the definition of infimum there exists a sequence $\{x_\nu \in K : \nu \in N\}$ such that $\lim J(x_\nu) = \gamma$, and since K is sequentially compact we can suppose sequence $\{x_\nu\}$ so chosen that it has a limit $x_* \in K$.

That J is lower semi-continuous on K means that, for every y_0 in K and every sequence $\{y_\nu\}$ in K with limit y_0, $\lim \inf J(y_\nu) \geq J(y_0)$, hence that $\lim \inf J(x_\nu) \geq J(x_*)$. These observations combine to say that

$$J(x_*) \leq \lim \inf J(x_\nu) = \lim J(x_\nu) = \gamma \leq J(x_*).$$

The outer terms being equal, all terms must be equal; consequently, $J(x_*) = \gamma$ and (7.1) holds. Thus $J(x_*)$ is the global minimum of $J(x)$ on K.

Exercise 7.1

1. Given the function $\phi: [-1/\pi, 1/\pi] \to R$ with values $\phi(x) = 1 + \sin 1/x$ or 0 according as $x \neq 0$ or $= 0$, point out that the hypotheses of Theorem 7.1 are satisfied and exhibit one sequence $\{x_\nu\}$ with the property of that sequence in the proof of Theorem 7.1. Do the same with $\phi(0) = -2$ in place of 0.

2. Formulate and prove a theorem similar to Theorem 7.1 on the existence of the global maximum of an upper semi-continuous function.

7.3   LENGTH OF A MAPPING

Given a compact interval $[a,b]$ of positive length and a continuous function $x: [a,b] \to R^n$, let π denote a *partition* of $[a,b]$, which means a finite subset $\{t_0, t_1, \ldots, t_k\}$ such that $a = t_0 < t_1 \cdots < t_k = b$. The length $\mathscr{L}(x)$ of x is, by definition,

$$(7.2) \qquad \mathscr{L}(x) \equiv \sup_{\pi} \sum_i |x(t_i) - x(t_{i-1})|.$$

The sum on i can be thought of as a sum of lengths of consecutive chords that join endpoints $x(a)$ and $x(b)$. The supremum is on the class of all partitions π of $[a,b]$. If and only if x maps the interval $[a,b]$ onto a single point of R^n, the sum in (7.2) is zero for every choice of the partition π and hence $\mathscr{L}(x) = 0$.

Clearly $\mathscr{L}(x) \geq 0$. If the path traced by $x(t)$ is sufficiently crinkly, $\mathscr{L}(x) = \infty$. If $\mathscr{L}(x) < \infty$, then x is called *rectifiable*.

The supremum in (7.2) is also called the *total variation* of the function x. It seems convenient in the present chapter, however, to follow a traditional restriction and to use the alternative term only in the case $n = 1$. Among several symbols for the total variation or length $\mathscr{L}(x)$ of a function x: $[a,b] \to R$ is $T(x; [a,b])$. If $T(x; [a,b]) < \infty$, then x is said to be of *bounded variation* on $[a,b]$, and this is abbreviated by saying that x is BV on $[a,b]$.

Theorem 7.2

A mapping x: $[a,b] \to R^n$ is rectifiable iff each component x^j of x is BV on $[a,b]$.

PROOF

Use the definitions and the inequalities

$$|x^j(t_i) - x^j(t_{i-1})| \leqslant |x(t_i) - x(t_{i-1})| \leqslant \sum_j |x^j(t_i) - x^j(t_{i-1})|,$$

together with the sums of the three expressions with respect to i.

7.4   LOWER SEMI-CONTINUITY OF LENGTH

Theorem 7.3

Given the compact interval $I = [a,b]$ of positive length and a sequence $\{x_\nu : \nu \in N\}$ of continuous functions $x_\nu : I \to R^n$ converging pointwise to $x_0 : I \to R^n$, then

$$(7.3) \qquad \liminf \mathscr{L}(x_\nu) \geqslant \mathscr{L}(x_0).$$

PROOF

CASE 1, $\mathscr{L}(x_0) < \infty$

Given $\epsilon > 0$, there exists by definition (7.2) a partition π_ϵ of I consisting of points t_i, $i = 0, \ldots, k$ such that

$$(7.4) \qquad \sum_i |x_0(t_i) - x_0(t_{i-1})| > \mathscr{L}(x_0) - \epsilon/2.$$

With ϵ and π_ϵ fixed, there exists, as a consequence of the convergence of $x_\nu(t_i)$ to $x_0(t_i)$, $i = 0, \ldots, k$, an integer M_ϵ such that if $\nu > M_\epsilon$, then $\left| |x_\nu(t_i) - x_\nu(t_{i-1})| - |x_0(t_i) - x_0(t_{i-1})| \right| < \epsilon/2k$ for every value of i. It follows that

$$(7.5) \qquad |x_\nu(t_i) - x_\nu(t_{i-1})| > |x_0(t_i) - x_0(t_{i-1})| - \epsilon/2k.$$

After summation on i we see with reference to definition (7.2) of length and (7.4) that

$$\mathscr{L}(x_\nu) > \mathscr{L}(x_0) - \epsilon \qquad if\, \nu > M_\epsilon,$$

from which conclusion (7.3) follows.

CASE 2, $\mathscr{L}(x_0) = \infty$

Given $\epsilon > 0$, there exists a partition π_ϵ of I such that

$$\sum_i |x_0(t_i) - x_0(t_{i-1})| > 1/\epsilon.$$

Inequality (7.5) again holds and consequently

$$\mathscr{L}(x_\nu) > 1/\epsilon - \epsilon/2 \qquad if \qquad \nu > M_\epsilon.$$

It follows that $\mathscr{L}(x_\nu) \to \infty$; hence the desired conclusion holds in the form $\lim \mathscr{L}(x_\nu) = \mathscr{L}(x_0)$.

We frequently wish to replace a given continuous mapping y: $[c,d]$ $\to R^n$ by a Fréchet-equivalent (defined in Section 6.3) continuous mapping x having an assigned domain $[a,b]$. That this can be done is easily seen with reference to definition (6.9) of the distance ρ. Let h be an SPH that maps $[a,b]$ onto $[c,d]$ and define x as the composition $y \circ h$. Thus $x(t) \equiv y[h(t)]$, $t \in [a,b]$, and the euclidean distance $|x(t) - y[h(t)]|$ is identically zero on $[a,b]$. Consequently, $\rho(x,y) = 0$.

Theorem 7.4

Given the compact intervals I_0, I_1, ..., of positive length, given continuous mappings y_ν: $I_\nu \to R^n$, $\nu = 0, 1, \ldots$, and given that $\rho(y_\nu, y_0) \to 0$ as $\nu \to \infty$, then

(7.6) $$\lim \inf \mathscr{L}(y_\nu) \geq \mathscr{L}(y_0).$$

PROOF

Let h_ν be an SPH that maps I_0 onto I_ν with the property that $\sup|y_0(t) - y_\nu[h_\nu(t)]| < \rho(y_0, y_\nu) + 1/\nu$ and define $x_\nu(t) \equiv y_\nu[h_\nu(t)]$, $t \in I_0$. Thus x_ν: $I_0 \to R^n$ is Fréchet-equivalent to y_ν: $I_\nu \to R^n$ and one verifies from definition (7.2) of $\mathscr{L}$ that $\mathscr{L}(x_\nu) = \mathscr{L}(y_\nu)$. The stated conclusion then follows from Theorem 7.3.

One defines lower semi-continuity of a functional, in particular of $\mathscr{L}$, by following the pattern of the corresponding definition (1.15) for a point-function with convergence of y_ν to y_0 now taken to mean that $\rho(y_\nu, y_0) \to 0$. If inequality (7.6) holds for every sequence $\{y_\nu: \nu \in N\}$ such that $\rho(y_\nu, y_0) \to 0$, the functional $\mathscr{L}$ is lower semi-continuous at y_0. This is what Theorem 7.4 asserts.

7.5 LENGTH OF A CURVE

Recall the definition of a Fréchet curve C in E_n in Section 6.3 and of Fréchet distance d between curves in Section 6.4.

Theorem 7.5

If x and y are any two representations of the same Fréchet curve C in E_n, then $\mathscr{L}(x) = \mathscr{L}(y)$.

PROOF

Since x and y both represent C, we have that $\rho(x,y) = 0$ by the definition of a Fréchet curve. Set $z_\nu \equiv y$, $\nu = 1, 2, \ldots$. Clearly $\mathscr{L}(z_\nu)$ has the limit $\mathscr{L}(y)$ as $\nu \to \infty$; consequently, by Theorem 7.4

$$\mathscr{L}(y) = \liminf \mathscr{L}(z_\nu) \geq \mathscr{L}(x).$$

Repeating the argument with the roles of x and y reversed, we obtain the complementary inequality $\mathscr{L}(x) \geq \mathscr{L}(y)$.

In view of this theorem we can define the length $L(C)$ of a Fréchet curve C as follows.

$$(7.7) \qquad L(C) \equiv \mathscr{L}(x), \qquad \text{where } x \text{ is an arbitrary representation of } C.$$

That $0 \leq L(C) \leq \infty$ follows from the corresponding property of $\mathscr{L}(x)$.

A curve-function J is called *lower semi-continuous at C_0* if, for every choice of a sequence $\{C_\nu : \nu \in N\}$ such that Fréchet distance $d(C_\nu, C_0) \to 0$,

$$\liminf J(C_\nu) \geq J(C_0).$$

Theorem 7.6

Given the class $\mathscr{C}$ of all continuous oriented Fréchet curves in E_n, the function $L: \mathscr{C} \to R^*$ is lower semi-continuous at each $C_0 \in \mathscr{C}$.

PROOF

Let $y_\nu: I_\nu \to R^n$ be a representation of C_ν, $\nu = 0, 1, 2, \ldots$. The desired conclusion is immediate from definition (7.7) and Theorem 7.4.

7.6 THE REPRESENTATION IN TERMS OF LENGTH

Given a Fréchet curve C represented by $x: [a,b] \to R^n$, let $C_{x,t}$ called a *subcurve of C relative to x* be the curve represented by the restriction of x

to a subinterval $[a,t]$ of $[a,b]$. All intervals I that are domains of representations of curves have been required thus far to be a positive length. It is now convenient to relax this condition and to include the degenerate representation x: $[a,a] \to R^n$ of the subcurve $C_{x,a}$. Clearly there is exactly one image point. If we modify the definition of partition preceding (7.2) so as to require only that $a = t_0 \leq t_1 \leq \cdots \leq t_k = b$ with weak inequalities, then $\mathscr{L}(x) = L(C_{x,a}) = 0$ for a degenerate curve.

We see from definitions (7.2) and (7.7) that $L(C_{x,t})$ is nondecreasing in t on $[a,b]$. The representation x is called *proper* if $L(C_{x,t})$ is strictly increasing, otherwise *improper*. The representations y and x of our Example 6.1 are proper and improper, respectively.

Theorem 7.7

Given a Fréchet curve C in E_n of positive finite length, then among the representations of C is a particular one, X: $[0,L(C)] \to R^n$, with the important property that, if $C_{X,s}$ is a subcurve of C relative to X, then

$$(7.8) \qquad L(C_{X,s}) = s, \qquad \forall s \in [0,L(C)].$$

This mapping X, traditionally called the parameterization of C in terms of arc length, is useful in vector analysis, the theory of functions of a complex variable, and in differential geometry as well as in variational theory. The existence of X is often taken for granted. It is easy to prove for a smooth curve C in the manner suggested under our problem 2, Exercise 6.2, provided one is given a proper representation of C to start with, but this is a substantial gift.

PROOF OF THEOREM 7.7

Let x: $[a,b] \to R^n$ be an arbitrary representation of C with $0 < b-a < \infty$. Let s: $[a,b] \to [0,L(C)]$ be the function with values $s(t) = L(C_{x,t})$. Intuition says loudly that since the function x is continuous and $L(C) = \mathscr{L}(x)$ is finite, then surely

(α)   *the function s is continuous on $[a,b]$.*

A proof of (α) follows the proof of the theorem.

Since s is clearly nondecreasing, it has a countable set, empty, finite, or denumerable as the case may be, of intervals of constancy. Define $t(s)$ as the maximal t such that $s(t) = s$. Thus if $s(t) = s_1$ for one and only one value t_1, then $t(s_1) = t_1$; if $s(t) = s_1$ for more than one value of t, then the totality of such values must, as a result of the continuity of s, comprise a closed interval $[t_1, t_1']$ and $t(s_1) = t_1'$. The choice of the maximal value t_1' rather than the minimal value t_1 or some other explicit value belonging to $[t_1, t_1']$ such as the midpoint represents an arbitrary step in the proof. The function t: $[0,L(C)] \to [a,b]$ that we have chosen as a tool is an (not the) inverse of the function s. The function t is strictly increasing. There is a one-one correspondence between intervals on which $s(t)$ is constant and possible discontinuities of $t(s)$.

Next define $X(s) \equiv x[t(s)]$, $0 \leqslant s \leqslant L(C)$ and verify that X is continuous on $[0,L(C)]$ even though $t(s)$ may not be. Consequently, X represents some continuous Fréchet curve. We hope that it is the same curve C with which we started. To establish that it is, we must show that $\rho(X,x) = 0$. To that end we shall show (in a manner reminiscent of Example 6.1) that the function s can be approximated arbitrarily closely by an SPH.

Define $h_e: [a,b] \to [0,L(C)]$ by assigning values

$$h_e(t) \equiv \frac{L(C)}{L(C)+e}\left[s(t) + \frac{e}{b-a}(t-a)\right], \qquad e > 0.$$

Then

$$h_e(t) - s(t) = \frac{-es(t)}{L(C)+e} + \frac{L(C)}{L(C)+e}\left[\frac{e}{b-a}(t-a)\right]$$

tends to zero uniformly in t as $e \to 0$. It follows that

$$\inf_e \sup_t |X[h_e(t)] - X[s(t)]| = 0$$

and, since $X[s(t)] = x(t)$, we have proved that $\rho(X,x) = 0$.

To complete the proof of (7.8) we must establish that, if $s = s(t)$, then $C_{X,s}$ and $C_{x,t}$ are one and the same curve. We have shown already that this is so if $t = b$ and $s(t) = L(C)$. It is easy to verify for the degenerate case, $t = a$ and $s(t) = 0$. If $t \in (a,b)$, the entire argument preceding this paragraph can be applied with $[a,t]$ in place of $[a,b]$ and with $C_{x,t}$ in place of C.

PROOF OF (α)

Given the continuous function $x: [a,t] \to R^n$, let $\mathscr{L}(x; [t,t'])$ denote the length of the restriction of x to the subinterval $[t,t']$ of $[a,b]$. Given also that $\epsilon > 0$, there exists a partition $\pi = \{t_0,t_1,\ldots,t_k\}$ depending on ϵ such that

$$\mathscr{L}(x; [a,b]) - \sum_{i=1}^k |x(t_i) - x(t_{i-1})| < \epsilon.$$

Let τ be a fixed point of $[a,b]$. We can suppose the partition π to have been so chosen that $\tau = t_j \in \pi$ for, if not, we can now adjoin τ to π and the preceding inequality still holds. Let h be a positive number so small that no partition point is between τ and $\tau + h$. We can then adjoin $\tau + h$ to the partition π and, after relabeling the partition points, $\tau + h$ becomes t_{j+1}. We continue to use k for the number of subintervals.

One can verify that s is additive, that is, that

$$\mathcal{L}(x; [a,t_{j+1}]) - \mathcal{L}(x; [a,t_j]) = \mathcal{L}(x; [t_j,t_{j+1}])$$

and

$$\mathcal{L}(x; [a,b]) = \sum_{i=1}^{k} \mathcal{L}(x; [t_{i-1},t_i]),$$

hence from the inequality on the preceding page that

$$\sum_{i=1}^{k} \{\mathcal{L}(x;[t_{i-1},t_i]) - |x(t_i) - x(t_{i-1})|\} < \epsilon.$$

Each term in this sum being nonnegative must be below ϵ and, since $t_j = \tau$ and $t_{j+1} = \tau + h$,

$$\mathcal{L}(x; [\tau,\tau+h]) < |x(\tau+h) - x(\tau)| + \epsilon.$$

We can finally suppose h to have been so chosen that, as a result of the continuity of x, the first term on the right is below ϵ. It follows that

$$s(\tau+h) - s(\tau) = \mathcal{L}(x; [\tau,\tau+h]) < 2\epsilon;$$

consequently s is right continuous on $[a,b)$. The proof of (α) is completed by a similar argument with $h < 0$ showing that s is left continuous on $(a,b]$.

The following is a useful corollary to Theorem 7.7.

Theorem 7.8

Every Fréchet curve C in E_n of finite length has a representation $\xi: [0,1] \to R^n$ with the property that if $C_{\xi,t}$ is a subcurve relative to ξ, then

(7.9) $$L(C_{\xi,t}) = tL(C), \quad \forall t \in [0,1].$$

PROOF

If $L(C) > 0$, then Theorem 7.7 is applicable and we define $\xi(t) \equiv X[tL(C)]$. The reader should verify from definition (6.9) that $\rho(\xi,X) = 0$ and hence that ξ represents C. To establish (7.9) we must verify next that the subcurves $C_{X,s}$ and $C_{\xi,t}$ of C are one and the same iff $s = tL(C)$. Then (7.9) follows from (7.8).

In the degenerate case $L(C) = 0$, the graph of C is a single point a of R^n. The mapping $\xi: [0,1] \to R^n$ such that $\xi(t) = a$ for all $t \in [0,1]$ then has property (7.9).

The representation ξ in Theorem 7.8 is sometimes called the *representation of C in terms of reduced length*.

Theorem 7.9

The representations X and ξ satisfy respective Lipschitz conditions

(7.10) $|X(s_2) - X(s_1)| \leq |s_2 - s_1|, \qquad \forall s_1, s_2 \in [0, L(C)]$

and

(7.11) $|\xi(t_2) - \xi(t_1)| \leq |t_2 - t_1| L(C), \qquad \forall t_1, t_2 \in [0,1].$

PROOF

If $s_1 \leq s_2$, denote by $C_{1,2}$ the subcurve of C represented by the restriction of X to $[s_1, s_2]$. Verify that $L(C_{1,2}) = L(C_{X,s_2}) - L(C_{X,s_1})$. Inequality (7.10), which says that the length of a chord is dominated by the length of the corresponding subcurve, then follows from the definitions of $\mathscr{L}$ and L and (7.8).

Given the left member of (7.10), set $s_i = t_i L(C)$, $i = 1, 2$. We then have from (7.10) and the definition of $\xi(t)$ for the case $L(C) > 0$ that (7.11) holds. In the case $L(C) = 0$, $\xi(t) = a$ and (7.11) is trivially true.

Exercise 7.2

1. Define the norm $\|\pi\|$ of a partition π of an interval as the largest of the differences $t_i - t_{i-1}$. Given a continuous function $x: [a,b] \to R^n$, prove that $\Sigma |x(t_i) - x(t_{i-1})|$ has the limit $\mathscr{L}(x)$ defined by (7.2) as $\|\pi\| \to 0$ and hence that this limit may be either finite or ∞.

7.7 THE HILBERT COMPACTNESS THEOREM

Theorem 7.10

The class $\mathscr{C}$ consisting of all Fréchet curves C whose graphs are subsets of a given sequentially compact (hence compact, hence bounded and closed) nonempty subset A of R^n with lengths $L(C)$ at most a nonnegative constant λ is a sequentially compact class of curves.

PROOF

If $\lambda = 0$, then the graph of each $C \in \mathscr{C}$ consists of a single point and a sequence $\{C_\nu: \nu \in N\}$ in $\mathscr{C}$ has a limit iff the sequence of graphs converges to a point. That an arbitrary sequence of points in a sequentially compact set A has at least one convergent subsequence is a classic theorem. Having disposed of the trivial case, we suppose in the remainder of the proof that $\lambda > 0$.

Let $\{C_\nu: \nu \in N\}$ be an arbitrary sequence in $\mathscr{C}$. We must show that at least one subsequence converges in terms of Fréchet distance d to a curve

C_0 in $\mathscr{C}$. Let $\{\xi_\nu\colon [0,1] \to R^n\colon \nu \in N\}$ be the sequence of reduced-length representations of the curves C_ν. Each ξ_ν satisfies a condition (7.11), hence

(7.12) $|\xi_\nu(t) - \xi_\nu(t')| \leqslant \lambda|t - t'|,\quad \forall t,t' \in [0,1], \nu = 1, 2, \ldots.$

Let $T \equiv \{t_1\colon i \in N\}$ be a fixed sequence in $[0,1]$ that is *dense in* $[0,1]$; that is, each point of $[0,1]$ is an accumulation point of points t_i. In particular, T can be a sequentialization of all rational reals in $[0,1]$.

The sequence $\{\xi_\nu(t_1)\}$ of values at t_1 is in general a nonconvergent sequence in A but, by the given sequential compactness of A, it has a subsequence $\{\xi_{1\nu}(t_1)\}$ converging to a point which we denote by $\xi_0(t_1)$. Points of the sequence are in A and A is closed; hence $\xi_0(t_1) \in A$.

Proceeding inductively, suppose that $\{\xi_{m\nu}\colon \nu \in N\}$ is a subsequence of the original sequence $\{\xi_\nu\}$ such that the point-sequences $\{\xi_{m\nu}(t_i)\colon \nu \in N\}$ have respective limits $\xi_0(t_i)$, $i = 1, \ldots, m$, in A. Sequence $\{\xi_{m\nu}(t_{m+1})\colon \nu \in N\}$ is in general divergent, but again there must be a subsequence $\{\xi_{(m+1)\nu}(t_{m+1})\}$ of this point sequence that converges to a point $\xi_0(t_{m+1})$ of A. The sequence $\{\xi_{(m+1)\nu}\colon \nu \in N\}$ of functions is then a subsequence of the original $\{\xi_\nu\}$ such that the sequence $\{\xi_{(m+1)\nu}(t_i)\colon \nu \in N\}$ converges to a point of A for $i = 1, \ldots, m+1$. The inductive step from m to $m+1$ is complete.

Accordingly, there is a *double sequence*,

$$\xi_{11}, \xi_{12}, \xi_{13}, \ldots$$

(7.13) $$\xi_{21}, \xi_{22}, \xi_{23}, \ldots$$

$$\vdots$$

$$\xi_{m1}, \xi_{m2}, \xi_{m3}, \ldots$$

$$\vdots$$

Each row of this infinite array is a subsequence of its predecessor and of $\{\xi_\nu\}$ and such that, for each m, the mth row converges for the first m terms $t_1, \ldots, t_m$ of T. The proof is completed with the aid of two lemmas, (α) and (β), the proofs of which are at the end of the main proof.

(α) *The diagonal sequence* $\{\xi_{\nu\nu}(t_i)\colon \nu \in N\}$ *converges to points* $\xi_0(t_i) \in A, i = 1, 2, \ldots.$

(β) $|\xi_0(t_i) - \xi_0(t_j)| \leqslant \lambda|t_i - t_j|,\quad \forall t_i, t_j \in T.$

Granted (α) and (β) we extend the domain of ξ_0 from T to $[0,1]$ as follows. Given $t \in [0,1] - T$, let $\{\tau_i\colon i \in N\}$ be a sequence in T converging to t. In view of the compactness of A, we can, moreover, suppose that τ_i has been so selected that the sequence $\{\xi_0(\tau_i): i \in N\}$ has a limit. Define

(7.14) $$\xi_0(t) \equiv \lim \xi_0(\tau_i), \qquad t \in [0,1] - T.$$

To see that the left member depends only on t and not on the sequence $\{\tau_i\}$, let $\{\tau_i^* \in T: i \in N\}$ be another sequence with t as limit. By the triangle property of euclidean distance,

$$|\xi_0(\tau_i^*) - \xi_0(t)| \leq |\xi_0(\tau_i^*) - \xi_0(\tau_j)| + |\xi_0(\tau_j) - \xi_0(t)|.$$

Respective terms on the right converge to zero as $i,j \to \infty$ by (β) and (7.14); hence $\lim \xi_0(\tau_i^*) = \xi_0(t)$.

We need the additional lemmas:

$(\beta+)$ $|\xi_0(t) - \xi_0(t')| \leq \lambda|t - t'|, \qquad \forall t,t' \in [0,1].$

$(\alpha+)$ *The diagonal sequence* $\{\xi_{\nu\nu}(t)\}$ *converges for all* $t \in [0,1]$.

$(\alpha++)$ *Indeed this convergence is uniform on* $[0,1]$.

It is immediate from $(\beta+)$ that ξ_0 is continuous on $[0,1]$ and hence is a representation of some curve C_0 whose graph is in A. It follows from $(\beta+)$ and definition (7.2) of $\mathscr{L}$ that $\mathscr{L}(\xi_0) \leq \lambda$ and hence from definition (7.7) of L that $L(C_0) \leq \lambda$, consequently $C_0 \in \mathscr{C}$. The uniform convergence $(\alpha++)$ of $\xi_{\nu\nu}$ to ξ_0 assures us that $\rho(\xi_{\nu\nu},\xi_0) \to 0$ as $\nu \to \infty$.

This completes the proof except for the lemmas, to which we now turn.

PROOF OF (α)

Select $t_m \in T$ and $\epsilon > 0$. The mth row of array (7.13) has been so selected that $\xi_{m\nu}(t_m) \to \xi_0(t_m)$ as $\nu \to \infty$; hence there is an $N_{\epsilon,m}$ depending on ϵ and t_m such that

$$\nu > N_{\epsilon,m} \Rightarrow |\xi_{m\nu}(t_m) - \xi_0(t_m)| < \epsilon.$$

All symbols $\xi_{\nu\nu}$ on the diagonal of the array with $\nu > N'_{\epsilon,m} \equiv \max(m,N_{\epsilon,m})$ are new labels for certain terms $\xi_{m\nu}$ in row m with $\nu > N_{\epsilon,m}$. Consequently,

$$\nu > N'_{\epsilon,m} \Rightarrow |\xi_{\nu\nu}(t_m) - \xi_0(t_m)| < \epsilon.$$

PROOF OF (β)

Suppose that, for some pair $t_i,t_j \in T$, $|\xi_0(t_i) - \xi_0(t_j)| > \lambda|t_i - t_j|$. With t_i,t_j fixed and hence $\xi_0(t_i)$, $\xi_0(t_j)$ fixed and given $\epsilon > 0$, select ν so large that

(7.15) $$\Big||\xi_{\nu\nu}(t_i) - \xi_{\nu\nu}(t_j)| - |\xi_0(t_i) - \xi_0(t_j)|\Big| < \epsilon.$$

With ϵ sufficiently small and the hypothesis on ξ_0 with which we started, (7.15) denies that $\xi_{\nu\nu}$ has property (7.12) and hence we must infer the truth of (β).

PROOF OF $(\beta+)$

Let $\{\tau_i\}$ and $\{\tau_i'\}$ be respective sequences in T with limits t and t'. If the stated condition $(\beta+)$ does not hold we can use sequences $\{\xi_0(\tau_i)\}$ and $\{\xi_0(\tau_i')\}$ to obtain the same kind of contradiction used in the last proof.

PROOF OF $(\alpha+)$

Given $t \in [0,1]$ and $\epsilon > 0$, there necessarily exists $t_i \in T$ such that $|t_i - t| < \epsilon/3\lambda$. It follows from (7.12) that $|\xi_{\nu\nu}(t) - \xi_{\nu\nu}(t_i)| < \epsilon/3$ and from $(\beta+)$ that $|\xi_0(t_i) - \xi_0(t)| < \epsilon/3$. By Lemma (α) there is an integer $N_{\epsilon,i}$ depending on ϵ and t_i such that

$$\nu > N_{\epsilon,i} \Rightarrow |\xi_{\nu\nu}(t_i) - \xi_0(t_i)| < \epsilon/3.$$

As a consequence of the three preceding inequalities and the triangle property of euclidean distance,

$$(7.16) \qquad\qquad |\xi_{\nu\nu}(t) - \xi_0(t)| < \epsilon$$

provided that $\nu > N_{\epsilon,i}$. But the left member is free of i and hence the inequality holds if $\nu > N_{\epsilon,t}'$ with

$$N_{\epsilon,t}' \equiv \min\{N_{\epsilon,i} : t_i \in T \text{ and } |t_i - t| < \epsilon/3\lambda\}.$$

This completes the proof.

PROOF OF $(\alpha++)$

To show that there exists an N_ϵ depending only on ϵ such that (7.16) holds for all $t \in [0,1]$ if $\nu > N_\epsilon$, suppose the contrary. Then, for some positive number ϵ_1 there is a strictly increasing sequence $\{\nu_i : i \in N\}$ and, corresponding to ν_i, a number $\tau_i \in [0,1]$ such that

$$|\xi_{\nu_i\nu_i}(\tau_i) - \xi_0(\tau_i)| \geq \epsilon_1.$$

The sequence $\{\tau_i : i \in N\}$ must have a subsequence converging to some $t \in [0,1]$; hence we can suppose that ν_i and τ_i were so selected that $\tau_i \to t$. By the triangle inequality

$$|\xi_{\nu_i\nu_i}(\tau_i) - \xi_0(\tau_i)| \leq |\xi_{\nu_i\nu_i}(\tau_i) - \xi_{\nu_i\nu_i}(t)| + |\xi_{\nu_i\nu_i}(t) - \xi_0(t)| + |\xi_0(t) - \xi_0(\tau_i)|.$$

The first and third terms on the right converge to zero as a consequence of the Lipschitz conditions (7.12) and $(\beta+)$, respectively, and the second term by $(\alpha+)$. We thus have a contradiction and infer that $(\alpha++)$ is valid.

7.8   THE ASCOLI–ARZELÀ THEOREM

The functions x: $[a,b] \to R^n$ of a class $\mathscr{X}$ of such functions with a common domain $[a,b]$ are called *equicontinuous on* $[a,b]$ if

$$(7.17) \qquad \forall \epsilon > 0, \exists \delta(\epsilon) > 0$$
$$such\ that\ x \in \mathscr{X}, t, t' \in [a,b]\ and\ |t-t'| < \delta(\epsilon),$$
$$\Rightarrow |x(t) - x(t')| \le \epsilon.$$

Observe that (7.17) is stated with $\le$ in the final inequality. Definitions of limits and of different kinds of continuity are, by custom, usually stated with strict inequalities. The content of these definitions remains the same, however, if weak inequalities are used. Doing so in (7.17) is a convenience in proving the next theorem.

If $\mathscr{X}$ consists of a single element x, this reduces to the definition of uniform continuity on $[a,b]$ of that function. In general, definition (7.17) embodies two kinds of uniformity, uniformity both with respect to $t \in [0,1]$ and to $x \in \mathscr{X}$. Alternatively stated, the $\delta(\epsilon)$ in (7.17) is free of both t and x.

Equicontinuity of an infinite class $\mathscr{X}$ is a strong property, and yet such classes are not difficult to find. Given an infinite class of Fréchet curves C, of lengths $L(C) < \lambda$, the reduced length representations ξ are equilipschitzian; that is, they satisfy a common Lipschitz condition (7.12). This is seen to imply equicontinuity but not conversely.

The functions x of a class $\mathscr{X}$ are said to be *equally* (uniformly) *bounded* on a common domain $[a,b]$ if there are real numbers m^j and M^j, $j = 1, \ldots, n$ such that

$$(7.18) \quad m^j \le x^j(t) \le M^j, \qquad \forall t \in [a,b], \qquad \forall x \in \mathscr{X}, j = 1, \ldots, n.$$

The following theorem is one of a number of versions of a theorem sometimes ascribed to Ascoli and again to Arzelà.

Theorem 7.11

Given a set of real numbers m^j and M^j and a function δ from the positive reals to the positive reals with values $\delta(\epsilon)$, the set $\mathscr{X}$ consisting of all functions x: $[a,b] \to R^n$ that satisfy (7.17) and (7.18) is sequentially compact in terms of uniform convergence.

PROOF

It must be shown that an arbitrary sequence $\{x_\nu \in \mathscr{X} : \nu \in N\}$ has a subsequence converging uniformly to a function x in $\mathscr{X}$. The proof is generally similar to that of Theorem 7.10. We give an outline for the case $n = 1$ and suggest that the reader supply the details.

Let $T \equiv \{t_i \in [a,b] : i \in N\}$ be a fixed sequence that is dense in $[a,b]$. By successive selection of subsequences we define a double sequence

$$x_{11}, x_{12}, x_{13}, \ldots$$

(7.19) $$x_{21}, x_{22}, x_{23}, \ldots$$

$$\vdots$$

$$x_{m1}, x_{m2}, x_{m3}, \ldots$$

$$\vdots$$

such that the simple sequence in the mth row converges for $t_1, t_2, \ldots, t_m$ and verify that the diagonal sequence $\{x_{\nu\nu}(t) : \nu \in N\}$ converges to a limit denoted by $x_0(t)$ for each $t \in T$. One then extends x_0 from T to $[a,b]$ in the manner of the preceding proof except that (7.17) must be used in place of the Lipschitz condition that we no longer have. It remains to show that x_0 is in the class $\mathscr{X}$, that $x_{\nu\nu}$ converges to x_0, and that this convergence is uniform.

Granted that the theorem is true for $n = 1$, let $x_\nu = (x_\nu^1, \ldots, x_\nu^n)$, $\nu \in N$, be a sequence in $\mathscr{X}$. We construct another array like (7.19), which terminates with the nth row in the following manner.

Let $\{x_{1\nu} : \nu \in N\}$ be a subsequence of $\{x_\nu\}$ such that the sequence $\{x_{1\nu}^1\}$ of first components converges uniformly on $[a,b]$ in accord with the case $n = 1$ to a limit x_0^1. Let $\{x_{2\nu}\}$ be a subsequence of $\{x_{1\nu}\}$ such that $\{x_{2\nu}^2\}$ converges uniformly to a limit x_0^2. Sequence $\{x_{2\nu}^1\}$ as a subsequence of $\{x_{1\nu}^1\}$ automatically converges to x_0^1; hence first and second components of the functions $x_{2\nu}$ both converge. Continuing thus, we arrive after n steps at a sequence $\{x_{n\nu} : \nu \in N\}$ such that the sequences $\{x_{n\nu}^j : \nu \in N\}$ of components converge uniformly to x_0^j, $j = 1, \ldots, n$. It follows that $\{x_{n\nu}\}$ converges uniformly to the function $x_0 \colon [a,b] \to R^n$ defined as the function whose n components are $x_0^1, \ldots, x_0^n$.

Exercise 7.3

1. Granted the Ascoli–Arzelà Theorem, obtain the Hilbert Compactness Theorem as an easy corollary with the aid of reduced-length representations of the curves.

2. Let $\mathscr{C}$ denote the class of all Fréchet curves C of lengths $L(C) \leq \lambda$ whose graphs are subsets of a given compact subset A of R^n with the initial and terminal points of *graph* C in respective nonempty disjoint closed subsets S_1 and S_2 of A. Point out with the aid of Theorem 7.10 that this class $\mathscr{C}$ is sequentially compact.

3. Let A be the two-sphere $\{(x,y,z) \in R^3 : x^2 + y^2 + z^2 = 25\}$. Given the set S_1 consisting of the single point $(0,0,5)$ and $S_2 \equiv \{(x,y,z) \in A :$

$z = -4$}, let $\mathscr{C}$ be the class of all Fréchet curves whose graphs are subsets of A and join S_1 to S_2. The last clause means that if x: $[a,b] \to R^3$ represents a curve $C \in \mathscr{C}$, then $x(a) \in S_1$ and $x(b) \in S_2$. With Theorem 7.1 as a guide, use Theorems 7.6 and 7.10 in proving that there exists a curve C_0 of least length in the class $\mathscr{C}$.

4. Supply a proof of Theorem 7.11 by filling in all the missing details in the given outline of a proof.

5. Let y_ν: $[0,2] \to R$ be the sawtooth function consisting of all points $[x, y_\nu(x)]$ on the oblique sides of ν consecutive equilateral triangles whose bases, each of length $1/\nu$, fill the interval $[0,1]$ together with all points $(x,0), x \in (1,2]$. Let $x_\nu(x) \equiv x, x \in [0,2]$, $\nu = 1, 2, \ldots$. Describe precisely the reduced-length representation (ξ_ν, η_ν) of the curve C_ν represented by $[x_\nu(x), y_\nu(x)]$. Verify that sequence (ξ_ν, η_ν), $\nu = 1, 2, \ldots$ has a limit pair (x_0, y_0) but that this pair is not the reduced-length representation of the curve C_0 to which C_ν converges.

6. Identify reasons why the Ascoli–Arzelà theorem does or does not (whichever is correct) apply to the sequence $\{y_\nu : \nu \in N\}$ described in problem 5.

7.9 THE HELLY COMPACTNESS THEOREM

Let $\mathscr{Y}$ denote the class of all functions y: $[a,b] \to R$, each of which is nondecreasing on the common interval $[a,b]$ of positive finite length. Let $\mathscr{Y}_{m,M}$ be the equally bounded subclass consisting of all $y \in \mathscr{Y}$ such that

$$(7.20) \qquad m \leqslant y(t) \leqslant M, \qquad \forall t \in [a,b].$$

Theorem 7.12

The class $\mathscr{Y}_{m,M}$ is sequentially compact in terms of pointwise convergence. Alternatively stated, every sequence in $\mathscr{Y}_{m,M}$ has at least one subsequence converging pointwise to a function in the class $\mathscr{Y}_{m,M}$.

PROOF

Let $\{y_\nu : \nu \in N\}$ be an arbitrary sequence in $\mathscr{Y}_{m,M}$ define T as in Sections 7.7 and 7.8 with points a and b both placed in T, and follow the proofs of Theorems 7.10 and 7.11 until an array like (7.19) is obtained. The general element of the array is now $y_{m\nu}$. Set $z_\nu \equiv y_{\nu\nu}$. Then, as before, $z_\nu(t)$ converges for each $t \in T$ to a limit $y_0(t)$. Thus far y_0 is a real-valued function with domain T. We wish to extend y_0 to the whole of $[a,b]$. Define

$$\lambda(t) \equiv \liminf z_\nu(t) \qquad and \qquad \Lambda(t) \equiv \limsup z_\nu(t).$$

We shall show following the main proof that

(α) λ and Λ are both nondecreasing on $[a,b]$.

Each of these functions is then necessarily continuous on $[a,b]$ except for a countable (empty, finite, or denumerable as the case may be) subset of $[a,b]$ at each point of which it has a finite jump. We have remarked that $z_\nu(t)$ converges on T, hence

$$(7.21) \qquad \lambda(t) = \Lambda(t) = y_0(t), \qquad \forall t \in T.$$

Moreover, every point of $[a,b]$ is an accumulation point of points of T. Given $t \in [a,b] - T$, then t is necessarily an interior point of $[a,b]$ because T was chosen to include both a and b. Since T is dense in $[a,b]$, there are points of T on either side of t and arbitrarily near to t and hence there is a sequence $\{t_i \in T : i \in N\}$ converging to t with terms t_i alternately to the left and to the right of t.

Both λ and Λ may be continuous at t. If so, then as a consequence of (7.21) $\lambda(t_i)$ and $\Lambda(t_i)$ necessarily have a common limit as $i \to \infty$ and we define

$$(7.22) \qquad y_0(t) \equiv \lim \lambda(t_i) = \lim \Lambda(t_i)$$

if λ and Λ are both continuous at t.

We shall show at the end of the proof that

$$(7.23) \qquad \lim z_\nu(t) = y_0(t) \qquad \textit{if λ and Λ are both continuous at t.}$$

In the event that neither λ nor Λ has any discontinuities, we shall be through.

If this is not the case, let $\tau_1, \tau_2, \ldots, \tau_k$ or $\tau_1, \tau_2, \ldots$ be the finite or infinite sequence in $[a,b]$, as the case may be, consisting of points at which λ or Λ or both or them are discontinuous. In the infinite case we apply the diagonal process again. Consider the array

$$(7.24) \qquad \begin{array}{l} z_{11}(t), z_{12}(t), z_{13}(t), \ldots \\ z_{21}(t), z_{22}(t), z_{23}(t), \ldots \\ \vdots \\ z_{k1}(t), z_{k2}(t), z_{k3}(t), \ldots \\ \vdots \end{array}$$

with the first row being a subsequence of the sequence $\{z_\nu(t) : \nu \in N\}$ that converges for τ_1, the second row being a subsequence of the first row that converges for τ_1 and τ_2, the kth row being a subsequence of each preceding row so chosen that it converges for $\tau_1, \tau_2, \ldots, \tau_k$, and so on. The diagonal sequence $z_{\nu\nu}(t)$ then converges for $\tau_1, \tau_2, \ldots$, and we complete the extension of y_0 by defining

$$(7.25) \qquad y_0(\tau_i) \equiv \lim z_{\nu\nu}(\tau_i), \qquad i = 1, 2, \ldots.$$

In the case of a finite sequence $\tau_1, \tau_2, \ldots, \tau_k$, the construction of the array terminates with the kth row and we define

$$(7.26) \qquad y_0(\tau_i) \equiv \lim z_{k\nu}(\tau_i), \qquad i = 1, 2, \ldots, k.$$

Thus, regardless of the case that may occur, we now have a function $y_0: [a,b] \to R$. Since each value $y_0(t)$ is the limit of a sequence of values $y(t)$ of functions satisfying (7.20), the limit $y_0(t)$ must satisfy these inequalities. It remains to show that y_0 is nondecreasing. If so, it is in the given class and the proof is complete.

Suppose that there exist $t_1, t_2 \in [a,b]$, $t_1 < t_2$ such that $y_0(t_1) - y_0(t_2) = p > 0$. Given $\epsilon > 0$, select ν so large that, in accord with (7.23) and the structure of array (7.24),

$$\left| z_{\nu\nu}(t_j) - y_0(t_j) \right| < p/3, \qquad j = 1, 2.$$

Then

$$z_{\nu\nu}(t_1) > y_0(t_1) - p/3 \qquad and \qquad -z_{\nu\nu}(t_2) > -y_0(t_2) - p/3.$$

By addition of these inequalities we deny that $z_{\nu\nu}$ is nondecreasing and hence must infer that y_0 is nondecreasing.

It remains to prove statements (α) and (7.23).

PROOF THAT λ IS NONDECREASING

Let $\{w_\nu(t): \nu \in N\}$ be a subsequence of $\{z_\nu(t): \nu \in N\}$ so chosen that $w_\nu(t_2)$, $t_2 \in (a,b]$, has, in accord with definition (1.9) of the limit inferior as applied to a sequence, the limit $\lambda(t_2)$. Select $t_1 \in [a,b]$, $t_1 < t_2$ and let $\{v_\nu(t)\}$ be a subsequence of $\{w_\nu(t)\}$ such that $v_\nu(t_1)$ has a limit. Since $v_\nu \in \mathcal{Y}$, we know that $v_\nu(t_2) \geqslant v_\nu(t_1)$, $\nu = 1, 2, \ldots$; therefore,

$$\lambda(t_2) = \lim v_\nu(t_2) \geqslant \lim v_\nu(t_1) \geqslant \lambda(t_1).$$

That Λ is nondecreasing can be similarly proved.

PROOF OF (7.23)

Given $\epsilon > 0$, let j be an integer so large that, for the value t of (7.22) and (7.23),

$$(7.27) \qquad \left| \lambda(t_j) - y_0(t) \right| < \epsilon/6 \qquad and \qquad \left| \lambda(t_{j+1}) - y_0(t) \right| < \epsilon/6,$$

and hence

(7.28) $$|\lambda(t_j) - \lambda(t_{j+1})| < \epsilon/3.$$

Next select ν so large that

(7.29) $|z_\nu(t_j) - \lambda(t_j)| < \epsilon/6$ and $|z_\nu(t_{j+1}) - \lambda(t_{j+1})| < \epsilon/6.$

By (7.28), (7.29), and the triangle property of absolute values,

(7.30) $\begin{aligned}|z_\nu(t_{j+1}) - z_\nu(t_j)| &\leq |z_\nu(t_{j+1}) - \lambda(t_{j+1})| \\ &+ |\lambda(t_{j+1}) - \lambda(t_j)| + |\lambda(t_j) - z_\nu(t_j)| < 2\epsilon/3.\end{aligned}$

Again by the triangle inequality

(7.31) $|z_\nu(t) - y_0(t)| \leq |z_\nu(t) - z_\nu(t_j)| + |z_\nu(t_j) - \lambda(t_j)| + |\lambda(t_j) - y_0(t)|.$

Recall that the sequence $\{t_i\}$ chosen preceding (7.22) consists of terms alternately below and above t. It follows from this and the monotonicity of z_ν that the first term on the right in (7.31) is dominated by the left member of (7.30) and hence by $2\epsilon/3$. The second and third terms on the right in (7.31) are below $\epsilon/6$ by the first inequalities (7.29) and (7.27), respectively; consequently $|z_\nu(t) - y_0(t)| < \epsilon$ provided that ν is sufficiently large.

The Hilbert and Ascoli–Arzelà theorems have been widely used in the proofs of theorems of the calculus of variations. The Helly theorem and its extensions (8b,c, 41a) are effective with certain problems having monotonicity restrictions on admissible functions. See, for example, (8d) or (12g).

Exercise 7.4

1. State an immediate corollary to the Helly Theorem 7.12 for a class of equally bounded nonincreasing functions and point out why it follows easily from Theorem 7.12.

2. State and prove, using Theorem 7.12, a similar theorem for the class $\mathcal{Y}$ of all nondecreasing functions $y: (a,b) \to R$. Observe at the outset that such functions need not be bounded.

3. Given a function $y: [a,b] \to R$ that is BV on $[a,b]$, then $y = p - n$, where p and n are both nondecreasing or both nonincreasing on $[a,b]$. Given a sequence $\{y_\nu: \nu \in N\}$ of such functions, each of total variation $\mathcal{L}(y_\nu) = T(y_\nu; [a,b]) \leq \lambda$ and such that $m \leq y_\nu(t) \leq M$, $t \in [a,b]$, $\nu = 1, 2, \ldots$, prove using Theorem 7.12 that there exists a subsequence converging to a limit $y_0: [a,b] \to R$. Also prove that $T(y_0; [a,b]) \leq \lambda$ and that $m \leq y_0 \leq M$.

4. Let $\{r_m:\ m \in N\}$ be a fixed sequentialization of all rational real numbers. Let u be the unit step-function (problem 8, Exercise 1.3). Precisely how much can be said of the sequence of partial sums of the infinite series $\Sigma\ (\tfrac{1}{2})^m u(t - r_m)$ on the basis of the Helly Theorem 7.12? On the basis of everything you may know about infinite series of functions?

5. Let $\alpha(x)$ be the piecewise linear function that joins (0,0) to (2,2) to (3,1) to (5,3). Establish that there exists a function $y_0 \colon [0,5] \to R$ furnishing a global minimum for the integral

$$J(y) \equiv \int_0^5 [y - \alpha(x)]^2\, dx$$

on the class $\mathscr{Y}$ of all nondecreasing functions $y \colon [0,5] \to R$. End values $y(0)$ and $y(5)$ are free.

6. Consider the classical nonparametric Bolza Problem of minimizing the integral of problem 5 subject to the side-condition $\dot{y} = \dot{z}^2$, a device for requiring that y be nondecreasing. Verify that $y_0(x) = x$, $\tfrac{3}{2}$, or $x - 2$ according as $x \in [0,\tfrac{3}{2}]$, $(\tfrac{3}{2},\tfrac{7}{2}]$, or $(\tfrac{7}{2},5]$, together with an appropriate z_0 and a multiplier, satisfies the Multiplier Rule of Chapter 5. Then try to devise, drawing from Chapter 5 or any other source, a demonstration that y_0 furnishes a smaller value for $J(y)$ than any other nondecreasing PWS function $y \colon [0,5] \to R$ and indeed that this conclusion holds without the restriction that y be PWS.

7.10   THE WEIERSTRASS INTEGRAL

The compactness theorems of the last three sections and problems 2 and 3, Exercise 7.3, lead one to hope that existence theorems for other curve-functions J or functionals J can be constructed along the lines of Theorem 7.1.

Such a proof requires a sequence $\{x_\nu\}$ of admissible functions converging to an admissible function x_0 such that $J(x_\nu)$ converges to the infimum γ of $J(x)$. If one is restricted to the Riemann integral and hence to functions x such that $F(x,\dot{x})$ is Riemann integrable, the composite function $F(x_0,\dot{x}_0)$ is not necessarily Riemann integrable. It becomes necessary to replace the Riemann integral by some other that is free of this defect if we are to proceed.

Weierstrass was aware of this need and defined such an integral around 1879. It was subsequently used by Hilbert and Osgood and until about 1912 by Tonelli. In the meantime the Lebesgue integral had appeared and Tonelli adapted it to the needs of variational theory in his subsequent work. It has become increasingly important for the calculus of variations as it has quite generally for the other parts of mathematics known collectively as analysis.

We include a brief account of the Weierstrass integral for several reasons. Because of its relative simplicity we can lead rather quickly to some typical existence theorems for variational problems. Proofs of certain theorems of the calculus of variations when made in terms of the Weierstrass integral are essentially simpler and exhibit more clearly the details of what is going on than do similar proofs in terms of the Lebesgue integral. Although the latter is a generally superior tool, there has been a continuing interest in the Weierstrass integral, as indicated by such papers as Aronszajn (1a), Cesari (9b,c), Ewing (12c,d,e), Menger (34a,b), Morse (38e, part III), and Pauc (43a,b).

In anticipation of the needs of certain proofs that use the Hilbert Compactness Theorem 7.10 we suppose given a nonempty subset A of R^n that is the closure of a bounded open subset (rather than the open set A of Section 6.6, which is not necessarily bounded). The set B is now the closure of a set with the properties of B in Section 6.6; namely, B is the closure of a nonempty open set in R^n that contains all points kr, $k \geq 0$, on every ray determined by the origin θ and a point $r \neq \theta$ of B.

Suppose given a parametric integrand $F: A \times B \to R$ with the following properties (i) and (ii).

(i) F is continuous on $A \times B$.

(ii) $F(x,kr) = kF(x,r)$, $\forall (x,r) \in A \times B$ and $\forall k \geq 0$.

We need to assume neither the continuity nor even the existence of any partial derivatives of F. However, conditions (i) and (ii) in terms of the closed sets A and B are rather strong and hence some functions that satisfy the conditions on F of Chapter 6 are now excluded.

In certain theorems we require that

(iii) B have the additional property of convexity and that $F(x,r)$ be convex in r for each fixed $x \in A$.

Definitions of a convex subset K of R^n and of a function that is convex on K are in Section 3.10.

At times we shall require that

(iv) $F(x,r) > 0$, $\forall x \in A$ and $\forall r \in B, r \neq \theta$.

A continuous rectifiable curve C is now called *admissible* for a given variational problem if it has at least one representation $x: [a,b] \to R^n$ satisfying the end-conditions and possible side-conditions of a given problem and such that

(7.32) $\begin{cases} x(t) \in A, & \forall t \in [a,b] \\[2mm] x(t') - x(t) \in B, & \forall t,t' \in [a,b] \text{ with } t < t'. \end{cases}$

If a sequence $\{x_\nu : \nu \in N\}$ of representations with properties (7.32) converges to a function $x_0 : [a,b] \rightarrow R^n$, one verifies from the closedness of sets A and B that x_0 also has properties (7.32). It follows from definition (6.9) of Fréchet distance that, if x has properties (7.32) and $\rho(x,y) = 0$, then y has properties (7.32).

Given that $x : [a,b] \rightarrow R^n$ is a representation with properties (7.32), let π be a partition of $[a,b]$, set $\Delta_i x \equiv x(t_i) - x(t_{i-1})$, and consider the *Weierstrass sum*

$$(7.33) \qquad S(x;F; [a,b]) \equiv \sum F[x(t_{i-1}), \Delta_i x].$$

That x is *Weierstrass integrable relative to F over $[a,b]$* means that the sum (7.33) has a finite limit as the norm of π, denoted by $\|\pi\|$, tends to zero. We denote this limit by $\mathcal{W}(x;F; [a,b])$ and call it the *Weierstrass integral of x relative to F over $[a,b]$*. Thus x is integrable in this sense if

$$(7.34) \qquad \forall \epsilon > 0, \exists \nu_\epsilon > 0 \text{ such that } \|\pi\| < \nu_\epsilon \Rightarrow |S(x;F) - \mathcal{W}(x;F)| < \epsilon.$$

The third argument $[a,b]$ of S and $\mathcal{W}$ has been suppressed.

The left endpoint t_{i-1} of the subinterval $[t_{i-1}, t_i]$ of $[a,b]$ under π can be replaced in the sum (7.33) by an arbitrary point of the subinterval without affecting either integrability or the integral.

Theorem 7.13

If F has properties (i) *and* (ii) *and $x : [a,b] \rightarrow R^n$ is a PWS function with properties* (7.32), *then the composite function $F(x,\dot{x})$ is Riemann integrable and*

$$(7.35) \qquad \mathcal{W}(x;F; [a,b]) = \int_a^b F(x,\dot{x}) \, dt \, (Riemann).$$

PROOF

Suppose initially that the derivative $\dot{x}$ is continuous. The sum (7.33) can be written in the notation

$$\sum F[x(t_{i-1}), \Delta_i x^1, \ldots, \Delta_i x^n]$$

After applying the Mean Value Theorem of the Differential Calculus to each difference $\Delta_i x^j$ and using the homogeneity (ii) of F, this becomes

$$(7.36) \qquad \sum F[x(t_{i-1}), \dot{x}^1(\tau_{i1}), \ldots, \dot{x}^n(\tau_{in})] (t_i - t_{i-1}).$$

If the $n+1$ values $t_{i-1}, \tau_{i1}, \ldots, \tau_{in}$ were all the same, the continuity (i) of F and the continuity of x and $\dot{x}$ would suffice for the stated conclusion. The difference between such a sum and the sum (7.36) can be shown to converge to zero with $\|\pi\|$ as a consequence of the properties of F and the uniform continuity of x and $\dot{x}$ on $[a,b]$.

For the general case in which x is only PWS, one can apply the preceding approach to each of the subintervals on which x is smooth.

This theorem and the one to follow show that the Weierstrass integral is an extension of the classical parametric integral discussed in Chapter 6 from PWS representations x to continuous representations of finite length.

Theorem 7.14

If F has properties (i) and (ii) and x: $[a,b] \to R^n$ is a continuous rectifiable function with the two properties (7.32), then x is Weierstrass integrable relative to F over $[a,b]$.

A proof under the present hypotheses can be constructed along the lines of Aronszajn (1a; III, p.235) or Ewing (12c,pp.677–678). The special case in which $F(x,\dot{x}) = |\dot{x}|$ and $\mathscr{W}$ is the length $\mathscr{L}$ is the relatively easy problem 1, Exercise 7.2. Since the corresponding result for Lebesgue integrals with a weaker hypothesis on F is given in Section 9.5, we omit the proof.

Theorem 7.15

If F has properties (i) and (ii) and if x: $[a,b] \to R^n$ and y: $[c,d] \to R^n$ both have properties (7.32) and both represent the same continuous rectifiable Fréchet curve C, then

$$\mathscr{W}(x;F;[a,b]) = \mathscr{W}(y;F;[c,d]).$$

We again omit the proof. In view of Theorem 7.13, one sees that Theorem 6.2 essentially covers the case in which x and y are PWS. Theorem 7.5 is the special case of Theorem 7.15 in which $F(x,\dot{x}) = |\dot{x}|$. For proof of the theorem as stated, one can follow that of (12c, Theorem 2.4) or see Section 9.5.

Granted Theorem 7.15, we define the *Weierstrass curvilinear integral* $W(C;F)$ *of the admissible curve C relative to F* by the statement that

$$(7.37) \qquad W(C;F) \equiv \mathscr{W}(x;F;[a,b]),$$

where x: $[a,b] \to R^n$ *is an arbitrary representation with properties (7.32) of the given curve C.*

Theorem 7.16

If F has properties (i), (ii), and (iii) and λ is an arbitrary positive real number, then $W(C;F)$ is lower semi-continuous on the class $\mathscr{C}$ of all continuous rectifiable curves C each of which has a representation with properties (7.32) and a length $L(C) \leqslant \lambda$.

PROOF

The theorem means that given $n \geq 1$ and given a curve $C_0 \in \mathscr{C}$ and an arbitrary sequence $\{C_\nu \in \mathscr{C} \colon \nu \in N\}$ such that $L(C_\nu) \leq \lambda$ and such that the Fréchet distance (Section 6.4) $d(C_\nu, C_0) \to 0$, then

$$(7.38) \qquad \liminf W(C_\nu; F) \geq W(C_0; F).$$

Let $x_0 \colon [a,b] \to R^n$ be a representation of C_0 with property (7.32). Given $\epsilon > 0$, let π be a partition of $[a,b]$ of norm so small that

$$(7.39) \qquad \sum F[x_0(t_{i-1}), \Delta_i x_0] > W(C_0; F) - \epsilon$$

and also so small that, if τ_1 and τ_2 are in any one of the closed subintervals $[t_{i-1}, t_i]$ and if u is any unit vector in B, then

$$(7.40) \qquad |F[x_0(\tau_1), u] - F[x_0(\tau_2), u]| < \epsilon.$$

Given $e > 0$, any curve $C \in \mathscr{C}$ at Fréchet distance below e from C_0 has a representation $x \colon [a,b] \to R^n$ satisfying the condition that

$$|x(t) - x_0(t)| < e, \qquad \forall t \in [a,b].$$

We can, moreover, require that e be so small that (7.40) holds with x_0 replaced by x and that

$$(7.41) \quad |F[x(t_{i-1}), \Delta_i x] - F[x_0(t_{i-1}), \Delta_i x_0]| < \epsilon/k, \qquad i = 1, \ldots, k,$$

in which k denotes the number of subintervals of $[a,b]$ under the partition π.

Let π_i, $t_{i-1} = \tau_0 < \tau_1 < \cdots < \tau_{p_i} = t_i$, be a partition of $[t_{i-1}, t_i]$ and set $\Delta_{ij} x \equiv x(\tau_j) - x(\tau_{j-1})$. Observe that

$$(7.42) \qquad \sum_{j=1}^{p_i} F[x(\tau_{j-1}), \Delta_{ij} x] = \sum F[x(\tau_0), \Delta_{ij} x] + R_i,$$

where

$$(7.43) \qquad R_i \equiv \sum \{F[x(\tau_{j-1}), \Delta_{ij} x] - F[x(\tau_0), \Delta_{ij} x]\}.$$

As a consequence of the homogeneity (ii) of F and the convexity (iii) of both the set B and the function F, we see that, if r_1 and r_2 are in B, then $(r_1 + r_2)/2$ and hence $r_1 + r_2$ are in B. It follows that

$$F(x, r_1 + r_2) \leq F(x, r_1) + F(x, r_2),$$

and this extends by induction to any finite sum of elements of B. From this, together with (7.42) and (7.43) and the fact that $\tau_0 = t_{i-1}$, we have that

(7.44) $$\sum F[x(\tau_{j-1}),\Delta_{ij}x] \geqslant F[x(t_{i-1}),\Delta_i x] + R_i.$$

The conditions imposed on e ensure via (7.40) for x that

$$|R_i| < \epsilon \sum_j |\Delta_{ij}x|$$

and after summing on i that

(7.45) $$\sum R_i \geqslant -\epsilon L(C) \geqslant -\epsilon\lambda.$$

We can suppose in view of Theorem 7.14 that the partitions π_i, $i = 1,\ldots,k$, have been chosen with norms so small that

(7.46) $$W(C;F) > \sum_i \sum_j F[x(\tau_{j-1}),\Delta_{ij}x] - \epsilon.$$

It follows from (7.41) that

(7.47) $$\sum F[x(t_{i-1}),\Delta_i x] > \sum F[x_0(t_{i-1}),\Delta_i x_0] - \epsilon,$$

and from (7.44) that

(7.48) $$\sum_i \sum_j F[x(\tau_{j-1}),\Delta_{ij}x] \geqslant \sum F[x(t_{i-1}),\Delta_i x] + \sum R_i.$$

By adding inequalities (7.39) and (7.45) through (7.48) we find that

$$W(C;F) > W(C_0;F) - \epsilon(3+\lambda).$$

Finally, let C run over any sequence C_ν, $\nu = 1, 2, \ldots$ such that $d(C_\nu, C_0) \to 0$ and obtain the stated conclusion (7.38).

Proofs concerning variational integrals of the Weierstrass type, although often straightforward, can require more space than similar proofs in terms of Lebesgue integrals, but the latter presuppose the content of Chapter 8 and parts of Chapter 9.

7.11 EXISTENCE THEOREMS FOR PARAMETRIC PROBLEMS

Proofs of the existence of a curve or function yielding the global minimum for a variational problem not only provide an important fact but also insight to the structure of the problem. Hilbert was a pioneer in this sector followed by Tonelli, whose many works over a span of some 40 years from around 1910 were a major inspiration to a considerable literature by his students and by others in a number of countries. Proofs in the Tonelli tradition are characterized by the use of semicontinuity of the functional. The variety of such theorems is too great to review here but we give a sample.

Theorem 7.17

If $F: A \times B \to R$ has properties (i), (ii), (iii), *and* (iv) *and $\mathscr{C}$ denotes the class of all admissible Fréchet curves C with graphs in A and having initial and terminal points in disjoint closed subsets S_1 and S_2 of A, then, if $\mathscr{C}$ is not empty, there exists $C_0 \in \mathscr{C}$ such that*

$$(7.49) \qquad W(C_0;F) \leqslant W(C;F), \qquad \forall C \in \mathscr{C}.$$

We begin by inserting two lemmas.

(α) *Properties* (i), (ii), *and* (iv) *of F and the properties of sets A and B stated in Section 7.10 imply that there exist positive constants m and M such that*

$$m|r| \leqslant F(x,r) \leqslant M|r|$$

if $(x,r) \in A \times B$ and $r \neq 0$.

PROOF OF (α)

Since B is closed, the subset of B consisting of points r at unit distance from the origin is bounded and closed, hence compact. Consequently, the set $S = A \times \{r \in B : |r| = 1\}$ is a compact (bounded and closed) subset of R^{2n}. By the continuity (i) of F there exist (x_*, r_*) and (x^*, r^*) $\in S$ at which $F(x,r)$ attains its infimum m and supremum M on S, respectively. That $m > 0$ follows from the continuity (i) of F and its positiveness when $r \neq 0$. Thus given $x \in A$ and $r \neq 0$, we have that

$$m \leqslant F(x,r/|r|) \leqslant M.$$

The stated conclusion then follows from the homogeneity (ii) of F.

(β) *If C is an admissible curve whose graph is a subset of A and if F has properties* (i), (ii), *and* (iv), *then*

$$mL(C) \leqslant W(C;F) \leqslant ML(C).$$

PROOF OF (β)

If $r \neq 0$, we use Lemma (α). If $r = 0$, it follows from (i) and (ii) that $F(x,0) = 0$ and hence that the inequalities

$$(7.50) \qquad m|r| \leqslant F(x,r) \leqslant M|r|$$

hold in the form $0 \leq 0 \leq 0$. Thus (7.50) holds for all $(x,r) \in A \times B$. The conclusion of Lemma (β) then follows from problem 1, Exercise 7.2, and the definitions (7.33) and (7.37).

PROOF OF THEOREM 7.17

Set $\gamma \equiv \inf\{W(C;F): C \in \mathscr{C}\}$. By property (iv) of F, $\gamma \geq 0$. Given any positive integer ν, there exists by the definition of infimum a curve $C_\nu \in \mathscr{C}$ such that $W(C_\nu;F) < \gamma + 1/\nu$; consequently $W(C_\nu;F)$ converges to γ. Such a sequence as $\{C_\nu \in \mathscr{C}: \nu \in N\}$ is called a *minimizing sequence*, a term which anticipates the truth of the theorem. By Lemma (β), $L(C_\nu) \leq (1/m) W(C_\nu;F)$; hence there is some real number λ such that $L(C_\nu) \leq \lambda, \nu = 1, 2, \ldots$. The Hilbert Compactness Theorem now applies to ensure that the sequence $\{C_\nu\}$ has a subsequence converging to a curve C_0 whose graph is necessarily a subset of the compact set A. We can avoid further notation by supposing that the original sequence $\{C_\nu\}$ has been so selected that it is already such a sequence. Let ξ_ν: $[0,1] \rightarrow A$ be the reduced-length representation of C_ν used in the proof of the Hilbert theorem. Since $\xi_\nu(0) \in S_1$ and $\xi_\nu(1) \in S_2$ with S_1 and S_2 closed by hypothesis and since Fréchet distance $d(C_\nu,C_0) \rightarrow 0$, it follows that the respective limits $\xi_0(0)$ and $\xi_0(1)$ are points of S_1 and S_2. Moreover, since C_ν is admissible, ξ_ν satisfies (7.32) and, since ξ_ν converges to ξ_0, ξ_0 satisfies (7.32) and C_0 is admissible.

Our present hypotheses include those of Theorem 7.16; therefore, (7.38) applies and

$$\gamma = \lim W(C_\nu;F) = \lim\inf W(C_\nu;F) \geq W(C_0;F) \geq \gamma.$$

The outer terms being equal, equality must hold at each step. Therefore, $W(C_0;F) = \gamma$ and this is equivalent to the desired conclusion (7.49).

Observe that this theorem includes the fixed-endpoint problem in which S_1 and S_2 are singleton sets and also various other problems. For instance, S_1 may be the graph of a continuous function $g: [a,b] \rightarrow R^n$ and, if $n > 2$, S_2 could be the graph of a continuous function $h: [a,b] \times [c,d] \rightarrow R^n$. Then S_1 is a "curve" and S_2 a "surface" in the sense of these terms in classical analytic geometry.

We comment further on the fact that the sets A and B introduced in Section 6.6 and used thereafter in Chapter 6 were open in order that certain functions y in a neighborhood of an admissible y_0 would also be admissible in the proofs of necessary conditions. In contrast with the needs of Chapter 6, proofs in the present discussion require that A be compact and that B be closed. Consequently, theorems of Chapters 6 and 7 may not both apply to the same problem.

A similar situation exists in the theory of extrema of a differentiable point-function ϕ. The extremizing values x_* and x^* mentioned at the beginning of Section 7.2 may be interior points of $[a,b]$ or they may be boundary points. We are reminded in Section 2.3 that the most familiar necessary conditions for a local extremum of ϕ are for the case in which the extremum occurs at an interior point of its domain, but there are also such results as (2.8) for the case of a minimizing boundary point. Similarly, the graph of the minimizing curve C_0 of Theorem 7.17 may or may not include boundary points of the set A. There is a literature on minimizing curves or functions that include portions of such a boundary, for example, Bolza (X,pp.41–43), (XI,pp.392–407) and Mancill (XV, years 1933–1937), but we do not discuss this topic in the present book.

Chapter 6 admits only PWS curves and representations but our admissible curves and representations are now merely rectifiable. Thus the minimizing curve C_0 of Theorem 7.17 is known only to be continuous and of finite length. It may happen to be PWS and its graph may happen to consist entirely of interior points of A. Granted these properties of C_0, then various results in Chapter 6 would be applicable, including the classical necessary conditions of Euler, Weierstrass, and Legendre, together with that of Jacobi, which we did not treat. Unfortunately there are no simple or general criteria for identifying those cases under Theorem 7.17, in which C_0 has these additional properties, but one finds that they often occur for examples that are simple enough so that all details can be checked out.

The next existence theorem will shed a little light on the preceding remarks. For simplicity we take $n = 2$.

Theorem 7.18

Given a function $F:R^2 \times R^2 \to R$ with properties (i), (ii), (iii), and (iv) and the class $\mathscr{C}$ of all continuous Fréchet curves of finite length whose graphs are in R^2 and join distinct fixed points of R^2, then there exists $C_0 \in \mathscr{C}$ such that

$$W(C_0;F) \leqslant W(C;F), \qquad \forall C \in \mathscr{C}.$$

PROOF

The present class $\mathscr{C}$ is clearly not empty. Set $\gamma \equiv \inf\{W(C;F): C \in \mathscr{C}\}$ and let $\{C_\nu \in \mathscr{C}: \nu \in N\}$ be a minimizing sequence. In view of Lemma (β) to the preceding theorem we can suppose C_ν to have been so chosen that the length

$$(7.51) \qquad L(C_\nu) \leqslant (1/m)W(C_\nu;F) \leqslant (\gamma + 1/\nu)/m, \qquad \nu = 1, 2, \ldots.$$

Let us now denote by A the compact subset of R^2 bounded by the ellipse consisting of all points of R^2 the sum of whose distance from the two fixed points in the theorem is $(\gamma + 2)/m$. By the Hilbert Compactness Theorem the sequence $\{C_\nu\}$ has a subsequence, again denoted by $\{C_\nu\}$, converging in terms of Fréchet distance to a curve C_0 that is necessarily in the class $\mathscr{C}$.

The semi-continuity of length L (Theorem 7.6) assures that

(7.52) $\liminf L(C_\nu) \geq L(C_0)$;

hence from (7.51) and our choice of the set A, $L(C_0) \leq (\gamma+1)/m$ and the graph of C_0 is interior to A.

By Theorem 7.16. $W(C;F)$ is lower semi-continuous on the class of curves of length at most $\lambda = (\gamma+1)/m$, and therefore we again have the inequalities

$$\gamma = \lim W(C_\nu;F) = \liminf W(C_\nu;F) \geq W(C_0;F) \geq \gamma,$$

which imply that $W(C_0;F) = \gamma$, and the proof is complete.

Although we know in this instance that the graph of C_0 is interior to A, we still cannot say in general whether C_0 is PWS. There is a literature on extensions of some of the classical necessary conditions including that of Euler to the case of a general rectifiable extremizing curve or function, but the usual approach, for example, in Tonelli (XXXV) or Reid (45e), has been in terms of the Lebesgue integral.

Existence theorems, whether in the calculus of variations or elsewhere, frequently provide no way in which to determine the thing that has been shown to exist. Generally speaking, one can obtain stronger conclusions if and only if he pays for them with stronger hypotheses. Hypotheses of the next theorem are so strong that it provides both the existence and identification of the minimizing curve. The theorem is suggested by such classical results as Example 2.1 and Theorem 3.9.

Theorem 7.19

If F is free of x and has properties (i), (ii), *and* (iii), *if the set A is convex, if $B = R^n$ and $\mathscr{C}$ is the class of all rectifiable Fréchet curves with graphs in A and joining fixed points h and k, then the curve $C_0 \in \mathscr{C}$ having the linear representation $x_0: [0,1] \to R^n$, $x_0(t) = h + t(k-h)$ furnishes a global minimum for $W(C;F)$. Moreover,*

$$W(C_0;F) = F(k-h).$$

PROOF

As a consequence of (iii) and (ii) we have the inequality

(7.53) $F\left(\sum r_i\right) \leq \sum F(r_i)$

explained following (7.43) in the proof of Theorem 7.16. Given the function x_0 mentioned in the theorem let π be a partition of $[0,1]$ that generates m abutting subintervals $[t_{i-1}, t_i]$ of equal length. By property (ii) of F,

$$F(k-h) = F[x_0(1) - x_0(0)] = mF\{[x_0(1) - x_0(0)]/m\}.$$

Each of the m vectors $x_0(t_i) - x_0(t_{i-1})$ equals $[x_0(1) - x_0(0)]/m$, and therefore

$$F(k-h) = \sum F[x_0(t_i) - x_0(t_{i-1})].$$

Let $\|\pi\| \to 0$ and use (7.34), Theorem 7.14, and definition (7.37) to conclude that

$$(7.54) \qquad\qquad F(k-h) = W(C_0; F).$$

If $x: [0,1] \to R^n$ represents an arbitrary curve $C \in \mathscr{C}$, it follows from (7.53) that

$$F(k-h) \leq \sum F[x(t_i) - x(t_{i-1})]$$

and by letting $\|\pi\| \to 0$ that

$$(7.55) \qquad\qquad F(k-h) \leq W(C; F).$$

The conclusions stated in the theorem follow from (7.54) and (7.55).

Recall the form of the parametric Problem of Bolza in Section 6.10. That a PWS representation y of an admissible curve satisfies a side-condition $\phi(y, \dot{y}) = 0$ on an interval $[a,b]$ means that $\phi[y(t), \dot{y}^-(t)]$ and $\phi[y(t), \dot{y}^+(t)]$ vanish on the half-open intervals $(a,b]$ and $[a,b)$, respectively. With $\phi(y, r)$ continuous on $A \times B$, one verifies that y satisfies the differential equation in this sense iff

$$(7.56) \qquad \int_a^t \phi[y(\tau), \dot{y}(\tau)] \, d\tau = 0, \qquad \forall t \in [a,b].$$

If ϕ is both nonnegative and continuous, then y satisfies the differential equation on $[a,b]$ iff

$$(7.57) \qquad \int_a^b \phi[y(\tau), \dot{y}(\tau)] \, d\tau = 0.$$

It follows from Theorem 7.13 that, if $\phi(y, r)$ has the stated properties and is also homogeneous in r, then these conditions can be stated in terms of the Weierstrass integral as

$$(7.58) \qquad \mathscr{W}(y; \phi; [a,t]) = 0, \qquad \forall t \in [a,b]$$

and

$$(7.59) \qquad \mathscr{W}(y; \phi; [a,b]) = 0.$$

The last conditions are meaningful, in the light of Theorem 7.14, if y is merely BV on $[a,b]$. Such a function y satisfying (7.58) or (7.59) will be called a *generalized solution* of the differential equation $\phi(y,\dot{y}) = 0$. Moreover, by Theorem 7.15 these conditions can be expressed by means of Weierstrass curvilinear integrals in the form

$$(7.58^*) \qquad\qquad W(C_{y,t};\phi) = 0, \qquad \forall t \in [a,b]$$

and

$$(7.59^*) \qquad\qquad W(C:\phi) = 0.$$

We wish to formulate a parametric Bolza Problem with the aid of the following functions:

$$
\begin{aligned}
&F: A \times B \to R &&\textit{with properties (i), (ii), (iii), and (iv),}\\
&\phi_\beta{:}A \times B \to R &&\textit{with properties (i), (ii), (iii), and (iv),}\\
&\chi_\alpha{:}A \to R &&\textit{continuous on } A,\\
&g: A \times A \to R &&\textit{continuous on } A \times A,\\
&\psi_\mu{:}A \times A \to R &&\textit{continuous on } A \times A.
\end{aligned}
$$

Consider the problem

$$(7.60) \qquad J(C) \equiv g[y(t_0),y(t_1)] + W(C;F) = minimum$$

on the class $\mathscr{C}$ consisting of all rectifiable Fréchet curves possibly satisfying side-conditions

$$W(C;\phi_\beta) = 0, \qquad \beta = 1,\ldots,m,$$

and/or

$$\chi_\alpha[y(t)] = 0, \qquad \forall t \in [t_0,t_1], \alpha = 1,\ldots,k,$$

and satisfying end-conditions

$$\psi_\mu[y(t_0),y(t_1)] = 0, \qquad \mu = 1,\ldots,p \geqslant 2.$$

The word "possibly," together with "and/or," conveys the qualification that there may be side-conditions of either, neither, or both kinds.

Clearly many types of problems are included under this formulation. For instance, if $g = 0$ and $W(C;F)$ is the length $L(C)$, if there is one side-condition $|y(t)| = a > 0$ of the second kind and end-conditions $y(t_0) = h_0 \in R^n$, $y(t_1) = h_1 \in R^n$ with $|h_0| = |h_1| = a$, then we have the problem of a curve of least length on the $(n-1)$-sphere joining two fixed points.

Exercise 7.5

In responding to problems 1 and 2, one can follow much of the proof of Theorem 7.17. Since the Hilbert Theorem 7.10 is involved, it is helpful to remember the use of reduced-length representations in the proof of that theorem.

1. Given functions F, ϕ_β, χ_α, g, and ψ_μ with the properties stated above and a problem (7.60) having at least one side-condition of each of the two types, and given further that the class of admissible curves is not empty, prove that there exists $C_0 \in \mathscr{C}$ such that $J(C_0) \leqslant J(C)$ for all $C \in \mathscr{C}$. Is it possible for $\mathscr{C}$ to be a finite class?

2. Let p^j: $A \to R$ be continuous on A, $j = 1, \ldots, n$. Given that B is convex, show that definition (3.20) of a convex function is satisfied in r for each fixed y by $\phi(y,r) \equiv |p^j(y)r^j|$ (summed on j). Then state and prove an existence theorem for the fixed-endpoint Lagrange Problem $W(C;F) = minimum$ subject to a single side condition $W(C;\phi) = 0$, where ϕ is of the form $|p^j(y)r^j|$.

3. Given that p^j: $A \to R$ is continuous on A, $j = 1, \ldots, n$, and that $\phi(y,r) = p^j(y)r^j$, point out why Theorem 7.16 applies to both of the integrals $W(C;\phi)$ and $W(C;-\phi)$. Given also that F has properties (i), (ii), (iii), and (iv), that $J(C) \equiv W(C;F)$, and that $\mathscr{C}$ is a nonempty class of rectifiable Fréchet curves satisfying (7.58*) with the given ϕ and satisfying the end-conditions $|y(t_0)| = 0$ and $|y(t_1)| = 5$, prove that there exists $C_0 \in \mathscr{C}$ such that $J(C_0)$ is the global minimum.

4. Given the Mayer Problem $J(C) \equiv g[y(t_0),y(t_1)] = minimum$ with one side condition $W(C;\phi) = 0$, where ϕ has properties (i), (ii), and (iii) but not (iv), let $\mathscr{C}$ be the class of all Fréchet curves satisfying the side condition and joining fixed endpoints. Point out why we cannot prove that there is a minimizing curve C_0 by simply following the proof of Theorem 7.17.

5. If $J(C)$ is the length $L(C)$ and we have fixed endpoints and one side-condition $W(C;\phi) = a \neq 0$, where ϕ has properties (i), (ii), (iii), and (iv), point out why there is a minimizing sequence $\{C_\nu\}$ converging to a curve C_0 and yet C_0 may not be in the class $\mathscr{C}$ of curves joining the fixed endpoints and satisfying the side-condition.

6. Formulate and prove an extension of Theorem 7.18 from $n = 2$ to a general n.

7. Devise a proof for the special case of Theorem 7.15 in which F is free of x and convex in r, drawing hints from Sections 7.4 and 7.5.

7.12   NONPARAMETRIC WEIERSTRASS INTEGRALS

In view of Theorem 7.15, the integral treated in Sections 7.10 and 7.11 is parametric, but we have remarked in Section 6.6 and elsewhere that many optimization questions lead to nonparametric variational problems and indeed that the mathematical model for an optimization problem from the sciences is usually of the nonparametric type. We have seen in Section 6.13, under classical hypotheses, that a parametric problem is equivalent to a nonparametric problem provided that the integrands are related as in (6.2) and that the class of admissible curves is restricted to curves C with representations (x,y): $[t_0,t_1] \to R^{m+1}$ such that x is strictly increasing.

We now point out one way in which these ideas can be extended to the setting of the present chapter and how the Weierstrass integral can be adapted to such circumstances.

Suppose given a nonparametric integrand $f: A \times R^m \to R$, where A is again the closure of a bounded open subset of the encompassing space, presently R^{m+1} with points $(x,y) = (x,y^1,\ldots,y^m)$. To simplify the discussion we restrict attention to a function f that is continuous and nonnegative on its domain and such that, for each choice of $(x,y) \in A$ and $q \in R^m$, $f(x,y,q/p)p$ has a limit, finite or ∞ as p tends to 0 through positive values. Then with suggestions from McShane (33c), Aronszajn, and Pauc (43a, pp. 66–76), we define an associated parametric integrand $F: A \times B \to R^*$ with $B \equiv \{(p,q) \in R^{m+1}: p \geq 0, q \in R^m\}$ by assigning the values

$$(7.61) \qquad F(x,y,p,q) \equiv \begin{cases} f(x,y,q/p)p & \text{if } p > 0, \\ \lim f(x,y,q/p)p \text{ as } p \to 0+ & \text{if } p = 0. \end{cases}$$

This parametric integrand inherits continuity from the given function f at every point of $A \times B$ having a positive third coordinate p. If $q \neq 0$, then usually $F(x,y,0,q) = \infty$, but $F(x,y,0,0) = 0$. For the special case in which f is linear in r; $f(x,y,r) = a^0(x,y) + a^j(x,y) r^j$, with summation on j, and the coefficients $a^j, j = 0,1,\ldots,m$, are continuous on A, the integrand (7.61) is continuous on $A \times B$. A similar conclusion is reached if f is the integrand for the nonparametric length integral. In such cases Sections 7.10 and 7.11 apply, but these are exceptional. For such a simple nonparametric integrand as $f(x,y,r) = r^2$, the square of r, one verifies from (7.61) that

$$(7.62) \qquad F(x,y,p,q) = \begin{cases} q^2/p & \text{if } p > 0, \\ 0 & \text{if } p = q = 0, \\ \infty & \text{if } p = 0, q \neq 0. \end{cases}$$

The continuity of F has been used throughout the last two sections. Although Pauc has pointed out (43a, pp. 66, 67) with reference to an unpublished work of Aronszajn a way in which the Weierstrass integral can be extended to such unbounded integrands as (7.62), the relative simplicity of the theory for a continuous integrand F is lost.

Among the advantages of the Lebesgue integral is the facility with which unbounded integrands or extended real-valued integrands are treated. Lebesgue integrals of the form encountered in variational theory are discussed in Chapter 9.

Chapter 8

MEASURE, INTEGRALS, AND DERIVATIVES

8.1 INTRODUCTION

Both parametric and nonparametric integrals, understood in the sense of Lebesgue's definition, are important ingredients in modern calculus of variations. In preparation for their introduction in Chapter 9 and in order that the book be reasonably self-contained, this chapter presents the basic theory of the Lebesgue integral as far as the Fundamental Theorem of the Integral Calculus. The reader should spend little time on the parts he already knows beyond observing what is there for possible later reference.

There are a number of approaches to this material. We elect a traditional one in which measure precedes the integral in the belief that this encourages insights that are more easily missed if one follows a streamline path to the convergence theorems. Although there are a number of books on measure and integration, few show any orientation to the special requirements of variational theory. It becomes a barrier to progress in the latter if one must identify and extract what is needed from several sources with differences in notation, in sequencing of concepts and results, and in level of generality and conciseness.

We are primarily interested in ordinary Lebesgue measure of subsets of the real numbers R and in integration with respect to this measure. However, much of what we do and say applies with little or no change to other important measures and integrals. Also needed is the related theory of differentiation of functions $y: [a,b] \to R^n$ of bounded variation so as to deal ultimately with integrals of composite integrands $F(y,\mathfrak{y})$ and $f(x,y,y')$ in problems of the calculus of variations.

8.2   LINEAR LEBESGUE OUTER MEASURE

Given R, the set of reals, and $A \subset R$, consider the family (or set) of all *countable open coverings of* A, that is, of all finite or denumerable unions $\cup (a_i, b_i)$ of open intervals that contain A. We adopt the convention that, when a union or summation symbol precedes a symbol with an index such as i or n, the range of that index is either the entire set N of positive integers or a finite subset $\{1, 2, \ldots, m\}$ of N unless some other index set is explicitly stated.

We do not require that the intervals (a_i, b_i) of a countable covering be disjoint and we regard the empty set $\emptyset$ as a particular open interval, namely $(a,a) = \{x \in R: a < x < a\}$. Consequently, a finite union, $\cup (a_i, b_i)$ is equivalent to a denumerable union for which all intervals beyond a certain one are empty.

Linear Lebesgue outer measure is a set function $\mu^*: \mathscr{P}(R) \to R^*$ whose domain $\mathscr{P}(R)$, called the *power set* of R, is the set of all subsets of R. Given $A \subset R$, $\mu^*(A)$ is defined as follows.

$$(8.1) \qquad \mu^*(A) \equiv \inf \left\{ \sum (b_i - a_i) : A \subset \cup (a_i, b_i) \right\},$$

the infimum being on the class of all countable open coverings of A. It is not difficult to verify that

$$(8.2) \qquad \begin{aligned} &\mu^*(\phi) = 0, \\ &0 \le \mu^*(A) \le \infty, \\ &\mu^*(A) \le \mu^*(B) \text{ if } A \subset B, \\ &\mu^*(\cup A_i) \le \sum \mu^*(A_i). \end{aligned}$$

Exercise 8.1

1. If I is an interval (a,b), $[a,b]$, $(a,b]$, or $[a,b)$, verify that $\mu^*(I) = b - a$.
2. If A is a countable subset of R cover element x_i with a single open interval of length $\epsilon/2^i$ and show that $\mu^*(A) = 0$.

3. Prove the third assertion (8.2).
4. Prove the fourth assertion (8.2). One may suspect that, if the sets A_i are disjoint, then equality must hold. This is not so, but an example would require nonmeasurable sets and this need not concern us.

8.3 LEBESGUE MEASURABILITY AND MEASURE

Let A be a fixed subset of R; let $\mathbf{C}(A)$, called the *complement of A*, denote the set $R - A$; and let T, called a *test set*, vary over the power set $\mathscr{P}(R)$. If

$$(8.3)\qquad \mu^*(T) = \mu^*(T \cap A) + \mu^*[T \cap \mathbf{C}(A)], \qquad \forall T \subset R,$$

then A is said to be *Lebesgue measurable*, and the *Lebesgue measure* $\mu(A)$ is by definition

$$(8.4)\qquad \mu(A) \equiv \mu^*(A) \qquad \textit{provided that } (8.3) \textit{ holds}.$$

Thus μ is a set-function whose domain is the subset of $\mathscr{P}(R)$ consisting of the measurable subsets of R.

One's initial reaction may be not only that (8.3) looks quite restrictive but that it cannot effectively be applied to a particular set. One certainly cannot check out (8.3) for each of uncountably many test sets T by taking each of them in turn. Ways around this impass must be found. It actually turns out that the class of sets satisfying (8.3) is very large.

Theorem 8.1

If A is measurable, then $\mathbf{C}(A)$ is measurable and

$$(8.5)\qquad \mu(T) = \mu(T \cap A) + \mu[T \cap \mathbf{C}(A)].$$

PROOF

Given that (8.3) holds, then, since $\mathbf{C}[\mathbf{C}(A)] = A$, (8.3) also holds with A replaced by $\mathbf{C}(A)$. Observe that such equations as (8.3) and (8.5) may on occasion hold in the form $\infty = \infty$, for example, when $T = R$.

Theorem 8.2

A sufficient condition for the measurability of A is that

$$(8.6)\qquad \mu^*(T) \geq \mu^*(T \cap A) + \mu^*[T \cap \mathbf{C}(A)] \qquad \textit{if } \mu^*(T) < \infty.$$

PROOF

$T = (T \cap A) \cup [T \cap \mathbf{C}(A)]$; hence by (8.2_4),

$$\mu^*(T) \leqslant \mu^*(T \cap A) + \mu^*[T \cap \mathbf{C}(A)],$$

which with (8.6) implies (8.3). Inequality (8.6) clearly holds if $\mu^*(T) = \infty$; hence it suffices to examine only those T such that $\mu^*(T) < \infty$.

Theorem 8.3

If $\mu^*(A) = 0$, then A is measurable and $\mu(A) = 0$.

PROOF

For an arbitrary test set T,

$$T \cap A \subset A \qquad and \qquad T \cap \mathbf{C}(A) \subset T;$$

hence, by (8.2_3),

$$\mu^*(T \cap A) \leqslant \mu^*(A) = 0 \qquad and \qquad \mu^*[T \cap \mathbf{C}(A)] \leqslant \mu^*(T).$$

It follows that

$$\mu^*(T \cap A) + \mu^*[T \cap \mathbf{C}(A)] \leqslant \mu^*(T),$$

which is (8.6). That $\mu(A) = 0$ then follows from our hypothesis that $\mu^*(A) = 0$ and definition (8.4).

Theorem 8.4

A countable set A is measurable and $\mu(A) = 0$.

PROOF

Use problem 2, Exercise 8.1, and Theorem 8.3.

Theorem 8.5

Every open subset G of the reals is the union of countably many disjoint open intervals. Moreover, there is only one such set of open intervals.

PROOF

The empty set $\emptyset$ is the empty open interval (a,a). Given $x \in G$, there necessarily exists an open interval $I_x \equiv (x - \delta_1, x + \delta_2)$ of maximal length such that $x \in I_x \subset G$. Each such interval contains a rational real number and the set of all such numbers is denumerable. The class of distinct such intervals I_x must therefore be countable (finite or denumerable), and the union of these intervals is G. That the decomposition of G into open intervals is unique and that they are disjoint is left as an exercise.

Theorem 8.6

Every interval I with endpoints a and b is measurable and $\mu(I) = b - a$.

PROOF

The degenerate cases $I = (a,a) = \emptyset$ and $I = [a,a]$ consisting of the single point a are covered by Theorems 8.3 and 8.4. We give the details for an open interval (a,b) of finite length.

Let T be an arbitrary open test set. If a or b is in T, such a point can be deleted with essentially no effect. We can therefore suppose that neither a nor b is in T. Let $\cup(a_n,b_n)$ be the decomposition of T into open intervals assured by Theorem 8.5. Let $\cup'(a_n,b_n)$ and $\cup''(a_n,b_n)$ be respective subunions of intervals (a_n,b_n) that are contained in the given interval (a,b) and in $\mathbf{C}[(a,b)] = R - (a,b)$.

By definition (8.1), $\mu^*(T) \leqslant \sum (b_n - a_n)$ but, since the intervals are disjoint and their union is T, equality must hold. Similarly,

$$\mu^*[T \cap (a,b)] = \sum{}' (b_n - a_n)$$

$$\text{and} \qquad \mu^*\{T \cap \mathbf{C}[(a,b)]\} = \sum{}'' (b_n - a_n);$$

hence (8.3) holds for any open test set T.

Given an arbitrary test set T and $\epsilon > 0$, there exists, by the definition of μ^*, an open set G_ϵ, namely, a union of open intervals, such that

$$(8.7) \qquad \mu^*(G_\epsilon) < \mu^*(T) + \epsilon.$$

Now $T \cap (a,b) \subset G_\epsilon \cap (a,b)$ and $T - (a,b) \subset G_\epsilon - (a,b)$; therefore, by (8.2_3),

$$\mu^*[T \cap (a,b)] \leqslant \mu^*[G_\epsilon \cap (a,b)]$$

$$\text{and} \qquad \mu^*[T - (a,b)] \leqslant \mu^*[G_\epsilon - (a,b)].$$

From these relations, with the aid of (8.7) and the conclusion above that (8.3) holds if T is open, now used with G_ϵ in place of T, we find that

$$\mu^*(T) + \epsilon > \mu^*[T \cap (a,b)] + \mu^*[T - (a,b)].$$

Since $\epsilon > 0$ but otherwise arbitrary, it follows that the right member is dominated by $\mu^*(T)$. This is (8.6) in view of the fact that $T - (a,b) =$

$T \cap \mathbf{C}[(a,b)]$ and the proof that (a,b) is measurable is complete. That $\mu(I) = b - a$ follows from problem 1, Exercise 8.1, and definition (8.4).

Theorem 8.7

If A and B are measurable, so also are $A \cup B$, $A - B$, and $A \cap B$.

PROOF

By the hypothesis on A,

(8.8) $\mu^*(T) = \mu^*(T \cap A) + \mu^*(T - A), \qquad \forall T \subset R.$

Since B is also measurable, (8.3) applies to B with $T - A$ as the test set, that is,

$$\mu^*(T - A) = \mu^*[(T - A) \cap B] + \mu^*[(T - A) - B]$$
$$= \mu^*[\qquad\qquad] + \mu^*[T - (A \cup B)].$$

Substitution of the last expression into (8.8) yields that

(8.9) $\mu^*(T) = \mu^*(T \cap A) + \mu^*[(T - A) \cap B] + \mu^*[T - (A \cup B)].$

One verifies that

$$T \cap (A \cup B) = (T \cap A) \cup [(T - A) \cap B].$$

From this, together with (8.2₄) and (8.9), follows that

(8.10) $\mu^*(T) \geqslant \mu^*[T \cap (A \cup B)] + \mu^*[T - (A \cup B)];$

therefore, $A \cup B$ is measurable in consequence of Theorem 8.2.

To prove that $A - B$ is measurable observe that $A - B = A \cap \mathbf{C}(B)$, hence that

$$\mathbf{C}(A - B) = [\mathbf{C}(A)] \cup B.$$

The right member is measurable by Theorem 8.1 and the preceding result. Hence $\mathbf{C}(A - B)$ is measurable and, by Theorem 8.1, $A - B$ is then measurable.

Finally, $A \cap B = B - \mathbf{C}(A)$ is measurable by Theorem 8.1 and the measurability of a difference.

Theorem 8.8

Every countable union $\cup A_i$ of measurable sets is measurable. Moreover,

(8.11) $\mu(\cup A_i) = \sum \mu(A_i) \qquad$ *if the sets A_i are disjoint.*

PROOF

Since $A_1 \cup A_2$ is measurable by Theorem 8.7 and $\cup_1^{n+1} A_i = (\cup_1^n A_i) \cup A_{n+1}$, we have by induction that every finite union is measurable.

To establish (8.11) for a finite union, set $S_n \equiv \cup_1^n A_i$ and, proceeding inductively, suppose that

$$(8.12) \qquad \mu^*(T \cap S_k) = \sum_1^k \mu^*(T \cap A_i).$$

Using $T \cap S_{k+1}$ as a test set, we know from the measurability of S_k, already established, that

$$\mu^*(T \cap S_{k+1}) = \mu^*[(T \cap S_{k+1}) \cap S_k] + \mu^*[T \cap S_{k+1} - S_k]$$
$$= \mu^*[T \cap S_k] + \mu^*(T \cap A_{k+1}).$$

After using expression (8.12) for the next to last term we have completed a proof by induction that

$$(8.13) \qquad \mu^*\left(T \cap \bigcup_1^n A_i\right) = \sum_1^n \mu^*(T \cap A_i).$$

Choosing $T = \cup_1^n A_i$ and using definition (8.4) yields (8.11), in case the union is finite.

For the proof of (8.11) in the denumerable case, suppose initially that the sets A_i are disjoint. By the measurability of a finite union, with (8.13) and (8.2_3),

$$\mu^*(T) = \mu^*\left(T \cap \bigcup_1^n A_i\right) + \mu^*\left(T - \bigcup_1^n A_i\right)$$
$$= \sum_1^n \mu^*(T \cap A_i) + \mu^*(\qquad\qquad)$$
$$\geq \sum_1^n \mu^*(T \cap A_i) + \mu^*\left(T - \bigcup_1^\infty A_i\right).$$

In view of Theorem 8.2 we can restrict attention to test sets of finite outer measure. Then

$$\mu^*\left(T \cap \bigcup_1^\infty A_i\right) < \infty.$$

Now $T \cap \cup_1^\infty A_i \supset T \cap \cup_1^n A_i$; consequently,

$$(8.14) \qquad \mu^*\left(T \cap \bigcup_1^\infty A_i\right) \geq \mu^*\left(T \cap \bigcup_1^n A_i\right) = \sum_1^n \mu^*(T \cap A_i).$$

Denote by λ the necessarily finite limit of this sum as $n \to \infty$. Observe that

$$T \cap \bigcup_1^\infty A_i = \left(T \cap \bigcup_1^n A_i\right) \cup \left(T \cap \bigcup_{n+1}^\infty A_i\right),$$

whence

(8.15) $\mu^*\left(T \cap \bigcup_1^\infty A_i\right) \leq \sum_1^n \mu^*(T \cap A_i) + \mu^*\left(T \cap \bigcup_{n+1}^\infty A_i\right).$

Moreover,

$$T \cap \bigcup_{n+1}^\infty A_i = \bigcup_{n+1}^\infty (T \cap A_i),$$

so that

(8.16) $\mu^*\left(T \cap \bigcup_{n+1}^\infty A_i\right) \leq \sum_{n+1}^\infty \mu^*(T \cap A_i).$

Because of the convergence of the sum in (8.14), the right member of (8.16) has the limit zero; consequently, from (8.15),

(8.17) $\mu^*\left(T \cap \bigcup_1^\infty A_i\right) \leq \lambda.$

Letting $n \to \infty$ in the inequality with $\mu^*(T)$ on the left that precedes (8.14), we see that

$$\mu^*(T) \geq \lambda + \mu^*\left(T - \bigcup_1^\infty A_i\right),$$

therefore, with the aid of (8.17), that

(8.18) $\mu^*(T) \geq \mu^*\left(T \cap \bigcup_1^\infty A_i\right) + \mu^*\left(T - \bigcup_1^\infty A_i\right).$

This says, by Theorem 8.2, that $\bigcup_1^\infty A_i$ is measurable. It follows from (8.14) and (8.17) that

$$\mu^*\left(T \cap \bigcup_1^\infty A_i\right) = \sum_1^\infty \mu^*(T \cap A_i).$$

The particular test set, $T = \bigcup_1^\infty A_i$ then yields (8.11).

If the sets A_i are not disjoint, define $B_1 \equiv A_1$, $B_2 \equiv A_2 - A_1, \ldots , B_i \equiv A_i - \bigcup_1^{n-1} A_j, \ldots$. The B_i are disjoint and

$$\bigcup_1^\infty B_i = \bigcup_1^\infty A_i.$$

The union on the left is measurable by (8.18) with the present B_i in the role of the A_i of (8.18); consequently the right member is measurable and the proof is complete.

Theorem 8.9

Every countable intersection $\cap \, A_i$ of measurable sets is measurable.

PROOF

Use Theorems 8.1 and 8.8 and the De Morgan "law" that $\mathbf{C}(\cap \, A_i)$ $= \cup \, \mathbf{C}(A_i)$.

Given a sequence $\{A_n : n \in N\}$ of sets, define

$$(8.19) \qquad \qquad \liminf A_n \equiv \bigcup_{n \geq 1} \bigcap_{m \geq n} A_m,$$

$$(8.20) \qquad \qquad \limsup A_n \equiv \bigcap_{n \geq 1} \bigcup_{m \geq n} A_m.$$

It can be verified that $\liminf A_n \subset \limsup A_n$. If and only if equality holds, the common set is, by definition, the limit of the sequence of sets. By Theorems 8.8 and 8.9, if the sets A_n are all measurable, then sets (8.19) and (8.20) together with $\lim A_n$, when it exists, are all measurable.

This section can be summarized as follows. Every open subset G of the reals is Lebesgue measurable by Theorems 8.5 and 8.6; hence every closed subset is measurable by Theorem 8.1. Every set that can be constructed by countable unions and intersections or a succession of such operations upon measurable sets is measurable; hence many sets that are neither open nor closed are measurable. Loosely stated, all subsets of R that one is likely to deal with are measurable. The only known examples of nonmeasurable sets are constructed with the aid of what is known as the Axiom of Choice.

Exercise 8.2

1. If $\{x\}$ is a singleton subset of R and A is any measurable subset of R, show that $A \cup \{x\}$ and $A - \{x\}$ are measurable and of the same measure as A. Replace $\{x\}$ by any finite set and extend the preceding results by induction.

2. Granted the results proved in this section, show that intervals $[a,b]$ of finite length and that intervals $[a,b)$ or $(a,b]$ of either finite or infinite length are measurable and that in each case the measure is $b - a$.

8.4   MEASURABLE FUNCTIONS

Given a measurable subset E or R and a function $f: E \to R^*$, we denote the set $\{x \in E : f(x) > a\}$ by the abbreviation $\{f(x) > a\}$ and use symbols $\{f(x) \leq a\}$, $\{f(x) = a\}$, $\{f(x) = \infty\}$, $\{a < f(x) \leq b\}$, etc., with similar meanings.

A function $f: E \to R^*$ is called *measurable on E* if the sets

(8.21) $\{f(x) \geq a\}, \{f(x) < a\}, \{f(x) \leq a\}, and \{f(x) > a\}$

are all measurable for every real number a. This definition may seem to be rather artificial. It is designed, however, to give us exactly the functions we shall want in defining the Lebesgue integral.

Observe that $\{f(x) < a\} = E - \{f(x) \geq a\}$; hence if every set of the first type (8.21) is measurable, so also is each set of the second type. Granted this, then the set

$$\{f(x) = a\} = \bigcap_{n \in N} \left(\{f(x) \geq a\} \cap \left\{ f(x) < a + \frac{1}{n} \right\} \right)$$

is measurable. Also

$$\{f(x) \leq a\} = \{f(x) < a\} \cup \{f(x) = a\}$$

is measurable, and finally

$$\{f(x) > a\} = E - \{f(x) \leq a\}$$

is measurable. Consequently it would suffice to use only sets of the first type (8.21) in the definition of a measurable function. It can be verified that any one of the other types (8.21) would also suffice in the definition.

Granted that f is measurable on E under definition (8.21), such sets as $\{f(x) = \infty\} = \cap \{f(x) > n\}$ and $\{a < f(x) \leq b\} = \{f(x) > a\} \cap \{f(x) \leq b\}$, $a, b \in R$, are all measurable.

Theorem 8.10

Given a measurable set E, a function $f: E \to R^*$ measurable on E, and a measurable subset S of E, then f restricted to S is measurable on S.

PROOF

$\{x \in S: f(x) \geq a\} = S \cap \{x \in E: f(x) \geq a\}$ and the set on the right is measurable by Theorem 8.7.

Theorem 8.11

If E is measurable and if $f: E \to R^*$ and $g: E \to R^*$ are both measurable on E, then each of the following functions is measurable on the subset of E consisting of all $x \in E$ for which the stated function is meaningful:

(i)	$c + f, \quad c \in R^*$	(iii)	$f + g$	(v)	fg
(ii)	$cg, \quad c \in R^*$	(iv)	$f - g$	(vi)	f/g

The purpose of the phrase "for which the stated function is meaningful" is to exclude meaningless expressions such as $\infty - \infty$ in conclusions (i), (iii), and (iv); or $0(\infty)$ in (ii) and (v); or $3/0$, ∞/∞, etc., in (vi).

PROOF OF (iii)

The residual set S on which we wish to prove that $f + g$ is measurable is

$$S = E - (\{f(x) = \infty\} \cap \{g(x) = -\infty\}) \cup (\{f(x) = -\infty\} \cap \{g(x) = \infty\}).$$

We now establish that the set

(8.22) $\{x \in S: f(x) + g(x) > a\}, \qquad a \in R$

is measurable. We suppose that conclusions (i) and (ii) have already been proved. Set S as a difference between measurable sets is measurable; hence g is measurable on S by Theorem 8.10. By (ii) with $c = -1$, $-g$ is measurable on S; hence $a - g$ is measurable on S by (i). The inequality $f(x) + g(x) > a$ is equivalent to $f(x) > a - g(x)$. We complete the investigation of set (8.22) by showing that the set $\{x \in S: f(x) > a - g(x)\}$ is measurable. Let $r_1, r_2, r_3, \ldots$ be a fixed sequentialization of all rational real numbers. Now

$$\{x \in S: f(x) > a - g(x)\}$$
$$= \bigcap_n (\{x \in S: f(x) > r_n\} \cap \{x \in S: a - g(x) < r_n\}).$$

The set on the right is measurable by Theorems 8.8 and 8.9; therefore, the set on the left and consequently set (8.22) is measurable.

Theorem 8.12

Given a measurable set E and a sequence $\{f_n: E \to R^*: n \in N\}$ of functions all measurable on E, then each of the following functions is measurable on E:

(i) $\sup f_n$	(iii) $\lim \sup f_n$	(v) $\lim f_n$, *if it exists*
(ii) $\inf f_n$	(iv) $\lim \inf f_n$	(vi) $\|f\|$

PROOF

To prove (i), set $G \equiv \sup f_n$. Then $G(x) = \sup f_n(x)$, $\forall x \in E$. The set $\{G(x) > a\} = \cup_n \{f_n(x) > a\}$. Each set in the union is measurable under the hypothesis that f_n is measurable on E; hence the union is measurable by Theorem 8.8. Conclusion (ii) follows from (i) and Theorem 8.11(ii) together with the fact that $\inf f_n = -\sup(-f_n)$. The index n in these proofs can either run over the entire set N of positive integers or over a finite subset of N, in particular over the set $\{1,2\}$ of two elements.

Let 0 denote the function whose value at every point of E is the real number 0. Function 0 is easily verified to be measurable on E. If f is also measurable on E, then $\sup(f,0)$ and $\inf(f,0)$ are measurable on E. Define

$$f^+ \equiv \sup(f,0) \qquad and \qquad f^- \equiv -\inf(f,0)$$

and observe that

(8.23) $$|f| = f^+ + f^-.$$

Conclusion (vi) then follows from Theorem 8.11(iii).

Conclusion (iii), (iv), and (v) of this theorem follow from (i) and (ii) and the respective definitions

$$\liminf f_n \equiv \sup_{n\geq 1} \inf_{m\geq n} f_m,$$

$$\limsup f_n \equiv \inf_{n\geq 1} \sup_{m\geq n} f_m,$$

which are equivalent to (1.9) and (1.10) for the special case in which the f of those definitions is a sequence and the a is ∞.

When we deal later with Lebesgue integrals of the form $\int_a^b f(t,y,\dot y)\,dt$ we shall wish to know that $\dot y$ is measurable.

Theorem 8.13

Given a function y: $[a,b] \to R$ that is measurable on $[a,b]$ and given that the derivative $\dot y(t)$ exists, finite, ∞, or $-\infty$, at each point of a measurable subset E of $[a,b]$, then the function $\dot y$: $E \to R^*$ is measurable on E.

PROOF

By the definition of a derivative (Section 1.7), the difference quotient

(8.24) $$\frac{y(x+h) - y(x)}{h}$$

has a finite or infinite limit at each point of E. In the light of problem 1, Exercise 8.2, we can suppose without loss of generality that neither a nor b is a point of E and avoid such details as restricting h to be positive if x in (8.24) is a or negative if x is b.

Now y is measurable on E by Theorem 8.10. If we define $y(x+h)$ to be $y(a)$ or $y(b)$, respectively, when $x+h < a$ or $> b$, it is easy to verify from definition (8.21) that the translation of y with values $y(x+h)$ is measurable on $[a,b]$ and hence, by Theorem 8.10, on E. The constant-valued function ϕ: $[a,b] \to R$ with values $\phi(x) \equiv h$ is clearly measurable on $[a,b]$ and hence, by Theorem 8.10, on E. Consequently, by Theorem 8.11(iv) and (vi), quotient (8.24) is a value of a function measurable on E provided $h \neq 0$. This is true, in particular, if $h = 1/n$, $n = 1, 2, \ldots$.

Let $n \to \infty$. The limit of (8.24) exists by hypothesis if $x \in E$. Hence if $f_n(x)$ now denotes quotient (8.24) with $h = 1/n$, the desired conclusion follows from Theorem 8.12(v).

Exercise 8.3

1. Given $f: [a,b] \to R$, if f is continuous on $[a,b]$, show that f is measurable on $[a,b]$ by showing that the set $\{x \in [a,b]: f(x) \geq c\}$ is closed. Alternatively, show that $\{x \in [a,b]: f(x) > c\}$ is open.
2. Given $f: [a,b] \to R$, if f is lower or upper semi-continuous on $[a,b]$, show that f is measurable on $[a,b]$.
3. Point out that the Dirichlet function, $f: R \to \{0,1\}$, $f(x) = 0$ or 1 according as x is rational or irrational, is measurable on R.
4. Given that the infinite series $\Sigma a_n x^n$ converges on $[-1,1]$, point out with the aid of theorems of this section that the function f defined by this series is measurable on $[-1,1]$.
5. Given a measurable set E, an extended real-valued function f that is measurable on E and a set Z of measure zero, show that an arbitrary extension f^* of f from E to $E \cup Z$ obtained by assigning extended real values $f^*(x)$, $x \in Z - E$, at pleasure, is measurable on $E \cup Z$.
6. Given that every set of the form $\{f(x) > a\}$ is measurable, prove that every set of the form $\{f(x) \geq a\}$ is measurable. With reference to the discussion following (8.21), point out why any one of the sets (8.21) would suffice in the definition of a measurable function.
7. Prove that the function (v) under Theorem 8.11 is measurable over a suitable subset S of E.
8. Show for the translated function in the proof of Theorem 8.13 that $\{y(x+h) \geq a\}$ is measurable for every real value of a.

The theorems and problems of this section show that all functions $f: E \subset R \to R^*$ that one is likely to encounter are measurable on E. The only examples of nonmeasurable functions are defined with the aid of nonmeasurable sets and are very complicated.

8.5 THE LEBESGUE INTEGRAL

Let E be a nonempty measurable subset of R, $0 \leq \mu(E) < \infty$. By a *measurable partition* π of E we mean a finite class $\{E_1, \ldots, E_n\}$, $n \geq 1$ of disjoint nonempty measurable subsets of E whose union is E.

Given a function $f: E \to R$ that is bounded and measurable on E, set

(8.25)
$$m_i \equiv \inf\{f(x): x \in E_i\}, \qquad m \equiv \inf\{f(x): x \in E\},$$
$$M_i \equiv \sup\{f(x): x \in E_i\}, \qquad M \equiv \sup\{f(x): x \in E\}.$$

These traditional symbols m and M must not be mistaken to suggest that $f(x)$ has minimum and maximum values on the respective sets. The sets E_i and E are in general not closed and f may be badly discontinuous. Let ξ_i be an arbitrary element of E_i and consider the inequalities

$$(8.26) \quad m\mu(E) \leqslant \sum m_i\mu(E_i) \leqslant \sum f(\xi_i)\mu(E_i)$$
$$\leqslant \sum M_i\mu(E_i) \leqslant M\mu(E).$$

The first and third sums (8.26), called *lower* and *upper sums*, respectively, will be denoted by $s(f;E;\pi)$ and $S(f;E;\pi)$.

Define *lower* and *upper integrals of f over E*,

$$(8.27) \quad I_*(f;E) \equiv \sup_\pi s(f;E;\pi) \quad and \quad I^*(f;E) \equiv \inf_\pi S(f;E;\pi).$$

For such a simple function f as one having a constant value on E, the upper and lower sums are independent of the choice of π. In general, $s(f;E;\pi)$, for fixed f and E, has infinitely many values corresponding to different choices of π. This is the set whose supremum defines the lower integral. Similar remarks apply to upper sums and the upper integral.

Given two measurable partitions π' and π'' of E, let π denote the class consisting of all nonempty intersections $E_i' \cap E_j''$ of respective sets from the classes π' and π''. One verifies that π is a measurable partition of E and that

$$s(f;E;\pi') \leqslant s(f;E;\pi) \leqslant S(f;E;\pi) \leqslant S(f;E;\pi''),$$

hence that $s(f;E;\pi') \leqslant S(f;E;\pi'')$. It follows that

$$(8.28) \quad I_*(f;E) \leqslant I^*(f;E).$$

In the event that $I_*(f;E) = I^*(f;E)$, then f is said to be *integrable over E in the broad sense* and the common value, denoted by $I(f;E)$, is the integral of f over E. If, in addition, the common value is finite, then f is said to be *integrable over E*. There is not a standard terminology. Some writers use the respective terms integrable over E and summable over E for the notions integrable over E in the broad sense and integrable over E as defined above.

Neither is there a standard notation. Among the several alternatives to symbol $I(f;E)$ are

$$\int_E f, \quad \int_E f\,d\mu, \quad and \quad \int_E f(x)\,dx.$$

When either of the last two forms is used, one must understand that nothing has been said about differentials or is to be inferred. The one and only purpose of symbol $d\mu$ is to remind us that the integral has been defined using a measure μ. We shall see in Theorem 8.15 that the integral $I(f;E)$ is an extension of the ordinary Riemann integral. The last symbol above is simply a slight modification of the traditional symbol for that integral. If one meets the more familiar symbol $\int_a^b f(x)\,dx$, he must often decide from the context whether it is intended as a Lebesgue or a Riemann integral over the set $E = [a,b]$.

Theorem 8.14

If f is bounded and measurable on the nonempty set E of finite measure, then f is integrable over E.

PROOF

We see from (8.26), (8.27), and (8.28) that

$$(8.29) \quad I^*(f;E) - I_*(f;E) \leq S(f;E;\pi) - s(f;E;\pi) \leq \sum (M_i - m_i)\mu(E_i).$$

This is valid for any measurable partition π. It suits the purpose of this proof to use partitions of a special form. Relabel the interval $[m,M]$, in which all values $f(x)$ fall, as $[y_0, y_{n-1}]$ and introduce $y_1, y_2, \ldots, y_{n-2}$ with $y_i - y_{i-1} = (M-m)/(n-1)$, $i = 1, \ldots, n-1$. Using the notation for sets from (8.21), set

$$E_i \equiv \{y_{i-1} \leq f(x) < y_i\}, \quad i = 1, \ldots, n-1$$

and

$$E_n \equiv \{f(x) = y_{n-1}\}.$$

For the partition π_n determined by these sets, $m_i = y_{i-1}$ and $M_i = y_i$, $i = 1, \ldots, n-1$, while $m_n = M_n = y_{n-1}$. It follows from (8.29) that

$$(8.30) \quad I^*(f;E) - I_*(f;E) \leq \frac{M-m}{n}\sum_1^{n-1}\mu(E_i) \leq \frac{M-m}{n}\mu(E).$$

Since the first member if free of n while the last member $\to 0$ as $n \to \infty$, the left member must be zero.

Theorem 8.15

If E is an interval $[a,b]$ of finite length and the bounded function $f\colon [a,b] \to R$ is Riemann integrable over $[a,b]$, then f is Lebesgue integrable over $[a,b]$ and

$$(8.31) \quad I(f;[a,b]) = \int_a^b f(x)\,dx\ (Riemann).$$

PROOF

The usual definition of the Riemann integral by way of upper and lower sums employs closed subintervals $[a_{i-1}, a_i]$ that are not overlapping but have common endpoints.

Define

$$k_i \equiv \inf\{f(x): x \in [a_{i-1}, a_i]\}, \qquad K_i \equiv \sup\{f(x): x \in [a_{i-1}, a_i]\}.$$

The disjoint intervals $E_i \equiv [a_{i-1}, a_i)$, $i = 1, \ldots, n-1$ and $E_n \equiv [a_{n-1}, a_n]$, with $a_0 = a$ and $a_n = b$, constitute a measurable partition of $[a,b]$ in the sense of this section. Let m_i and M_i be defined by (8.25) for this particular type of partition. Then

$$m_i \geqslant k_i \qquad \text{and} \qquad M_i \leqslant K_i, \qquad i = 1, \ldots, n.$$

Moreover,

$$\sum k_i(a_i - a_{i-1}) \leqslant \sum m_i(a_i - a_{i-1}) \leqslant I_*(f;[a,b])$$
$$\leqslant I^*(f;[a,b]) \leqslant \sum M_i(a_i - a_{i-1}) \leqslant \sum K_i(a_i - a_{i-1}).$$

By hypothesis, f is Riemann integrable; hence as $\max(a_i - a_{i-1}) \to 0$ the outer terms both converge to the Riemann integral. This implies that the third and fourth terms, being constants, must be equal. Therefore, f is Lebesgue integrable and (8.31) holds.

That $I(f;[a,b])$ exists for functions f that are not Riemann integrable over $[a,b]$ is shown by examples. Let $f: [0,1] \to R$ be the Drichlet function with values 0 or 1 according as $x \in [0,1]$ is rational or irrational. For every partition of $[0,1]$ into subintervals the lower and upper (Darboux) sums associated with the Riemann integral have respective values 0 and 1. This function accordingly is not Riemann integrable. However, if we use the measurable partition $\pi \equiv \{E_1, E_2\}$, where E_1 and E_2 consist of all rationals and all irrationals in $[0,1]$, respectively, then $\mu(E_1) = 0$ and $\mu(E_2) = 1$, the lower and upper sums $s(f;[0,1];\pi)$ and $S(f;[0,1];\pi)$ are both unity; consequently, $I_*(f;[0,1]) = I^*(f;[0,1]) = 1$ Function f is Lebesgue integrable over $[0,1]$ and $I(f;[0,1]) = 1$.

The Lebesgue definition also extends the Riemann concept significantly in another way; namely, the integral $I(f;E)$ replaces a closed interval $[a,b]$ by a general nonempty measurable set E. Thus far we have excluded the rather trivial case in which E is empty. Given a function $f: \emptyset \to R$ that is otherwise arbitrary define

$$(8.32) \qquad I(f;\emptyset) \equiv 0.$$

A function $s: E \to R$ is called *simple* if there is a measurable partition π of the measurable set E into disjoint measurable subsets E_i, $i = 1, \ldots, n$, with $s(t)$ a constant c_i on E_i. If we examine (8.25), (8.26), and the proof of Theorem 8.14, we see that when f is bounded and measurable and $\mu(E) < \infty$, there are two sequences $\{s_\nu(t): \nu \in N\}$ and $\{S_\nu(t): \nu \in N\}$ of simple functions with

$$s_\nu(t) = m_i \quad and \quad S_\nu(t) = M_i \quad on\ E_i,\ i = 1, \ldots, n$$

and with

$$s_\nu(t) \leqslant f(t) \leqslant S_\nu(t)\ on\ E.$$

Each of the functions s_ν and S_ν is bounded and measurable on E so that integrals $I(s_\nu;E)$ and $I(S_\nu;E)$ both exist by Theorem 8.14 and, moreover, from the nature of s_ν and S_ν,

$$I(s_\nu;E) = \sum m_i\mu(E_i) \quad and \quad I(S_\nu;E) = \sum M_i\mu(E_i).$$

Consequently, Theorem 8.14 shows that

(8.33) $$\lim I(s_\nu;E) = \lim I(S_\nu;E) = I(f;E).$$

Let $f: E \to R^*$ be nonnegative and measurable on E and define

(8.34) $$I(f;E) \equiv \sup\{I(s;E): s\ is\ simple, 0 \leqslant s \leqslant f\}.$$

If f happens to be bounded, then $I(f;E)$ is the same real number already defined and which exists by Theorem 8.14.

As an immediate consequence of the meaning of supremum, we see from (8.34) that there must exist a sequence $\{s_n: n \in N\}$ of nonnegative simple functions all below f such that

(8.34*) $$I(f;E) = \lim I(s_n;E), \quad finite\ or\ \infty$$

as the case may be. Moreover, we can suppose that $s_n \leqslant s_{n+1}, n = 1,2, \ldots,$ for if this is not so for the sequence in (8.34*), we can define $s_n^*(x) \equiv \max\{s_m(x): m \leqslant n\}$ and use s_n^* in (8.34*).

If $f: E \to R^*$ is measurable on E but not otherwise restricted, then $f = f^+ - f^-$, where f^+ and f^- are the nonnegative functions defined preceding (8.23). Extended real numbers $I(f^+;E)$ and $I(f^-;E)$ are then given by (8.34) and we can define

(8.35) $$I(f;E) \equiv I(f^+;E) - I(f^-;E)$$

provided that at least one term on the right is finite.

When the right member is the meaningless expression $\infty - \infty$, f is not integrable over E even in the broad sense.

Consistently with the terminology introduced following (8.28), we say that f is integrable over E in the broad sense if $I(f;E)$ exists and is finite, ∞, or $-\infty$ and understand by the shorter statement that f is integrable over E that $I(f;E)$ exists and is finite. Clearly f is integrable over E in the general case (8.35) if and only if both f^+ and f^- are integrable over E.

The set E has been understood to be of finite measure, as stated at the beginning of the section. We shall continue to understand that any measurable set E that may be mentioned is of finite measure unless there is an explicit statement to the contrary. That $\mu(E) < \infty$ does not imply that the set E is bounded.

However, if $\mu(E) = \infty$, $I(f;E)$ can be defined as follows. Set $E_n \equiv [-n,n] \cap E$. Then $I(f^+;E_n)$ and $I(f^-;E_n)$, both defined by (8.34), are both nondecreasing in n. Define

(8.36) $$I(f;E) \equiv \lim_{n\to\infty} I(f^+;E_n) - \lim_{n\to\infty} I(f^-;E_n),$$

provided that at least one of these limits is finite.

Continuing the terminology already introduced, we say that f is integrable over the set E of infinite measure iff both limits on the right are finite and that f is integrable in the broad sense if at least one of these limits is finite. If both of these limits are ∞, f is simply not integrable over E even in the broad sense. Henceforth when we say that a function f is integrable over a set E we shall understand without saying so that E is a measurable set and that f is measurable on E. Integrals have been defined only over measurable sets and for functions measurable on those sets.

Exercise 8.4

1. Given measurable partitions $\pi' = \{E_1', \ldots, E_m'\}$ and $\pi'' = \{E_1'', \ldots, E_n''\}$ and that $E_i' \cap E_j'' \neq \emptyset$, define m_{ij} and M_{ij} in the manner of (8.25) and let m_i', m_j'', M_i', M_j'' be the similarly defined extended real numbers for the sets E_i' and E_j''. Identify the reasons for the inequalities

$$m_i' \leq m_{ij} \quad and \quad M_{ij} \leq M_j''.$$

2. (a) Given that $f(x) = \infty$ for all points x of a set Z of measure zero. Verify directly from definition (8.34) that $I(f;Z) = 0$.
   (b) Given that $f(x) = 0$ for all points x of a set A such that $\mu(A) = \infty$. Verify directly from definition (8.36) that $I(f;A) = 0$.
   (c) Refer to the last sentence in Section 1.3 and formulate a special convention on the interpretation of $0(\pm\infty)$ and $(\pm\infty)0$ when these combinations occur in a discussion of Lebesgue integrals.

3. Given that $f: E \to R^*$ is integrable over E and that $g(x) = f(x)$ on E except for the points x of a subset Z of E of measure zero, prove that g is integrable over E and that $I(g;E) = I(f;E)$.

4. Construct an example of an unbounded set E of finite measure and a function $f: E \to R^*$ that is simple enough so that the value of $I(f;E)$ can be determined on the basis of Section 8.5.

5. If the function $f: E \to R^*$ is integrable over E and A is a measurable subset of E, prove that f is integrable over A.

6. If f and g are both integrable over E, prove that $f + g$ and αf, $\alpha \in R$, are both integrable over E and that

   (i) $I(f+g; E) = I(f;E) + I(g;E)$,
   (ii) $I(\alpha f;E) = \alpha I(f;E)$.

7. Given functions f, g, and h from E to R^*, if f and h are integrable over E, if g is measurable on E and $f \le g \le h$ on E, prove

   (i) g is integrable over E,
   (ii) $I(f;E) \le I(g;E) \le I(h;E)$.

8. Given $f: A \cup B \to R^*$ with A and B disjoint and that f is integrable over A and also over B, prove that

   (i) f is integrable over $A \cup B$,
   (ii) $I(f;A \cup B) = I(f;A) + I(f;B)$.
   (iii) Extend this result to any finite union by induction.

9. Prove that if f is integrable over E, then $|f|$ is integrable over E. Given that A is a nonmeasurable subset of E and that $f(x) = -1$ or 1 according as $x \in A$ or $x \in E - A$, then $|f(x)| \equiv 1$ is integrable over E but f is not. Prove that if $|f|$ is integrable over E and f is measurable on E, then f is integrable over E.

10. Construct an example of a function $f: [0,1] \to R$ that is unbounded and not of fixed sign, whose improper Riemann integral over $[0,1]$ is finite but such that the improper integral of $|f|$ over $[0,1]$ is ∞. Then point out that your function f is not Lebesgue integrable over $[0,1]$.

11. With reference to problem 5, Exercise 8.3, show that if f is integrable over E and Z is a set of measure zero, then an arbitrary extension f^* of f from E to $E \cup Z$ is integrable over $E \cup Z$ and $I(f^*;E \cup Z) = I(f;E)$.

8.6   CONVERGENCE THEOREMS

The theorems of this section are among the most useful results in the Lebesgue theory. They exhibit some of the reasons for the importance of the Lebesgue integral for modern analysis.

Theorem 8.16. (Monotone Convergence Theorem)

Given a sequence $\{f_n: E \to R^*: n \in N\}$ of functions integrable in the broad sense on the common measurable domain E, if $0 \leqslant f_n \leqslant f_{n+1}, n = 1, 2, \ldots,$ and if $f_n(t)$ has a limit $f_0(t) \in R^*$ as $n \to \infty$ for each $t \in E$, then

$$(i) \quad I(f_n;E) \text{ has a limit, finite or } \infty,$$

and

$$(ii) \quad \lim I(f_n;E) = I(f_0;E).$$

PROOF

Function f_0 is measurable on E by Theorem 8.12(v). Integrals $I(f_n;E)$, $n = 1, 2, \ldots$ and $I(f_0;E)$ exist in the broad sense under definition (8.34) and it follows from (8.34) and the monotonicity of f_n in n that

$$I(f_n;E) \leqslant I(f_{n+1};E) \leqslant I(f_0;E).$$

The monotone sequence $\{I(f_n;E)\}$ of extended reals necessarily has a limit, finite or ∞, and

$$(8.37) \qquad \lim I(f_n;E) \leqslant I(f_0;E).$$

We wish to complete the proof by establishing the complementary inequality.

Given $\alpha \in (0,1)$ and a simple function $s, 0 \leqslant s < f_0$, define

$$E_n \equiv \{t \in E; f_n(t) \geqslant \alpha s(t)\}$$

and observe that $E_n \subset E_{n+1}$ and that $\cup E_n = E$. It can be verified that

$$(8.38) \qquad I(\alpha s;E_n) = \alpha I(s;E_n) \leqslant I(f_n;E_n) \leqslant I(f_n;E),$$

and from the nature of a simple function s and combinatorial properties (Section 8.3) of measurable sets that

$$I(s;E) = \lim I(s;E_n).$$

Letting $n \to \infty$, we then see that

$$\alpha I(s;E) \leqslant \lim I(f_n;E)$$

and, since this holds for all $\alpha \in (0,1)$, that

$$I(s;E) \leqslant \lim I(f_n;E).$$

But this relation holds for all simple functions $s, 0 \leqslant s \leqslant f_0$; hence, by definition (8.34) applied to f_0, $I(f_0;E) \leqslant \lim I(f_n;E)$, and the proof is complete.

Theorem 8.17 (Fatou's Lemma)

Given a sequence $\{f_n : E \to R^* : n \in N\}$ of functions measurable on the common measurable domain E, if there exists a function $g : E \to R^*$ that is integrable over E and such that $f_n \geqslant g$, $n = 1, 2, \ldots$, then

$$(8.39) \qquad \lim \inf I(f_n; E) \geqslant I(\lim \inf f_n; E).$$

PROOF

We first prove the theorem for the special case $g = 0$. Define $g_n : E \to R^*$ as the function with values $g_n(t) \equiv \inf \{f_i(t) : i \geqslant n\}$. Each g_n is measurable on E by Theorem 8.12(ii) and

$$(8.40) \qquad 0 \leqslant g_1(t) \leqslant g_2(t) \leqslant \cdots.$$

By the definition of limit inferior [following (8.23)]

$$(8.41) \qquad \lim \inf f_n(x) = \lim g_n(x) = \sup\{g_n(x) : n \in N\}.$$

Now $f_n \geqslant g_n$, $n = 1, 2, \ldots$, from which follows that

$$I(f_n; E) \geqslant I(g_n; E).$$

From the monotonicity (8.40) and the Monotone Convergence Theorem, the right member has a limit, finite or ∞, and

$$\lim \inf I(f_n; E) \geqslant \lim I(g_n; E) = I(\lim g_n; E).$$

By the first equality (8.41), we then have the stated conclusion (8.39) for the special case in which $g = 0$.

With g now only required to be integrable, set

$$(8.42) \qquad h_n(x) \equiv f_n(x) - g(x)$$

and define $A \equiv E - \{g(x) = \infty\} \cup \{g(x) = -\infty\}$. Now g is integrable over the measurable set $\{g(x) = \infty\}$ by problem 5, Exercise 8.4. If this set is of positive measure, it follows from definition (8.34) that $I(g; \{g(x) = \infty\}) = \infty$, contrary to the integrability of g over that set. Consequently, $\mu(\{g(x) = \infty\}) = 0$ and similarly for the set $\{g(x) = -\infty\}$. Relation (8.42) is meaningful on A and $h_n \geqslant 0$ on A; consequently, by the special case of the present theorem that is already proved,

$$(8.43) \qquad \lim \inf I(f_n - g; A) \geqslant I[\lim \inf (f_n - g); A].$$

Since g is free of n,

$$\lim \inf (f_n - g) = \lim \inf f_n - g.$$

Similarly and with the aid of problem 6, Exercise 8.4,

$$\liminf I(f_n - g;A) = \liminf [I(f_n;A) - I(g;A)]$$

$$= \liminf I(f_n;A) - I(g;A).$$

The stated conclusion (8.39) then follows from (8.43).

Theorem 8.18   (Dominated Convergence Theorem)

Given a sequence $\{f_n: E \to R^*: n \in N\}$ of functions measurable on the common measurable domain E, if $f_n(t)$ has a limit $f_0(t) \in R^*$ for all $t \in E$ and if there exists a function $g: E \to R^*$ that is integrable over E and such that $|f_n| \leq g, n = 1, 2, \ldots,$ then

and
(i) $\lim f_n$ *is integrable over E*

(ii) $\lim I(f_n;E) = I(\lim f_n;E)$.

PROOF

By Theorem 8.12(v), $\lim f_n$ is measurable on E. Since $I(g;E) < \infty$ and $|f_n| < g$ by hypothesis, one verifies that $I(f_n^+;E)$ and $I(f_n^-;E)$ are finite and obtains conclusion (i).

To prove conclusion (ii), observe that

$$-g \leq \pm f_n \leq g;$$

hence by Fatou's Lemma applied first to f_n and then to $-f_n$, we have that

$$\liminf I(f_n;E) \geq I(f_0;E)$$

and

$$\liminf I(-f_n;E) \geq I(-f_0;E).$$

The last statement is equivalent to the inequality

$$\limsup I(f_n;E) \leq I(f_0;E),$$

and since a limit inferior is always dominated by the corresponding limit superior, conclusion (ii) of the theorem follows.

Two functions $f: E \to R^*$ and $g: E \to R^*$ are said to be *equal almost everywhere* (abbreviated a.e.) if $f(x) = g(x)$ on E except for the points x of a subset of E of measure zero. Problem 3, Exercise 8.4, calls attention to the fact that equality a.e. suffices for equality of the integrals. Hypotheses can be weakened in Theorems 8.16 and 8.18 by requiring only that f_n converge to f_0 almost everywhere, in Theorem 8.17 only that $f_n \geq g$ almost everywhere.

Exercise 8.5

1. Given $f(x) = 1/\sqrt{x}$, $x \neq 0$, distinguish conceptually between the Lebesgue integral $I(f;(0,1])$ and the improper or Cauchy–Riemann integral

$$\int_0^1 f(x) \, dx \equiv \lim_{\epsilon \to 0} \int_\epsilon^1 f(x) \, dx.$$

   Discuss the relation between the limit on the right and the Monotone Convergence Theorem.

2. Let $\{r_n : n \in N\}$ be a fixed sequentialization of all rational real numbers on $[0,1]$ and define $f_n(x) \equiv 0$ if $x = r_1, r_2, \ldots$ or $r_n, f_n(x) \equiv 1$ if x is any other number in $[0,1]$. Point out that the Dominated Convergence Theorem (indeed its special case known as the Bounded Convergence Theorem) applies. Observe that each f_n is Riemann integrable but that $\lim f_n$ is not.

3. The *characteristic function* $\chi_E(x) = 1$ or 0 according as $x \in E$ or $x \in \mathbf{C}(E)$. Given that E is measurable, point out that χ_E is integrable over E and that this integral is $\mu(E)$.

4. Recall the meaning of $\sum_1^\infty u_n(x)$ in terms of the sequence of partial sums. Formulate corollaries to the Monotone and Bounded Convergence Theorems having as conclusions that

   (i) $\sum u_n$ *is integrable over* E,

   (ii) $I\left(\sum u_n; E\right) = \sum I(u_n; E).$

8.7   OTHER PROPERTIES OF INTEGRALS

Theorem 8.19   (Mean Value Theorem)

If f is integrable over E and $\mu(E) < \infty$, there exists $\alpha \in R$ such that

$$\inf \{f(x) : x \in E\} \leqslant \alpha \leqslant \sup \{f(x) : x \in E\}$$

and

(8.44) $I(f;E) = \alpha\mu(E).$

PROOF

   If f is bounded, m and M in (8.26) are both finite (that is, real), and it follows from (8.26) and the given integrability of f that

$$m\mu(E) \leqslant I(f;E) \leqslant M\mu(E).$$

Clearly (8.44) holds for some $\alpha \in [m,M]$.

If $\mu(E) = 0$, then, whether or not f is bounded, $I(f;E) = 0$ and (8.44) holds for an arbitrary α. In the remainder of the proof suppose that $\mu(E) > 0$.

If f is nonnegative and not necessarily bounded, let $\{s_n : n \in N\}$ be a sequence of simple functions, nondecreasing in n and satisfying (8.34*). Then

$$(8.45) \qquad\qquad I(s_n;E) = \alpha_n \mu(E)$$

by the proof for the bounded case. By our choice of s_n, $I(s_n;E)$ is nondecreasing in n; hence so also is α_n, as a result of our restriction that $\mu(E) > 0$. Let $n \to \infty$. It follows from (8.34*) and (8.45) that

$$I(f;E) = \alpha\mu(E) \qquad where\; \alpha = \lim \alpha_n.$$

The reader should verify that α satisfies the inequalities stated in the theorem.

In the general case apply the preceding case to both f^+ and f^- and find that

$$I(f^+;E) = \alpha\mu(E), \qquad 0 \leq \alpha \leq \sup \{f(x) : f(x) \geq 0\}$$

and

$$-I(f^-;E) = -\beta\mu(E), \qquad \inf \{f(x) : f(x) < 0\} \leq -\beta \leq 0.$$

By addition and definition (8.35),

$$I(f;E) = (\alpha - \beta)\mu(E),$$

where $(\alpha - \beta)$ plays the role of the α in (8.44).

Observe that the restriction $\mu(E) < \infty$ in Theorem 8.19 is essential. If $\mu(E) = \infty$ and f is integrable, the left member of (8.44) is a real number, hence is general not $\alpha(\infty)$.

Theorem 8.20

Given a countable class of disjoint measurable sets A_i and a nonnegative function $f: \cup A_i \to R^*$ that is integrable over each of the sets A_i, then f is integrable in the broad sense over $\cup A_i$ and

$$(8.46) \qquad\qquad I(f; \cup A_i) = \sum I(f; A_i) \leq \infty.$$

Moreover, f is integrable over $\cup A_i$ iff either the number of sets A_i is finite or the infinite series in (8.46) has a finite sum.

PROOF

The case in which the number of sets A_i is finite is problem 8, Exercise 8.4. We assume this result and address the present proof to the denumerable case.

Set $S_n = \cup_1^n A_i$ and define $f_n(x)$ as $f(x)$ or 0 according as $x \in S_n$ or not. The sequence $\{f_n : n \in N\}$ is nondecreasing in n and, by the Monotone Convergence Theorem.

(8.47) $$\lim I(f_n; \cup A_i) = I(f; \cup A_i),$$

in which the union is over all i.

From the definition of f_n follows that

$$I(f_n; \cup A_i) = I(f_n; \bigcup_1^n A_i) = I(f; \bigcup_1^n A_i)$$

and, by problem 8(iii), Exercise 8.4,

(8.48) $$I(f; \bigcup_1^n A_i) = \sum_1^n I(f; A_i).$$

The left member of (8.47) is thus the limit as $n \to \infty$ of the right member of (8.48), and this is the second member of (8.46).

Theorem 8.21

If f is integrable over the union $\cup A_i$ of a countable class of disjoint measurable sets, then f is integrable over each set A_i and

(8.49) $$I(f; \cup A_i) = \sum I(f; A_i).$$

PROOF

By problem 5, Exercise 8.4, f is integrable over each of the sets A_i; hence so also are f^+ and f^-.

Denote the given union by S. Since f is integrable over S, we have by definition (8.35) that

$$I(f; S) = I(f^+; S) - I(f^-; S).$$

By Theorem 8.20, the right member equals

$$\sum I(f^+; A_i) - \sum I(f^-; A_i) = \sum [I(f^+; A_i) - I(f^-; A_i)] = \sum I(f; A_i),$$

and the proof is complete.

The results of this section with problem 4, Exercise 8.5, can be described by saying that under suitable hypotheses $I(f; E)$ is countably additive both in f and in E.

8.8   FUNCTIONS OF BOUNDED VARIATION

In preparation for the theory of differentiation and the Fundamental Theorem of the Integral Calculus, we turn to functions $\phi\colon [a,b] \to R$ of bounded variation, a concept already defined in Section 7.3. It is convenient in the first part of this section to use I as an alternative symbol for the interval $[a,b]$.

Given $\phi\colon I \to R$ and a partition π of I, define

(8.50) $$V(\phi;I;\pi) \equiv \sum |\phi(t_i) - \phi(t_{i-1})|,$$

called the *variation of ϕ on I relative to π*. Then

(8.51) $$T(\phi;I) \equiv \sup \{V(\phi;I;\pi)\colon \pi \text{ a partition of } I\}$$

is the *total variation* of ϕ on I. Clearly $0 \leqslant T(\phi;I) \leqslant \infty$.

Theorem 8.22

If c is an interior point of $[a,b]$, then

(8.52) $$T(\phi;[a,b]) = T(\phi;[a,c]) + T(\phi;[c,b]).$$

PROOF

There necessarily exists a sequence $\{\pi_\nu\colon \nu \in N\}$ of partitions of an interval I such that $V(\phi;I;\pi_\nu) \to T(\phi;I)$ as $\nu \to \infty$. Let $\{\pi_\nu\}$ now be such a sequence for $[a,b]$ and define

$$\pi_{1\nu} \equiv \pi_\nu \cap [a,c] \quad and \quad \pi_{2\nu} \equiv \pi_\nu \cap [c,b].$$

In view of the triangle inequality,

$$|\phi(t_i) - \phi(t_{i-1})| \leqslant |\phi(t_i) - \phi(c)| + |\phi(c) - \phi(t_{i-1})|,$$

we can and do suppose that c is a point of each partition π_ν and hence that the sets $\pi_{1\nu}$, $\pi_{2\nu}$ both include c and therefore are respective partitions of $[a,c]$ and $[b,c]$. One verifies that

(8.53) $$V(\phi;[a,b];\pi_\nu) = V(\phi;[a,c];\pi_{1\nu}) + V(\phi;[c,b];\pi_{2\nu}).$$

Sequence π_ν was chosen so that the left member converges but terms on the right may not converge. Nevertheless, if $\nu \to \infty$, it follows that

$$T(\phi;[a,b]) \leqslant \limsup V(\phi;[a,c];\pi_{1\nu}) + \limsup V(\phi;[c,b];\pi_{2\nu})$$
$$\leqslant T(\phi;[a,c]) + T(\phi;[c,b]).$$

To obtain the complementary inequality redefine $\pi_{1\nu}$ and $\pi_{2\nu}$ as general terms of sequences of partitions of $[a,c]$ and $[c,b]$ such that terms on the right in (8.53) converge to the respective total variations. Redefine π_ν as $\pi_{1\nu} \cup \pi_{2\nu}$. Then let $\nu \to \infty$ in (8.53) and see that

$$T(\phi;[a,c]) + T(\phi;[c,b]) \leqslant \lim \sup V(\phi;[a,b],\pi_\nu) \leqslant T(\phi;[a,b]).$$

This theorem extends by mathematical induction to any finite union of abutting intervals.

Theorem 8.23

A function $\phi:[a,b] \to R$ is BV on $[a,b]$ iff $\phi = p - n$, where p and n are monotone functions on $[a,b]$ of the same type.

PROOF

Suppose first that $f = p - n$, where p and n are both monotone, but in this half of the theorem one may be nondecreasing, the other nonincreasing. Then

$$V(\phi;[a,b];\pi) = \sum |[p(t_i) - n(t_i)] - [p(t_{i-1}) - n(t_{i-1})]|$$

$$\leqslant \sum |p(t_i) - p(t_{i-1})| + \sum |n(t_i) - n(t_{i-1})|$$

$$= |p(b) - p(a)| + |n(b) - n(a)| < \infty.$$

The last expression, being independent of π, is an upper bound for $T(\phi;[a,b])$.

Conversely, if ϕ is BV on $[a,b]$, define functions p and n from $[a,b]$ to R as follows.

$$p(t) \equiv \tfrac{1}{2}[T(\phi;[a,t]) + \phi(t)] \quad and \quad n(t) \equiv \tfrac{1}{2}[T(\phi;[a,t]) - \phi(t)].$$

By subtraction $p - n = \phi$. Verify with the aid of Theorem 8.22 that for any $t_1, t_2 \in [a,b]$, $t_1 < t_2$,

$$p(t_2) - p(t_1) = \tfrac{1}{2}[T(\phi;[t_1,t_2]) + \phi(t_2) - \phi(t_1)].$$

It follows from definition (8.51) that the right member is nonnegative. That $n(t_2) - n(t_1) \geqslant 0$ is shown similarly. Clearly $-p$ and $-n$ are both nonincreasing. Moreover, $p + x$ and $n + x$ are strictly increasing while $-(p+x)$ and $-(n+x)$ are strictly decreasing. By elementary algebra,

$$\phi(t) = -n(t) - [-p(t)] = [p(t) + t] - [n(t) + t] = -[n(t) + t] - [-p(t) - t].$$

Given an interval $[a,b]$ of positive length and a function $\phi\colon [a,b] \to R$, ϕ is called AC (*absolutely continuous*) on $[a,b]$ if

(8.54) $\forall \epsilon > 0, \exists \delta_\epsilon > 0$ such that $\displaystyle\sum_1^n |b_i - a_i| < \delta_\epsilon \Rightarrow$

$$\sum_1^n |\phi(b_i) - \phi(a_i)| < \epsilon,$$

in which a_i and b_i are ends of $n \geqslant 1$ nonoverlapping but possibly abutting subintervals $[a_i,b_i]$ of $[a,b]$ and n is an arbitrary positive integer.

Theorem 8.24

If $\phi\colon [a,b] \to R$ is AC on $[a,b]$, then ϕ is BV on $[a,b]$.

PROOF

A contrapositive argument is convenient. Suppose that ϕ is not BV on $[a,b]$. Let π_m denote the particular partition comprising $m+1$ uniformly spaced points with $(b-a)/m$ as the length of each subinterval. For at least one subinterval of $[a,b]$ under π_m — call such a subinterval $[\alpha,\beta]$ — we have that $T(\phi;[\alpha,\beta]) = \infty$; otherwise $T(\phi;[a,b])$ would be finite by the extension of Theorem 8.22 to m intervals. Let $\pi = \{\alpha = \tau_0, \tau_1, \ldots, \tau_k = \beta\}$ be an arbitrary partition of $[\alpha,\beta]$. Then

$$\sum |\tau_i - \tau_{i-1}| = \beta - \alpha = (b-a)/m.$$

This can be made arbitrarily near zero by choosing m to be sufficiently large. But $T(\phi;[\alpha,\beta]) = \infty$ and, in accord with (8.51), π can be so chosen that $\sum |\phi(\tau_\nu) - \phi(\tau_{i-1})|$ is arbitrarily large. Consequently, there exists no δ_ϵ with the property required by (8.54); indeed (8.54) fails with a vengeance.

A function $\phi\colon [a,b] \to R$ is called *lipschitzian* on $[a,b]$ if there is a real number k such that

(8.55) $|\phi(t) - \phi(t')| \leqslant k|t - t'|$, $\forall t,t' \in [a,b]$.

Theorem 8.25

If ϕ is lipschitzian on $[a,b]$, then ϕ is AC on $[a,b]$.

PROOF

If $k = 0$, then $\phi(t)$ is constant and (8.54) holds with an arbitrary real value of δ_ϵ. If $k > 0$, choose $\delta_\epsilon = \epsilon/k$.

Theorem 8.26

If $\phi\colon [a,b] \to R$ is PWS on $[a,b]$, then ϕ is lipschitzian on $[a,b]$.

PROOF

We discuss the case in which ϕ is smooth on each of two subintervals $[a,c]$ and $[c,b]$ of $[a,b]$. Set $k \equiv \sup \{|\dot{\phi}(t)|: t \in [a,b]\}$ with symbol $\dot{\phi}$ interpreted under convention (2.17). If t and t' are both in $[a,c]$ or both in $[c,b]$, (8.55) is immediate from the Mean Value Theorem of the Differential Calculus and our definition of k. If $t \in [a,c]$ and $t' \in [c,b]$, then

(8.56) $$\phi(t') - \phi(t) = [\phi(t') - \phi(c)] + [\phi(c) - \phi(t)].$$

By the Mean Value Theorem, there exist ξ_1 and ξ_2 such that

$$\phi(c) - \phi(t) = \dot{\phi}(\xi_1)(c - t) \quad and \quad \phi(t') - \phi(c) = \dot{\phi}(\xi_2)(t' - c).$$

It follows from (8.56) that ϕ satisfies the Lipschitz condition

$$|\phi(t') - \phi(t)| \leq k[(t' - c) + (c - t)] = k(t' - t) = k|t' - t|.$$

Exercise 8.6

1. The following properties of a function ϕ: $[a,b] \rightarrow R^*$ are successively more restrictive. For each of the stated properties find an example of a function ϕ having that property but not having the next property and demonstrate that the examples meet these specifications. The properties are measurable on $[a,b]$, integrable over $[a,b]$, BV on $[a,b]$, AC on $[a,b]$, lipschitzian on $[a,b]$, and PWS on $[a,b]$.

2. Consider the sequence of equilateral triangles having as bases the intervals $[(\frac{1}{2})^n, (\frac{1}{2})^{n-1}]$, $n = 1, 2, \ldots$. Let ϕ be the function from $[0,1]$ to R consisting of points (x,y) on the oblique sides of all these triangles together with the point $(0,0)$. Show that ϕ is lipschitzian on $[0,1]$ and determine the smallest constant k for which ϕ satisfies a Lipschitz condition.

3. Show that the function ϕ: $[-1,1] \rightarrow R$, $\phi(x) \equiv x^{1/3}$ is AC on $[-1,1]$.

4. Prove that if ϕ and ψ are both AC on $[a,b]$, then $\phi + \psi$ and $\phi - \psi$ are AC on $[a,b]$.

5. Given the Cantor set E defined as the subset of $[0,1]$ obtained by deleting the open middle third $(\frac{1}{3}, \frac{2}{3})$, then deleting the open third of each of the remaining closed thirds $[0, \frac{1}{3}], [\frac{2}{3}, 1]$, then deleting the open third of each of the remaining closed intervals of length $\frac{1}{9}$, and so on, show that the Cantor set is of measure zero. Define the Cantor–Lebesgue function ω: $[0,1] \rightarrow R$ as follows. $\omega(0) \equiv 0$, $\omega(1) \equiv 1$, $\omega(t) \equiv \frac{1}{2}$ on $(\frac{1}{3}, \frac{2}{3})$, then $\frac{1}{4}$ and $\frac{3}{4}$ on $(\frac{1}{9}, \frac{2}{9})$ and $(\frac{7}{9}, \frac{8}{9})$, respectively. On each deleted open interval assign to $\omega(t)$ the constant value that is midway between those already assigned on

adjacent intervals deleted at the preceding step in the construction. Finally, for points $t \in E$ define $\omega(t) \equiv \sup \{\omega(\tau): \tau \in [0,1] - E, \tau < t\}$. Show that ω does not satisfy the definition of absolute continuity on $[0,1]$.

8.9   THE VITALI COVERING THEOREM

The following theorem is basic to the theory of differentiation.

Theorem 8.27   (Vitali)

Given a subset E of the reals and a class $K \equiv \{I_\alpha: \alpha \in A$, an index set$\}$ of compact intervals I_α of positive length such that

$$(8.57) \qquad \mu^*(\cup I_\alpha) < \infty, \qquad \text{suppose}$$

$$(8.58) \qquad \text{that, for every open interval } I \text{ and point } x \in I \cap E,$$
$$\text{there exists } I_\alpha \in K \text{ such that } x \in I_\alpha \subset I.$$

Then there exists a countable subset of K consisting of disjoint intervals whose union covers almost all of E.

PROOF

If $E = \emptyset$ the theorem is trivially true. We now consider the case $E \neq \emptyset$.

Let $\mathscr{I}_1$ be an arbitrary interval in K. If $E \subset \mathscr{I}_1$ the theorem is true. If $E \not\subset \mathscr{I}_1$, there necessarily exists $\mathscr{I}_2 \in K$ such that

$$(8.59) \qquad \mathscr{I}_2 \subset \mathbf{C}(\mathscr{I}_1) \qquad and$$
$$\mu(\mathscr{I}_2) > \tfrac{1}{2} \sup \{\mu(I_\alpha): \mathbf{C}(\mathscr{I}_1) \supset I_\alpha \in K\}.$$

Proceeding inductively, suppose given the disjoint intervals

$$\mathscr{I}_1, \mathscr{I}_2, \ldots, \mathscr{I}_m \in K$$

such that if $j = 1, 2, \ldots, m-1$, then $\mathscr{I}_{j+1} \subset \mathbf{C}(\cup_1^j \mathscr{I}_\nu)$ and

$$\mathbf{C}\left(\overset{j}{\underset{1}{\cup}} \mathscr{I}_\nu\right) \supset I_\alpha \in K \Rightarrow \mu(I_\alpha) < 2\mu(\mathscr{I}_{j+1}).$$

Then either $E \subset \cup_1^m \mathscr{I}_\nu$ and the theorem is true or, as in the case $m = 1$ above, there exists, under hypothesis (8.58), $\mathscr{I}_{m+1}$ in K such that

$$(8.60) \qquad \mathbf{C}\left(\overset{m}{\underset{1}{\cup}} \mathscr{I}_\nu\right) \supset \mathscr{I}_{m+1} \qquad and$$
$$\mu(\mathscr{I}_{m+1}) > \tfrac{1}{2} \sup \left\{\mu(I_\alpha): \mathbf{C}\left(\overset{m}{\underset{1}{\cup}} \mathscr{I}_\nu\right) \supset I_\alpha \in K\right\}.$$

The sequence with general term $\mathscr{I}_\nu$ that is thus inductively defined may terminate with a certain $\mathscr{I}_n$. Then $E \subset \cup_1^n \mathscr{I}_\nu$ and the theorem is true. If this does not occur, we have an infinite sequence $\{\mathscr{I}_\nu\}$ with (8.60) holding for $m = 1, 2, \ldots$. In the event that $E \subset \cup_1^\infty \mathscr{I}_\nu$, the theorem is true. It remains to investigate the case in which this does not occur.

As a consequence of hypothesis (8.57) and the disjointness of the intervals $\mathscr{I}_\nu$, the infinite series $\Sigma \mu(\mathscr{I}_\nu)$ has a finite sum. As necessary conditions for this,

(8.61) $$\mu(\mathscr{I}_\nu) \to 0 \ as \ \nu \to \infty$$

and

(8.62) $$\forall \epsilon > 0, \exists N_\epsilon \ such \ that \ \sum_{N_\epsilon}^\infty \mu(\mathscr{I}_\nu) < \epsilon.$$

Now $\cup_1^{N_\epsilon} \mathscr{I}_\nu \subset \cup_1^\infty \mathscr{I}_\nu$; hence $\mathbf{C}(\cup_1^\infty \mathscr{I}_\nu) \subset \mathbf{C}(\cup_1^{N_\epsilon} \mathscr{I}_\nu)$. Given $x \in E \cap \mathbf{C}(\cup_1^\infty \mathscr{I}_\nu)$, then $x \in \mathbf{C}(\cup_1^{N_\epsilon} \mathscr{I}_\nu)$ and, by hypothesis (8.58), there exists $\mathscr{I} \in K$ such that

$$x \in \mathscr{I} \subset \mathbf{C}\left(\bigcup_1^{N_\epsilon} \mathscr{I}_\nu\right).$$

With x, ϵ, and N_ϵ fixed, if it were true for all positive integers m that $\mathscr{I} \subset \mathbf{C}(\cup_1^m \mathscr{I}_\nu)$, then, from (8.60),

$$\mu(\mathscr{I}) < 2\mu(\mathscr{I}_{m+1}), \qquad m = 1, 2, \ldots,$$

in contradiction with (8.61). We infer that there is an integer n such that

(8.63) $$\mathscr{I} \subset \mathbf{C}\left(\bigcup_1^{n-1} \mathscr{I}_\nu\right) \quad but \quad \mathscr{I} \not\subset \mathbf{C}\left(\bigcup_1^n \mathscr{I}_\nu\right).$$

Since $\cup_1^m \mathscr{I}_\nu$ is an expanding sequence of sets, $\mathbf{C}(\cup_1^m \mathscr{I}_\nu)$ is a contracting sequence. It follows that

(8.64) $$n > N_\epsilon.$$

Moreover, from (8.63),

(8.65) $$\mathscr{I} \cap \mathscr{I}_n \neq \emptyset.$$

By the second statement (8.60), $\mu(\mathscr{I}) < 2\mu(\mathscr{I}_n)$; hence, as a consequence of (8.65), $\mathscr{I} \subset \mathscr{I}_n^*$, an interval concentric with $\mathscr{I}_n$ and five times as long.

In summary, we have now shown that

$$x \in E \cap \mathbf{C}\left(\bigcup_1^\infty \mathscr{I}_\nu\right) \Rightarrow \exists n > N_\epsilon \ such \ that \ x \in \mathscr{I}_n^*,$$

from which, with reference to (8.64), we see that

$$E \cap \mathbf{C}\Big(\overset{\infty}{\underset{1}{\cup}}\,\mathscr{I}_\nu\Big) \subset \overset{\infty}{\underset{N_\epsilon+1}{\cup}}\,\mathscr{I}_\nu^*.$$

Finally, by the definition of Lebesgue outer measure μ^* and (8.62), we have that

$$\mu^*\Big[E \cap \mathbf{C}\Big(\overset{\infty}{\underset{1}{\cup}}\,\mathscr{I}_\nu\Big)\Big] < \mu\Big(\overset{\infty}{\underset{N_\epsilon+1}{\cup}}\,\mathscr{I}_\nu^*\Big) < \overset{\infty}{\underset{N_\epsilon}{\sum}}\,\mu(\mathscr{I}_\nu^*) < 5\epsilon.$$

Since $\epsilon > 0$ but otherwise arbitrary, it follows with the aid of Theorem 8.3 that the set $E \cap \mathbf{C}(\cup\,\mathscr{I}_\nu)$ is measurable and of measure zero, completing the proof.

The following corollary is frequently useful.

Theorem 8.28

Given a bounded subset E of the reals, a class $K \equiv \{I_\alpha : \alpha \in A\}$ of compact intervals of positive length with property (8.58) and a positive number ϵ, there exists a finite subset of K consisting of disjoint intervals $\mathscr{I}_\nu$, $\nu = 1, \ldots, n_\epsilon$ such that

(i) $\overset{n_\epsilon}{\underset{1}{\cup}}\,\mathscr{I}_\nu$ *covers all of E except a subset E_0, $\mu^*(E_0) < \epsilon$*

and

(ii) $\overset{n_\epsilon}{\underset{1}{\sum}}\,\mu(\mathscr{I}_\nu) - \epsilon < \mu^*(E) < \overset{n_\epsilon}{\underset{1}{\sum}}\,\mu(\mathscr{I}_\nu) + \epsilon.$

PROOF

Hypothesis (8.57) of the Vitali Theorem is now a consequence of the boundedness of E. Hence there exists, in accord with the definition of μ^*, an open subset G of R such that

(8.66) $E \subset G$ and $\mu(G) < \mu^*(E) + \epsilon.$

If we examine hypothesis (8.58) and refer to Theorem 8.5 we see that it suffices to use the subset K_G of K consisting of only those intervals in K each of which is a subset of G.

By the Vitali Theorem with K_G in place of K, there is a finite or infinite sequence $\{\mathscr{I}_\nu \in K_G\}$ of disjoint intervals such that $\cup\,\mathscr{I}_\nu \supset E - E_1$, $\mu(E_1) = 0$. Thus

$$E = E_1 \cup [\cup (E \cap \mathscr{I}_\nu)]$$

and, consequently,

(8.67) $\mu^*(E) \leqslant \mu^*(\cup\,\mathscr{I}_\nu).$

Since $\cup \, \mathscr{I}_\nu \subset G$ and the intervals $\mathscr{I}_\nu$ are disjoint,

(8.68) $\sum \mu(\mathscr{I}_\nu) = \mu(\cup \, \mathscr{I}_\nu) \leqslant \mu(G) < \mu^*(E) + \epsilon \leqslant \sum \mu(\mathscr{I}_\nu) + \epsilon.$

CASE 1. $\{\mathscr{I}_\nu\}$ IS A FINITE SEQUENCE

The number n of these intervals serves as the n_ϵ in the theorem. Delete the second and third members of (8.68), subtract ϵ from each member that remains, and obtain inequalities that imply conclusion (ii).

CASE 2. $\{\mathscr{I}_\nu\}$ IS AN INFINITE SEQUENCE

Since $\cup \, \mathscr{I}_\nu \subset G$ and $\mu(G) < \infty$ by (8.66), the infinite series $\Sigma\mu(\mathscr{I}_\nu)$ has a finite sum. Therefore corresponding to the given ϵ is a number n_ϵ such that

(8.69) $\sum_{n_\epsilon+1}^{\infty} \mu(\mathscr{I}_\nu) < \epsilon.$

Using (8.67), we then see that

$$\mu^*(E) \leqslant \sum \mu(\mathscr{I}_\nu) < \sum_{1}^{n_\epsilon} \mu(\mathscr{I}_\nu) + \epsilon.$$

From this and (8.68),

$$\sum_{1}^{n_\epsilon} \mu(\mathscr{I}_\nu) < \mu^*(E) + \epsilon < \sum_{1}^{n_\epsilon} \mu(\mathscr{I}_\nu) + 2\epsilon,$$

which yields conclusion (ii) by subtraction of ϵ. In case 2, the set $E_1 \cup [\cup_{n_\epsilon+1}^{\infty} (E \cap \mathscr{I}_\nu)]$ plays the role of set E_0 in the theorem.

8.10 DERIVATIVES OF FUNCTIONS OF BOUNDED VARIATION

Refer to Section 1.7 and also define the *upper derivate* $\bar{D}\phi$ as the function such that

$$(\bar{D}\phi)(c) \equiv \begin{cases} (D^+\phi)(c) & if\ c = a \\ \max\,[(D^+\phi)(c), (D^-\phi)(c)] & if\ a < c < b, \\ (D^-\phi)(c) & if\ c = b. \end{cases}$$

The *lower derivate* $\underline{D}\phi$ is similarly defined.

Theorem 8.29

Given a compact interval $[a,b]$ of positive length and a function $\phi\colon [a,b] \to R$ that is nondecreasing on $[a,b]$, if $\bar{D}f \geqslant k$ on a subset E of $[a,b]$, then

(8.70) $\phi(b) - \phi(a) \geqslant k\mu^*(E).$

PROOF

If E is empty, (8.70) is obvious. Given $x \in E$, then either $(D^{+}\phi)(x) \geq k$ or $(D^{-}\phi)(x) \geq k$, with conjunctive "or"; hence x is the right or left endpoint of closed subintervals $[\alpha,\beta]$ of $[a,b]$ of arbitrarily small positive length such that

(8.71) $$\phi(\beta) - \phi(\alpha) \geq k(\beta - \alpha).$$

The class of all such subintervals $[\alpha,\beta]$ of $[a,b]$ has properties (8.57) and (8.58). Accordingly, given $\epsilon > 0$, there exists, by Theorem 8.28, a finite set of disjoint intervals $[\alpha_i,\beta_i]$, the number of which is denoted by n such that

(8.72) $$\mu^*(E) < \sum_{1}^{n} (\beta_i - \alpha_i) + \epsilon.$$

We can suppose the notation so chosen that

$$a \leq \alpha_1 < \beta_1 < \alpha_2 < \beta_2 < \cdots < \alpha_n < \beta_n \leq b.$$

It follows from (8.71) and (8.72) that

$$\sum_{1}^{n} [\phi(\beta_i) - \phi(\alpha_i)] \geq k \sum_{1}^{n} (\beta_i - \alpha_i) > k[\mu^*(E) - \epsilon].$$

Since ϕ is nondecreasing, $\phi(b) - \phi(a)$ dominates the first member. Conclusion (8.70) follows.

Theorem 8.30

Given a compact interval $[a,b]$ of positive length and a nondecreasing function $\phi: [a,b] \to R$, if, for each $x \in E \subset [a,b]$,

$$(\underline{D}\phi)(x) < h < k < (\bar{D}\phi)(x),$$

then E is Lebesgue measurable and $\mu(E) = 0$.

PROOF

Given $x \in E$, then $(D_-\phi)(x) < h$ or $(D_+\phi)(x) < h$; hence x is an endpoint of arbitrarily short subintervals $[\alpha,\beta]$ of $[a,b]$ such that

(8.73) $$\phi(\beta) - \phi(\alpha) < h(\beta - a).$$

Given $\epsilon > 0$, there is by Theorem 8.28 a finite set of such intervals $[\alpha_i,\beta_i]$, $i = 1, \ldots, n_\epsilon$ (we henceforth suppress the subscript ϵ), whose union covers all of E except a subset of measure below ϵ, and such that

$$\sum_{1}^{n} (\beta_i - \alpha_i) - \epsilon < \mu^*(E).$$

Hence, by (8.73),

$$(8.74) \qquad \sum_1^n [\phi(\beta_i) - \phi(\alpha_i)] < h\mu^*(E) + h\epsilon.$$

Theorem 8.29, applied separately to each set $E \cap [\alpha_i, \beta_i]$, yields the relations

$$\phi(\beta_i) - \phi(\alpha_i) \geq k\mu^*(E \cap [\alpha_i, \beta_i]), \qquad i = 1, \ldots, n;$$

therefore, with the aid of Theorem 8.28 and the monotonicity (8.2_3) of μ^*,

$$(8.75) \quad \sum_1^n [\phi(\beta_i) - \phi(\alpha_i)] \geq k \sum_1^n \mu^*(E \cap [\alpha_i, \beta_i]$$
$$\geq k\mu^*\left(\bigcup_1^n E \cap [\alpha_i, \beta_i]\right) > k\mu^*(E) - k\epsilon.$$

As a consequence of (8.74), (8.75), and the fact that $h < k$, we verify that

$$\mu^*(E) < (h+k)\epsilon/(k-h)$$

and, from the nature of ϵ and Theorem 8.3, that the theorem is true.

Theorem 8.31

If $[a,b]$ is a compact interval of positive length, a nondecreasing function $\phi: [a,b] \to R$ has a finite derivative $\dot\phi(t)$ a.e. on $[a,b]$.

PROOF

Define

$$E \equiv \{t \in [a,b]: (\underline{D}\phi)(t) < (\bar{D}\phi)(t)\}.$$

Let $\{r_\nu: \nu \in N\}$ be a fixed sequentialization of all nonnegative rational real numbers and define

$$E_{mn} \equiv \{t \in [a,b]: (\underline{D}\phi)(t) < r_m < r_n < (\bar{D}\phi)(t)\}.$$

Clearly E_{mn} is empty if $r_m \geq r_n$. Verify that $E = \cup E_{mn}$. By Theorem 8.30, $\mu(E_{mn}) = 0$ if $r_m < r_n$. It then follows from the countability of the class $\{E_{m,n}: m,n \in N\}$ and Theorem 8.8 that $\mu(E) = 0$ and consequently that $\dot\phi(t)$ exists, finite or ∞, if $t \in [a,b] - E$. To complete the proof it remains only to see that the set $S \equiv \{t \in [a,b]: \dot\phi(t) = \infty\}$ is of measure zero. By Theorem 8.29,

$$\phi(b) - \phi(a) \geq k\mu^*(S), \qquad \forall k > 0;$$

hence, if $\mu^*(S) > 0$, it would follow from consideration of arbitrarily large values of k that $\phi(b) - \phi(a) = \infty$, in contradiction with the hypothesis that ϕ is real-valued.

The following is an immediate corollary to Theorems 8.31 and 8.23.

Theorem 8.32

If $[a,b]$ is a compact interval of positive length, a function $\phi: [a,b] \to R$ that is BV on $[a,b]$ has a finite derivative $\dot{\phi}(t)$ a.e. on $[a,b]$.

Exercise 8.7

1. Construct an example of a function ϕ that is monotone on an interval and fails to have a finite derivative at each point of an infinite set of points where it is discontinuous. Construct a second example in which ϕ is continuous on its interval and $\dot{\phi}(t) = \infty$ at infinitely many points t of that interval.

8.11 INDEFINITE INTEGRALS

Given a compact interval $[a,b]$ and a function $\phi: [a,b] \to R^*$ that is integrable over $[a,b]$ and hence by problem 5, Exercise 8.4, over every subinterval $[a,t]$ of $[a,b]$, the function $\Phi: [a,b] \to R$ with values

$$\Phi(t) \equiv I(\phi;[a,t]) + C, \quad C \equiv \text{const.}$$

is called *an indefinite integral* or simply *an integral of* ϕ.

Theorem 8.33

If Φ is an integral of ϕ, then Φ is AC.

PROOF

It can be verified that given $\epsilon > 0$, there is a positive δ_ϵ such that if E is a measurable subset of $[a,b]$, then

(8.76) $I(|\phi|;E) < \epsilon$ *provided that* $\mu(E) < \delta_\epsilon$.

With reference to definition (8.54) of absolute continuity, let $[a_1,b_1], \ldots,$ $[a_n,b_n]$ be nonoverlapping subintervals of $[a,b]$. By the definition of Φ and Theorem 8.21,

$$\sum |\Phi(b_i) - \Phi(a_i)| \leq \sum I(|\phi|;[a_i,b_i]) = I(|\phi|; \cup [a_i,b_i]).$$

The desired conclusion then follows from (8.76).

Theorem 8.34

If ϕ: $[a,b] \to R$ is AC on $[a,b]$ and if $\dot{\phi}(t) = 0$ a.e. on $[a,b]$, then $\phi(t)$ = const. on $[a,b]$; hence $\dot{\phi}(t)$ must actually vanish everywhere on $[a,b]$.

PROOF

Define $E \equiv \{t \in [a,b]: \dot{\phi}(t) = 0\}$. Given $\epsilon > 0$ and $t \in E$, then t is the right or left end of arbitrarily short subintervals $[\alpha,\beta]$ of $[a,b]$ such that

$$(8.77) \qquad |\phi(\beta) - \phi(\alpha)| < \epsilon(\beta - \alpha).$$

By Theorem 8.28, there is an integer, which we denote here by $n_\epsilon - 1$ rather than n_ϵ and then suppress ϵ, together with disjoint intervals $[\alpha_1,\beta_1], \ldots, [\alpha_{n-1},\beta_{n-1}]$ with property (8.77) and such that

$$(8.78) \qquad \sum_1^{n-1} (\beta_i - \alpha_i) - \epsilon < \mu^*(E) < \sum_1^{n-1} (\beta_i - \alpha_i) + \epsilon.$$

We drop the asterisk on the middle term since E is measurable by hypothesis, $\mu(E) = b - a$.

Clearly $b - a \geqslant \Sigma(\beta_i - \alpha_i)$; hence by the second inequality (8.78)

$$(8.79) \qquad b - a \geqslant \sum (\beta_i - \alpha_i) > b - a - \epsilon.$$

Labels can be so chosen that

$$\alpha_1 < \beta_1 < \alpha_2 < \beta_2 < \cdots < \alpha_{n-1} < \beta_{n-1}.$$

Define $\beta_0 \equiv a$ and $\alpha_n \equiv b$. Then

$$(8.80) \quad |\phi(b) - \phi(a)| \leqslant \sum_1^{n-1} |\phi(\beta_i) - \phi(\alpha_i)| + \sum_1^n |\phi(\alpha_i) - \phi(\beta_{i-1})|.$$

It follows from (8.77) that the first sum is dominated by $\epsilon(b - a)$ and from (8.79) that $\Sigma_1^n(\alpha_i - \beta_{i-1}) < \epsilon$, consequently, from the absolute continuity of ϕ, that the second sum tends to zero with ϵ. In this last step ϵ plays the role of δ_ϵ in the definition of absolute continuity. The left member of (8.80) is free of ϵ and therefore must be zero.

The preceding argument, with $[a,t]$ in place of $[a,b]$, proves that $\phi(t) = \phi(a) = $ const. and, by elementary calculus, $\dot{\phi}(t)$ must vanish identically on $[a,b]$.

Theorem 8.35

Given a compact interval $[a,b]$ of positive length and a nondecreasing function $\phi: [a,b] \to R$,

(i)   the derivative $\dot{\phi}$ of ϕ is integrable over $[a,b]$

and

(ii) $I(\dot{\phi};[a,b]) \leq \phi(b-) - \phi(a+)$.

PROOF

As a consequence of Theorem 8.31, $\dot{\phi}(t)$ exists and is finite on a subset E of $[a,b]$ of measure $b-a$. Since ϕ is monotone, the sets $\{t \in [a,b]: \phi(t) < a\}$ are intervals, possibly empty for some values of a; therefore, ϕ is measurable on $[a,b]$ and, by Theorem 8.10, measurable on E.

Denote by ϕ^* the function from R to R that coincides with ϕ on the open interval (a,b) and set $\phi^*(t) \equiv \phi(a+)$ if $t \leq a$, $\phi^*(t) \equiv \phi(b-)$ if $t \geq b$. Define $\psi: R \to R$ by the relation

$$\psi(t) \equiv \phi^*(t+h).$$

Neither the measurability nor the Lebesgue measure of a set is affected by translation; hence ψ like ϕ and ϕ^* is measurable on $[a,b]$. By Theorem 8.11(iv), (vi), the difference quotient $Q: R \to R$,

$$Q(t) \equiv \frac{\psi(t) - \phi^*(t)}{h},$$

is then measurable on $[a,b]$, hence on E, while, by Theorem 8.13, the derivative $\dot{\phi}^*$ is measurable on E.

With h restricted to values $1/n$, we see by means of Fatou's Lemma that

(8.81) $\liminf\limits_{n \to \infty} I(Q;E) \geq I(\dot{\phi}^*;E) = I(\dot{\phi};[a,b])$.

The asterisk is dropped in the last integral since $\dot{\phi}(t) = \dot{\phi}^*(t)$ on $[a,b]$ except possibly at a and b. We can, moreover, interpret $\dot{\phi}$, consistently with problem 11, Exercise 8.4, as an arbitrary extension of the original $\dot{\phi}$ from $E \cap (a,b)$ to $[a,b]$ and integrate over $[a,b]$ rather than E.

One can verify that

$$I(Q;[a,b]) = \frac{1}{h} I(\phi^*;[a+h, b+h]) - \frac{1}{h} I(\phi^*;[a,b])$$

$$= \frac{1}{h} I(\phi^*;[b, b+h]) - \frac{1}{h} I(\phi^*;[a, a+h]).$$

By application of the Mean Value Theorem, Theorem 8.19, to each of the last two terms there are respective real numbers β and α_h,

$$\phi^*(b) = \beta = \phi^*(b+h) \qquad and \qquad \phi^*(a) \leq \alpha_h \leq \phi^*(a+h)$$

such that

$$I(Q;[a,b]) = \beta - \alpha_h = \phi(b-) - \alpha_h.$$

As $h \to 0$, $\alpha_h \to \phi(a+)$. Thus $I(Q;[a,b])$ has the limit $\phi(b-) - \phi(a+)$ and stated conclusions (i) and (ii) now follow from (8.81). We observe in retrospect that the limit inferior in (8.81) is actually a limit.

Theorem 8.36

Given the compact interval $[a,b]$, *if* ϕ: $[a,b] \to R^*$ *is integrable over* $[a,b]$, *hence over subintervals* $[a,t]$ *of* $[a,b]$, *and if* $I(\phi;[a,t]) = 0$ *for all* $t \in [a,b]$, *then* $\phi(t) = 0$ *a.e. on* $[a,b]$.

PROOF

Define $E \equiv \{t \in [a,b]: \phi(t) > k > 0\}$. Function ϕ being integrable is measurable on $[a,b]$; hence E is a measurable set.

If $\mu(E) > 0$, there is necessarily a closed set $F \subset E$, with $\mu(F) > 0$ and

$$(8.82) \qquad I(\phi;F) > I(k;F) = k\mu(F) > 0.$$

By the additivity of the integral as a set-function.

$$(8.83) \quad I(\phi;[a,b]) = I(\phi;[a,b]-F) + I(\phi;F) = I(\phi;(a,b)-F) + I(\phi;F),$$

the last equality being a consequence of problem 1, Exercise 8.2.

The open set $(a,b) - F$ is the union of a countable set of disjoint open intervals I_n by Theorem 8.5. Let (α,β) be any one of these. Then, under our hypothesis that $I(\phi;[a,t]) = 0$ and with another application or problem 1, Exercise 8.2,

$$I(\phi;(\alpha,\beta)) = I(\phi;[a,\beta]) - I(\phi;[a,\alpha]) = 0 - 0 = 0.$$

Therefore, $I(\phi; \cup I_n) = \Sigma I(\phi;I_n) = 0$ and, from (8.83) and (8.82),

$$I(\phi;[a,b]) = I(\phi;F) > 0,$$

contrary to the hypotheses that $I(\phi;[a,t]) = 0$ on $[a,b]$. Having reached this by supposing that $\mu(E) > 0$, we must infer that $\mu(E) = 0$ for any choice of the positive k.

Clearly,

$$\{t \in [a,b]: 1/n \geqslant \phi(t) > 1/(n+1)\} \subset \{t \in [a,b]: \phi(t) > 1/(n+1)\}.$$

The set on the right has measure zero; hence so also has the set on the left. Moreover,

$$\{t \in [a,b]: \phi(t) > 0\} = \{t \in [a,b]: \phi(t) > 1\} \cup$$
$$\left[\bigcup_1^\infty \left\{ t \in [a,b]: \frac{1}{n} \geqslant \phi(t) > \frac{1}{n+1} \right\} \right].$$

Each set on the right being of measure zero, so also is the set on the left.

The preceding argument applied to $-\phi$ shows that the set $\{t \in [a,b]: \phi(t) < 0\}$ is of measure zero, completing the proof of the theorem.

Theorem 8.37

If $\phi: [a,b] \to R^*$ is integrable over the compact interval $[a,b]$ of positive length and Φ is an integral of ϕ, then $\overset{\bullet}{\Phi}(t) = \phi(t)$ a.e. on $[a,b]$.

PROOF

Φ is AC by Theorem 8.33; hence, by Theorem 8.24, Φ is BV and, by Theorem 8.32, $\overset{\bullet}{\Phi}(t)$ exists and is finite a.e. on $[a,b]$.

CASE 1. ϕ IS BOUNDED ON $[a,b]$

This means that there is a real number M such that $|\phi(t)| \leqslant M$ on $[a,b]$, whence, with the aid of the Mean Value Theorem, Theorem 8.19,

$$(8.84) \qquad \frac{\Phi(t+h) - \Phi(t)}{h} = \frac{1}{h} I(\phi; [t, t+h]) \leqslant M, \qquad h > 0.$$

We understand without a shift in notation that ϕ has been extended beyond b, as was the function ϕ in the proof of Theorem 8.35. Let E_t denote the subset of $[a,t]$ on which $\overset{\bullet}{\Phi}(t)$ exists and is finite. With $h = 1/n$, let $n \to \infty$. In the light of the bound M in (8.84), Lebesgue's Bounded Convergence Theorem applies and the integral over E_t of the left member of (8.84) converges to $I(\overset{\bullet}{\Phi}; E_t)$. Since $\mu([a,t] - E_t) = 0$, we can then replace E_t by $[a,t]$ and, by the same steps used near the end of the proof of Theorem 8.35, the right member of the equation

$$nI\{\Phi(t+1/n) - \Phi(t); [a,b]\} = nI(\Phi; [t, t+1/n]) - nI(\Phi; [a, a+1/n])$$

converges to $\Phi(t) - \Phi(a)$.

Since Φ is given as an integral of ϕ, we now see that

$$I(\overset{\bullet}{\Phi}; [a,t]) = \Phi(t) - \Phi(a) = I(\phi; [a,t]).$$

It follows that $I(\dot{\Phi}-\phi;[a,t])=0$ and from Theorem 8.36 that $\dot{\Phi}(t)$ $=\phi(t)$ a.e. on $[a,b]$.

CASE 2. ϕ IS NONNEGATIVE ON $[a,b]$

Let $\{s_n : n \in N\}$ be a sequence of nonnegative simple functions, nondecreasing in n with the property (8.34*). The integrals

$$(8.85) \qquad I(\phi-s_n;[a,t]), \; I(\phi;[a,t]), and \; I(s_n;[a,t])$$

are all bounded for $t \in [a,b]$ and are nondecreasing in t.
   Define

$$\Phi_n(t) \equiv \Phi(t)-I(s_n;[a,t]).$$

Derivatives of both terms on the right exist and are finite a.e. on $[a,b]$ by Theorem 8.31; hence Φ_n has the same property. Moreover, the derivative of the integral is s_n a.e. on $[a,b]$ by case 1 of the present proof; consequently,

$$(8.86) \qquad \dot{\Phi}_n(t) = \dot{\Phi}(t) - s_n(t) \quad a.e. \; on \; [a,b].$$

The left member is nonnegative when it exists by the monotonicity of the first integral (8.85); hence

$$(8.87) \qquad \dot{\Phi}(t) \geqslant s_n(t) \quad a.e. \; on \; [a,b], \qquad n = 1, 2, \ldots ..$$

There is a subset A_n of $[a,b]$ of measure 0 (possibly empty) on which (8.87) fails. However, (8.87) is both meaningful [that is, $\dot{\Phi}(t)$ exists] and valid independently of n on the set $[a,b] - \cup A_n$, a set of measure $b-a$. If we let $n \to \infty$, it follows from (8.87) and the convergence of s_n to ϕ that

$$\dot{\Phi}(t) \geqslant \phi(t) \quad a.e. \; on \; [a,b],$$

and therefore that

$$(8.88) \qquad I(\dot{\Phi}-\phi;[a,t]) \geqslant 0.$$

To obtain the complementary inequality we remark from Theorem 8.35 and the fact that Φ is an integral of ϕ that

$$I(\dot{\Phi};[a,t]) \leqslant \Phi(t)-\Phi(a) = I(\phi;[a,t])$$

and hence that $I(\dot{\Phi}-\phi;[a,t]) \leqslant 0$. This with (8.88) and Theorem 8.36 yields the desired conclusion.

CASE 3.   THE GENERAL CASE

Express ϕ as $\phi^+ - \phi^-$ and define

$$\Phi^+(t) \equiv I(\phi^+;[a,t]) \qquad and \qquad \Phi^-(t) \equiv I(\phi^-;[a,t]).$$

Clearly $\Phi = \Phi^+ - \Phi^-$. Apply case 2 separately to Φ^+ and Φ^- and complete the proof.

Theorem 8.38   (The Fundamental Theorem of the Integral Calculus)

Given a compact interval $[a,b]$ of positive length and a function $\Phi;[a,b] \to R$, the statement that

$$(8.89) \qquad I(\dot{\Phi};[a,t]) = \Phi(t) - \Phi(a), \qquad \forall t \in [a,b]$$

is meaningful and valid if and only if Φ is AC on $[a,b]$.

Since the derivative of an AC function can fail to exist on some set Z of measure zero, symbol $\dot{\Phi}$ is to be understood as an arbitrary extension without shift in notation of an original $\dot{\Phi}$ from $[a,b] - Z$ to $[a,b]$.

PROOF

Suppose first that (8.89) holds as stated. It then holds for $t = b$ and hence (8.89) says among other things that $\dot{\Phi}$ is integrable over $[a,b]$. By Theorem 8.33, Φ as an integral is then AC on $[a,b]$.

If conversely Φ is AC on $[a,b]$, then it is BV on $[a,b]$ by Theorem 8.24 and $\dot{\Phi}(t)$ exists and is finite a.e. on $[a,b]$ by Theorems 8.23 and 8.31. Moreover, $\dot{\Phi}$ is integrable over $[a,b]$ by Theorems 8.35, 8.23, and the additivity of the integral.

As a consequence of Theorem 8.33, $I(\dot{\Phi};[a,t])$ is then AC on $[a,b]$. Define

$$(8.90) \qquad g(t) \equiv \Phi(t) - I(\dot{\Phi};[a,t]).$$

The derivative $\dot{g}(t)$ is seen to vanish a.e. on $[a,b]$ with the aid of Theorem 8.37. By an application of Theorem 8.34, we have that $g(t)$ is constant on $[a,b]$ and, setting $t = a$ in (8.90), we find that this constant is $\Phi(a)$. This completes the proof of (8.89).

Exercise 8.8

1. With reference to problem 5, Exercise 8.6, for the nature of the Cantor–Lebesgue function ω, verify that an arbitrary extension $\dot{\omega}^*$ of its derivative $\dot{\omega}$ from $[0,1] - E$ to $[0,1]$ is integrable over $[0,1]$ and that $I(\dot{\omega}^*;[0,1]) = 0 \neq \omega(1) - \omega(0)$.
2. Establish that the function $\phi:[-1,1] \to R$, $\phi(x) \equiv x^{1/3}$, is AC on

$[-1,1]$ by demonstrating that an indefinite integral Φ of this function has property (8.89).

3. Formulate and prove as a corollary to Theorem 8.35 a companion theorem for a nonincreasing function.

Chapter 9

VARIATIONAL
THEORY IN TERMS OF
LEBESGUE INTEGRALS

9.1 INTRODUCTION

A need in the calculus of variations for an integral with better convergence properties than that of Riemann has been described in Section 7.10. One answer is the Weierstrass integral, which continues to find a limited usefulness, but the Lebesgue integral is usually a superior tool.

In this chapter we show that, if y is BV on its interval and F is a semicontinuous parametric integrand, then the composite function $F(y,\dot{y})$ is Lebesgue measurable. Its integral is generally not useful, however, unless y is restricted to be AC, as will be pointed out. Similar remarks apply to nonparametric integrals. Also included are certain theorems concerning such integrals, a brief introduction to L_p-spaces, and some typical theorems on the existence of global extrema.

9.2   VARIATIONAL INTEGRALS OF THE LEBESGUE TYPE

The much used Theorem 8.5 on the decomposition of an open subset of the reals into disjoint open intervals does not extend to higher dimensions. For example, an L-shaped open subset of the plane is not a countable union of disjoint open two-dimensional intervals $I \equiv (a,b) \times (c,d)$. The following theorem, which is independent of the dimension p, will serve our purposes.

A *half-open interval in* R^p is the cartesian product (Section 1.2) of p one-dimensional intervals, either all of the type $(a,b]$ or all of the type $[a,b)$. Open p-dimensional intervals and closed p-dimensional intervals are similarly defined.

Theorem 9.1

Every open subset G *of* R^p *is a denumerable union of disjoint half-open* p-dimensional intervals.

PROOF

If $p = 1$, let (α,β) be an open interval and let $\{a_n: n \in N\}$ be a strictly decreasing sequence in (α,β) with α as limit. Then $[a_1,\beta)$ together with the union $\cup [a_{i+1},a_i)$ provides a denumerable decomposition of (α,β). A similar decomposition into disjoint intervals $(a,b]$ is obtained with the aid of a sequence $\{b_n: n \in N\}$ in (α,β) that converges to β. Since an open subset G of R is a countable union of disjoint intervals (α,β) by Theorem 8.5, the stated conclusion follows.

Given $p = 2$ and any positive integer k, consider the families of lines $x = \mu/2^k$ and $y = \nu/2^k$, $\mu, \nu = 0, \pm 1, \pm 2, \ldots$. For each k the entire plane R^2 is the denumerable union of all half-open intervals

$$(9.1) \qquad [\mu/2^k, (\mu+1)/2^k) \times [\nu/2^k, (\nu+1)/2^k).$$

For $k = 1$, a subset, possibly empty, of the intervals (9.1) is contained in G. There is then a subset of the intervals (9.1) with $k = 2$ consisting of such intervals as are contained in G but in none of the intervals (9.1) selected at the preceding step. Proceeding inductively in this manner we obtain a denumerable set of countable sets of intervals (9.1), which are disjoint and the union of which can be verified to be the given set G. The totality of all these intervals is clearly a countable set. One sees easily that no finite union of such intervals can be an open set, hence that the union is a denumerable union as stated in the theorem.

The same type of proof with minor changes applies if $p > 2$. Clearly the decomposition mentioned in the theorem is not unique.

Theorem 9.2

Given functions ϕ^j: $[a,b] \to R$, $j = 1, \ldots, p$, that are measurable on $[a,b]$, if G is an open subset of R^p containing all points $[\phi^1(t), \ldots, \phi^p(t)]$, $t \in [a,b]$, and if $F: G \to R^*$ is semi-continuous on G, then the composite function $F \circ \phi$ is measurable on $[a,b]$.

PROOF

Suppose that F is lower semi-continuous on G. Consider the mapping ϕ: $[a,b] \to R^p$, $\phi = (\phi^1, \ldots, \phi^p)$, and, given $S \subset R^p$, let $\phi^{-1}(S)$ denote the set $\{t \in [a,b]: \phi(t) \in S\}$. Let I be a generic symbol for a half-open interval of the type $[c^1,d^1) \times \cdots \times [c^p,d^p)$. Then $\phi^{-1}(I)$ consists of those $t \in [a,b]$ such that

$$(9.2) \qquad c^j \leqslant \phi^j(t) < d^j, \quad j = 1, \ldots, p.$$

Since ϕ^j is measurable by hypothesis, each subset A_j of $[a,b]$ consisting of those t satisfying (9.2) for a fixed j is measurable; hence the set $\cap_1^p A_j$ is measurable by Theorem 8.9, and this is precisely the set $\phi^{-1}(I)$. The open set G is a denumerable union of disjoint half-open intervals I by Theorem 9.1; consequently, the set $\phi^{-1}(G)$ is a denumerable union of disjoint measurable subsets of $[a,b]$ and is measurable by Theorem 8.8. A similar remark applies to any open subset of G.

We point out next that, for each choice of the real number k, the set

$$(9.3) \qquad S_k \equiv \{y \in G: F(y) > k\}$$

is an open subset of G. If S_k is not open and hence not empty, there must exist $y_0 \in S_k$ and a sequence $\{y_\nu: \nu \in N\}$ converging to y_0 and such that $F(y_\nu) \leqslant k$. It follows from the lower semi-continuity of F at y_0 that

$$F(y_0) \leqslant \lim \inf F(y_\nu) \leqslant k,$$

in contradiction with the choice of y_0 in S_k; hence we must infer that S_k is open as stated.

Observe finally that the subset

$$(9.4) \qquad \{t \in [a,b]: F[\phi(t)] > k\}$$

of $[a,b]$ is the image under ϕ^{-1} of the set S_k in R^p. That the set (9.3) in R^p is open implies that the set (9.4) is a measurable subset of $[a,b]$. Since k is real but otherwise arbitrary, the composite function $F \circ \phi$ is measurable under the discussion of Section 8.4.

If F is upper semi-continuous, apply the preceding proof to $-F$ and use Theorem 8.11(ii).

Let $f:[a,b] \times R^m \times R^m \to R$ be a nonparametric integrand that is lower semi-continuous at each point of the given domain. As a consequence of Theorems 8.13 and 8.32, if $y: [a,b] \to R^m$ is of finite length, hence if each component y^j of y is BV on $[a,b]$, there is a subset E_j of $[a,b]$ of measure $b-a$ such that the derivative $\dot{y}^j$ exists and is finite on E_j and is moreover measurable on E_j. It follows that all components of $\dot{y}$ are finite and measurable on the set

$$E \equiv [a,b] - \bigcup_j \{[a,b] - E_j\},$$

which differs from $[a,b]$ by a set of measure zero. In accord with problem 5, Exercise 8.3, any extension of $\dot{y}$ from E to $[a,b]$ obtained by assigning extended real values to the components of $\dot{y}(t)$ at points of $[a,b] - E$ yields a function $\dot{y}: [a,b] \to R^m$ with components measurable on $[a,b]$. Symbol $\dot{y}$ will now mean such an extension. Each component y^j of y is BV, and hence, as the difference between monotone functions, is seen to be measurable on $[a,b]$.

Identify $2m+1$ with the p of Theorem 9.2 and consider the function ϕ with components

$$\phi^1(t) \equiv t, \qquad \phi^{i+1}(t) \equiv y^i(t), \qquad \phi^{i+m+1}(t) \equiv \dot{y}^i(t), \qquad i = 1, \ldots, m.$$

Extend f from the stated domain to R^{2m+1} by setting $f(t,y,r) \equiv f(a,y,r)$ or $f(b,y,r)$, respectively, when $t < a$ or $t > b$, so that f now has an open domain and can be identified with the function F of Theorem 9.2. It follows from this theorem that the composite function $f(t,y,\dot{y})$ is measurable on $[a,b]$. Similar remarks apply to a parametric integrand.

A first requirement on a class $\mathcal{Y}$ of admissible functions $y: [a,b] \to R^m$ or R^n, $n = m+1$, is that the derivative of y exists a.e. on $[a,b]$. With $\dot{y}$ denoting the extension described above, we require further that $f(t,y,\dot{y})$ or $F(y,\dot{y})$, as the case may be, be integrable over $[a,b]$, that is, that the integral exist and be finite under (8.34) or (8.36). In view of the differentiability (Theorem 8.32) of functions y with components that are BV and the measurability of $\dot{y}$ on $[a,b]$, a consequence of Theorem 8.13 and problem 5, Exercise 8.3, a profusion of functions $f(t,y,\dot{y})$ and $F(y,\dot{y})$ will be integrable.

We often need to know that a change of variable is permissible.

Theorem 9.3

If $g: [a,b] \to R^*$ *is integrable over* $[a,b]$ *and* $h: [\alpha,\beta] \to [a,b]$ *is nondecreasing and* AC *on* $[\alpha,\beta]$ *with* $h(\alpha) = a$ *and* $h(\beta) = b$, *then*

(i)   *the product* $(g \circ h)\dot{h}$ *is integrable over* $[\alpha,\beta]$

and

(ii) $I(g; [a,b]) = I((g \circ h)\dot{h}; [\alpha,\beta])$.

PROOF

Define G as an integral (Section 8.11), namely,

(9.5) $G(t) \equiv I(g;[a,t]),   t \in [a,b].$

CASE 1

g is bounded and measurable on $[a,b]$ and $\dot{h}(\tau)$ is bounded away from 0 on the subset of $[\alpha,\beta]$ consisting of those τ at each of which $\dot{h}(\tau)$ exists and is finite. In this case there exists a positive real number M such that $|g(t)| \leq M$. We see from (8.26) that

$$-M(t_2-t_1) \leq I(g;[t_1,t_2]) \leq M(t_2-t_1),   \forall [t_1,t_2] \subset [a,b],$$

hence, by definition (9.5) of G, that

(9.6) $|G(t_2)-G(t_1)| \leq M|t_2-t_1|.$

Thus G is lipschitzian and, by Theorem 8.25, is AC on $[a,b]$. Moreover, by (9.6), given any finite set of nonoverlapping subintervals $[\alpha_i,\beta_i]$ of $[\alpha,\beta]$,

$$\sum |G[h(\beta_i)]-G[h(\alpha_i)]| \leq M \sum |h(\beta_i)-h(\alpha_i)|.$$

Since h is AC on $[\alpha,\beta]$, it follows that $G \circ h$ is AC on $[\alpha,\beta]$.

In view of Theorems 8.24 and 8.32, the derivatives $(G \circ h)'(\tau)$ and $\dot{h}(\tau)$ both exist and are both finite on a subset A of $[\alpha,\beta]$ of Lebesgue measure $\mu(A) = \beta-\alpha$. The prime and the dot both denote differentiation with respect to τ in this proof. It follows from the Fundamental Theorem 8.38 and definition (9.5) that

$$I((G \circ h)'; [\alpha,\beta]) = G[h(\beta)]-G[h(\alpha)] = G(b)-G(a) = I(g; [a,b]).$$

To complete the proof of case 1, we show that

(9.7) $(G \circ h)'(\tau) = g[h(\tau)]\dot{h}(\tau)$ a.e. on $[\alpha,\beta]$.

One verifies by examining the respective difference quotients that, if $\tau \in A$, then

$$(G \circ h)'(\tau) = \dot{G}[h(\tau)]\dot{h}(\tau).$$

From (9.5) and Theorem 8.37 we know that $\dot{G}(t) = g(t)$ a.e. on $[a,b]$, consequently that $\dot{G}[h(\tau)] = g[h(\tau)]$ except for points τ in a subset of $[\alpha,\beta]$ that map into the points $h(\tau)$ of a subset Z of measure zero. Now $\dot{h}(\tau)$ is bounded from zero by hypothesis, hence $\dot{h}(\tau) \geq \delta > 0$ on A. It follows that the points τ that mapped into Z constitute a subset of $[\alpha,\beta]$ of measure zero. This completes the proof of (9.7).

CASE 2. g IS BOUNDED AND MEASURABLE ON $[a,b]$.

If $h(\beta) = h(\alpha)$, then $\dot{h}(\tau) = 0$ on $[\alpha,\beta]$, hence $\dot{h}(g \circ h)$ is integrable and conclusion (ii) holds in the form $0 = 0$. Having disposed of the trivial subcase suppose that $h(\beta) > h(\alpha)$.

Next define a function $h_n : [\alpha,\beta] \to R$,

$$h_n(\tau) \equiv \frac{n}{n+1} h(\tau) + \frac{1}{n+1} \left\{ h(\alpha) + \frac{\tau-\alpha}{\beta-\alpha} [h(\beta) - h(\alpha)] \right\},$$

and observe that $h_n(\alpha) = h(\alpha)$, that $h_n(\beta) = h(\beta)$, and that

$$\dot{h}_n(\tau) = \frac{n}{n+1} \dot{h}(\tau) + \frac{1}{n+1} \frac{h(\beta) - h(\alpha)}{\beta-\alpha},$$

hence that $\dot{h}_n(\tau)$ is bounded from 0. Since h is AC on $[\alpha,\beta]$ so also is h_n. It follows from case 1 that $(g \circ h_n)\dot{h}_n$ is integrable over $[\alpha,\beta]$ and that

$$I(g; [a,b]) = I((g \circ h_n)\dot{h}_n; [\alpha,\beta]).$$

Moreover $|g[h_n(\tau)]\dot{h}_n(\tau)| \leq M\dot{h}_n(\tau)$ and hence from the form of $\dot{h}_n$ Lebesgue's Dominated Convergence Theorem applies to the last equation to show that

$$I(g; [a,b]) = I((g \circ h)\dot{h}; [\alpha,\beta]).$$

CASE 3. $g \geq 0$ ON $[a,b]$ AND MEASURABLE ON $[a,b]$

Let $g_n \equiv \inf(g,n)$ denote the truncation of g at the level n. By case 2 we have the stated conclusions (i) and (ii) for g_n. Let $n \to \infty$ and use the Monotone Convergence Theorem 8.16.

CASE 4. THE GENERAL CASE

Apply case 3 separately to the functions g^+ and g^-.

Exercise 9.1

1. Given that $F : R^n \times R^n \to R$ is semi-continuous on its domain and has the homogeneity property (6.20), let $\xi : [0,1] \to R^n$ be the reduced-length representation of a given curve C. Identify theorems that suffice to ensure the integrability of $F(\xi,\dot{\xi})$ over $[0,1]$. Do the same for the integrability of $F(X,\dot{X})$ over $[0,L(C)]$, where X is the representation in terms of length.

9.3   THE LEBESGUE LENGTH-INTEGRAL

Given a continuous Fréchet curve C in E_n of finite length having $x : [a,b] \to R^n$ as a representation, let $C_{x,t}$ denote the subcurve represented by the restriction of x to $[a,t] \subset [a,b]$ and set

(9.8) $$ s(t) \equiv L(C_{x,t}), \qquad t \in [a,b]. $$

Theorem 9.4

The derivative $\dot{s}(t)$ exists and is finite a.e. on $[a,b]$. Moreover, $\dot{x}(t)$ also exists and is finite and $|\dot{x}(t)| = \dot{s}(t)$ a.e. on $[a,b]$.

PROOF

Each component x^j of x is BV as a consequence of Theorem 7.2; hence the derivative $\dot{x}^j(t)$ exists and is finite on $[a,b]$ except at the points of a subset Z_j of $[a,b]$ of measure zero. It follows that all components of the vector $\dot{x}(t)$ and hence the length $|\dot{x}(t)|$ of that vector exist and are finite on $[a,b] - \cup Z_j$, which is almost all of $[a,b]$.

Clearly s is nondecreasing and hence finitely differentiable a.e. by Theorem 8.31. Let A denote the subset of $[a,b]$ of measure $b-a$ on which both $\dot{s}(t)$ and $|\dot{x}(t)|$ exist and are both finite. The length of any subcurve dominates the length of the corresponding chord; that is,

$$ s(t+h) - s(t) \geqslant |x(t+h) - x(t)|, \qquad t, t+h \in [a,b]. $$

If we choose t as a point of A, divide by h, and let $h \to 0$, we see that

(9.9) $$ \dot{s}(t) \geqslant |\dot{x}(t)|, \qquad t \in A. $$

Let B denote the subset of A at points of which the inequality holds in (9.9). The proof of the theorem is completed by showing that $\mu(B) = 0$. Define

$$ B_n \equiv \left\{ t_0 \in B : t \in B \text{ and } |t - t_0| < 1/n \Rightarrow \frac{s(t) - s(t_0)}{t - t_0} > \frac{|x(t) - x(t_0)|}{|t - t_0|} + 1/n \right\}. $$

It follows that

$$ t_0 \in B_n \Rightarrow \dot{s}(t_0) \geqslant |\dot{x}(t_0)| + 1/n \Rightarrow t_0 \in B_n $$

and therefore that $\cup B_n \subset B$. To establish the complementary inclusion, suppose given $t_0 \in B$. Then $\dot{s}(t_0) > |\dot{x}(t_0)|$ and, for a sufficiently large n_1,

$$ \dot{s}(t_0) > |\dot{x}(t_0)| + 1/n_1. $$

Moreover, since $\dot{s}(t_0)$ and $|\dot{x}(t_0)|$ are limits of respective difference quotients, we know that if $|t-t_0|$ is sufficiently small—say below $1/n_2$, then

$$\frac{s(t)-s(t_0)}{t-t_0} > \frac{|x(t)-x(t_0)|}{|t-t_0|} + 1/n_1.$$

With $n \equiv \max(n_1, n_2)$, we verify that $t_0 \in B_n$; consequently, $B \subset \cup B_n$.

Let n now be a fixed integer. The set B_n may be empty and hence of measure zero. If $B_n \neq \emptyset$ and $\epsilon > 0$, let π be any partition of $[a,b]$ of norm below $1/n$ and also so small that

$$(9.10) \qquad \mathscr{L}(x) - \sum |x(\tau_i) - x(\tau_{i-1})| < \epsilon .$$

Since the length $\mathscr{L}(x) = s(b) = \sum[s(\tau_i) - x(\tau_{i-1})]$, inequality (9.10) is equivalent to the relation

$$(9.11) \qquad \sum [s(\tau_i) - s(\tau_{i-1})] - \sum |x(\tau_i) - x(\tau_{i-1})| < \epsilon .$$

Since B_n is not empty, $[\tau_{i+1}, \tau_i] \cap B_n \neq \emptyset$ for at least one value of i. Let T be a point of such a subinterval. If T is τ_{i-1} or τ_i, then by the definition of B_n and our choice of π,

$$(9.12) \qquad s(\tau_i) - s(\tau_{i-1}) > |x(\tau_i) - x(\tau_{i-1})| + (\tau_i - \tau_{i+1})/n.$$

If T is an interior point of $[\tau_{i-1}, \tau_i]$, then

$$\tau_i - T < 1/n \quad \text{and} \quad T - \tau_{i-1} < 1/n,$$

whence

$$s(\tau_i) - s(T) > |x(\tau_i) - x(T)| + (\tau_i - T)/n$$

and

$$s(T) - s(\tau_{i-1}) > |x(T) - x(\tau_{i-1})| + (T - \tau_{i-1})/n.$$

By addition of these two inequalities and the triangle property of absolute values we again get (9.12).

Let Σ' denote summation over those i such that $[\tau_{i-1}, \tau_i] \cap B_n \neq \emptyset$. By (9.12) with reference to the definition of Lebesgue outer measure we verify that

$$\mu^*(B_n) \leqslant \sum {}' (\tau_i - \tau_{i-1}) < n \sum {}' [s(\tau_i) - s(\tau_{i-1}) - |x(\tau_i) - x(\tau_{i-1})|],$$

hence by (9.11) that $\mu^*(B_n) < n\epsilon$. But n is fixed and ϵ is arbitrary; consequently, $\mu^*(B_n) = 0$ and, by Theorem 8.3, B_n is measurable of measure zero. Finally, $\mu(\cup B_n) \leqslant \Sigma \mu(B_n) = 0$.

Recall definitions (8.51) and (8.54) of total variation and absolute continuity.

Theorem 9.5

A function ϕ: $[a,b] \to R$ is AC on $[a,b]$ iff the total variation $T(\phi;[a,t])$ on the subinterval $[a,t]$ is AC on $[a,b]$.

PROOF

Suppose first that ϕ is AC on $[a,b]$. Then, given $\epsilon > 0$ and non-overlapping subintervals $[a_i,b_i]$ of $[a,b]$ with length-sum below the δ_ϵ of definition (8.54) of absolute continuity, we know that

$$(9.13) \qquad \sum |\phi(b_i) - \phi(a_i)| < \epsilon .$$

If the intervals $[a_i,b_i]$ are subdivided, the new sum (9.13) remains below ϵ. It follows that $\sum T(\phi;[a_i,b_i]) \le \epsilon$, hence as a result of Theorem 8.22 that

$$(9.14) \qquad \sum \{T(\phi;[a,b_i]) - T(\phi;[a,a_i])\} \le \epsilon < 2\epsilon$$

provided that $\sum |b_i - a_i| < \delta_\epsilon$.

If, conversely, $T(\phi;[a,t])$ is AC on $[a,b]$, this means that there is a positive δ_ϵ^* such that the left member of (9.14) is below ϵ if $\sum |b_i - a_i| < \delta_\epsilon^*$. The left member of (9.14) dominates the left member of (9.13); hence (9.13) holds and the proof is complete.

A vector-valued function x is said to have a given property—for example, absolute continuity, bounded variation, etc.—if each component of x has the stated property.

Theorem 9.6

Given a rectifiable Fréchet curve C in E_n, a representation x: $[a,b] \to R^n$ of C and the function s defined by (9.8), then s is AC on $[a,b]$ if and only if x is AC on $[a,b]$.

PROOF

Let π be a partition of the subinterval $[a,t]$ of $[a,b]$; let x^j be any component of x and consider the inequalities

$$(9.15) \qquad \sum_i |x^j(t_i) - x^j(t_{i-1})| \le \sum_i |x(t_i) - x(t_{i-1})| \le \sum_j \sum_i |x^j(t_i) - x^j(t_{i-1})|.$$

If the norm, $\|\pi\| \equiv \min|t_i - t_{i-1}|$, tends to zero, the respective members of (9.15) have limits satisfying the relations

$$T(x^j;[a,t]) \le s(t) \le \sum_j T(x^j;[a,t]).$$

By the same procedure applied to an arbitrary subinterval $[a_i,b_i]$ of $[a,b]$,

$$(9.17) \qquad T(x^j;[a_i,b_i]) \le s(b_i) - s(a_i) \le \sum_j T(x^j;[a_i,b_i]).$$

The first inequality (9.17) with Theorem 9.5 and the fact that

$$(9.18) \qquad T(x^j;[a_i,b_i]) = T(x^j;[a,b_i]) - T(x^j;[a,a_i])$$

shows that, if s is AC on $[a,b]$, then so also is $x^j, j = 1, \ldots, n$, and hence x is AC on $[a,b]$.

If, conversely, x is AC on $[a,b]$, this means that each component x^j is AC; whence, by Theorem 9.5, the total variation of each x^j is AC on $[a,b]$. It follows from the second inequality (9.17) with the aid of (9.18) that s is necessarily AC on $[a,b]$.

Theorem 9.7   (Fundamental Theorem on the Lebesgue Length-Integral)

  If $x: [a,b] \to R^n$ represents a Fréchet curve C of finite length, then

$$\text{(i)} \quad I(|\dot{x}|;[a,t]) \leq L(C_{x,t}), \qquad \forall t \in [a,b]$$

and

$$\text{(ii)} \quad \textit{equality holds in (i) iff } x \textit{ is AC on } [a,b].$$

PROOF

Since s is monotone, its derivative $\dot{s}$ is integrable over $[a,t]$ by Theorem 8.35 and problem 5, Exercise 8.4; hence, by Theorems 9.4 and 8.35, $|\dot{x}|$ is integrable and

$$(9.19) \qquad I(\dot{s};[a,t]) = I(|\dot{x}|;[a,t]) \leq s(t) - s(a) = L(C_{x,t}).$$

To establish conclusion (ii) suppose first that x is AC. Then s is AC by Theorem 9.6 and equality holds in (9.19) by the Fundamental Theorem 8.38 of the Integral Calculus. Given conversely that such equality holds for all $t \in [a,b]$, then $s(t) = I(\dot{s};[a,t])$ and s is AC on $[a,b]$ by Theorem 8.38. Then Theorem 9.6 ensures that x is AC on $[a,b]$.

In contrast with the Weierstrass integral, the Lebesgue length-integral usually does not give the length of the curve unless we provide an AC representation. Every rectifiable curve C has such representations, of which the special representations in terms of length and reduced length are two. We mention the representation in terms of so-called μ-length of Marston Morse (38d) together with the ν-length (50b, Sec. 2) of Edward Silverman. It is not known, insofar as the author is aware, for which rectifiable curves C these representations are AC. Without such information one must avoid using these representations in Lebesgue integrals.

Exercise 9.2

1. Given $x:[0,1] \to R^2$ with $x^j = \omega$, the Cantor–Lebesgue function, $j = 1,2$ (see problem 5, Exercise 8.6), discover a simple AC function

y that is Fréchet-equivalent to x. Given $F(x,y,\dot{x},\dot{y}) \equiv xy\dot{x} + y\dot{y}$ in traditional notation, calculate the Lebesgue integral, $\int F$, for each of the representations mentioned above.

2. Show with the aid of Theorem 9.3 and ideas from the proof of Theorem 6.2 that if $x: [a,b] \rightarrow R^n$ and $y: [c,d] \rightarrow R^n$ are AC representations of the same curve C and F is a continuous parametric integrand, then the Lebesgue integrals $I(F(x,\dot{x});[a,b])$ and $I(F(y,\dot{y});[c,d])$ are equal.

3. Prove as a corollary to Theorem 9.4 that the ratio $|x(t+h) - x(t)|/[s(t+h) - s(t)]$ of the length of a chord to that of the corresponding subcurve has the limit unity as $h \rightarrow 0$ for almost all t.

4. Let $x: [0,1] \rightarrow R$ be the function consisting of the origin $(0,0)$ and all points $[t,x(t)]$ on a sequence of semi-circles having as diameters the intervals $[(\frac{1}{2})^n,(\frac{1}{2})^{n-1}]$, $n = 1,2,\ldots$. Investigate the behavior of $|x(h) - x(0)|/[s(h) - s(0)]$ as $h \rightarrow 0$.

9.4 CONVERGENCE IN THE MEAN AND IN LENGTH

A sequence $x_\nu: [a,b] \rightarrow R^n$, $\nu = 1, 2, \ldots$, is said to *converge in the mean-p* to $x_0: [a,b] \rightarrow R^n$ if

$$\int_a^b |x_\nu - x_0|^p \, dt \rightarrow 0 \quad \text{as} \quad \nu \rightarrow \infty.$$

We are concerned at present only with the case $p = 1$.

A sequence C_ν, $\nu = 1, 2, \ldots$, of Fréchet curves is said to converge in length to a curve C_0 if both the Fréchet distance $d(C_\nu,C_0)$ and the difference $L(C_\nu) - L(C_0)$ converge to zero. If we define

(9.20) $$l(C_1,C_2) \equiv d(C_1,C_2) + |L(C_1) - L(C_2)|,$$

it is easy to verify that l has properties (1.29) required of a distance, hence that convergence in length is equivalent to convergence in the metric l. See Ayer and Radò (2a,b) and McShane (33f) for further information and references.

Let $\xi: [0,1] \rightarrow R^n$ be the reduced-length representation introduced in Theorem 7.8.

Theorem 9.8

Given a sequence C_ν, $\nu = 1, 2, \ldots$, of Fréchet curves and a Fréchet curve C_0, all of finite length and such that $d(C_\nu,C_0) \rightarrow 0$ as $\nu \rightarrow \infty$, then C_ν converges in length to C_0 if and only if the sequence $\dot{\xi}_\nu$ converges in the mean to $\dot{\xi}_0$.

PROOF

The reduced-length representation ξ_ν of C_ν is AC on $[0,1]$ as a consequence of Theorems 7.9 and 8.25; therefore, by Theorem 9.7, $L(C_\nu) = I(|\dot{\xi}_\nu|;[0,1])$, $\nu = 0, 1, 2, \ldots$. It follows with the aid of elementary inequalities that

$$|L(C_\nu) - L(C_0)| \leq \int_0^1 \left||\dot{\xi}_\nu| - |\dot{\xi}_0|\right| dt \leq \int_0^1 |\dot{\xi}_\nu - \dot{\xi}_0| \, dt$$

and hence that convergence of the third member to zero implies convergence in length.

To prove the converse, that convergence of the first member to zero implies similar convergence of the third, requires a longer and more delicate argument. Proofs are to be found in McShane (33f, pp. 51–54) Radò (XXXIV, p. 247), and Tonelli (XXXV, Vol. 1, p. 186). It can be done with the aid of Theorems 9.17 and 9.19.

9.5   INTEGRABILITY OF PARAMETRIC AND NONPARAMETRIC INTEGRANDS; WEIERSTRASS INTEGRALS

Let C be a continuous Fréchet curve of positive finite length and let $X: [0,L(C)] \to R^n$ be its representation in terms of length discussed in Section 7.6. If $F: A \times B \to R$ is a parametric integrand that is lower semi-continuous on its domain and bounded when r is bounded, then, since $|\dot{X}(s)| = 1$ a.e., the composite function $F(X,\dot{X})$ is bounded on a subset of $[0,L(C)]$ of measure $L(C)$. The function $F(X,\dot{X})$ is measurable as a consequence of Theorems 9.2 and 8.10 and hence is Lebesgue integrable over $[0,L(C)]$ by Theorem 8.14 and problem 5, Exercise 8.3.

It then follows from the homogeneity of F and Theorems 9.3 and 9.6 with s in the latter playing the role of h in the former that, if $x:[a,b] \to R^n$ is any other AC representation of C, then $F(x,\dot{x})$ is integrable over $[a,b]$, although not in general bounded, and that the Lebesgue integrals $\int F(X,\dot{X}) \, ds$ and $\int F(x,\dot{x}) \, dt$ over their respective intervals are equal. This generalizes Theorem 6.2.

Given a piecewise linear representation $x: [a,b] \to R^n$, satisfying conditions (7.32) and that F is continuous in (x,r) and homogeneous in r, one can verify directly from the definitions or by Theorem 7.13 that the integrals

$$\int_a^b F(x,\dot{x}) \, dt \,\text{(Riemann)} \qquad and \qquad \mathscr{W}(x;F;[a,b]) \quad \text{(Weierstrass)}$$

both exist and are equal.

If x is merely AC on $[a,b]$, $F(x,\dot{x})$ may not be Riemann integrable, but, as remarked above, this function is Lebesgue integrable. Let

$$x_\nu: [a,b] \to R^n, \nu = 1, 2, \ldots$$

be a sequence of piecewise linear functions converging in length to x. The graph of x_ν can in particular be a suitable polygonal line inscribed in the graph of x. A theorem of Aronszajn [see Pauc (43a, p. 51) or Ewing (12c, p. 684)] ensures that

$$(9.21) \qquad \mathscr{W}(x_\nu;F;[a,b]) \to \mathscr{W}(x;F;[a,b]) \quad as\ \nu \to \infty.$$

A similar result can be obtained for Lebesgue integrals. It is proved in (38g, pp. 348–349) under the additional hypothesis that $F(x,r)$ be convex in r, but this hypothesis can be eliminated by a device of Tonelli used in (12c, p. 684). We conclude from these results that if $x: [a,b] \to R^n$ is AC, then the integrals

$$(9.22) \qquad \int_a^b F(x,\dot{x})\, dt \quad \text{(Lebesgue)} \qquad and \qquad \mathscr{W}(x;F;[a,b]) \quad \text{(Weierstrass)}$$

are equal. Whenever $F(x,\dot{x})$ happens to be Riemann integrable, we see from Theorem 8.15 that the first of these integrals also can be understood in the sense of Riemann's definition.

Statement (9.21) remains meaningful and valid if x is merely BV and the cited proofs are for this case. The Lebesgue integral is a bit snobbish. Unless we provide it with an AC function x it usually gives us an irrelevant value, as illustrated by the case of the length integral in Theorem 9.7. This fact is, however, seldom a handicap. We are generally able to choose AC representations, either that in terms of length s or another obtainable from this one by substituting $s = h(t)$, where h is an AC sense-preserving homeomorphism.

If C is restricted further to have at least one representation (x,y): $[a,b] \to R^{m+1}$ such that x is strictly increasing, the various integrals, whichever of them may apply, are now nonparametric in the sense discussed in Section 6.13.

We have commented in Section 7.12 on the difficulty of including unbounded integrands under the theory of the Weierstrass integral. With F as the nonnegative function defined by (7.61), let F_ν denote its truncation at the level ν; that is,

$$F_\nu(x,y,p,q) \equiv \begin{cases} F(x,y,p,q) & \textit{if}\, F(x,y,p,q) \leqslant \nu, \\ \\ \nu & \textit{if}\, F(x,y,p,q) > \nu\,. \end{cases}$$

Since F_ν is nondecreaseing in ν, we have, by the Monotone Convergence Theorem 8.16, that

$$\int_a^b F(x,\dot x)\, dt = \lim \int_a^b F_\nu(x,\dot x)\, dt.$$

If the nonparametric integrand f and its associated parametric integrand F were not required to be nonnegative, we could write f as $f^+ - f^-$, apply the limit on ν to the corresponding F^+ and F^- separately, and then combine the results provided at least one is finite. Granted the Lebesgue integral, these and other moves encountered in transferring back and forth between a nonparametric functional and the corresponding curve-function J become routine in contrast with analogous or substitute moves restricted to the spirit of Chapter 7.

Frequently one wishes to work directly with a nonparametric formulation. We have remarked following Theorem 9.2 that if f is semicontinuous and $y: [a,b] \to R^m$ is BV, then the composite integrand $f(x,y,\dot y)$ is measurable on $[a,b]$. Under various further restrictions ($f > 0$, f bounded, etc.), $f(x,y,\dot y)$ is integrable over $[a,b]$, but again the integral has a generally irrelevant value unless we restrict y to be AC.

Exercise 9.3

1. Given the sequence of curves C_ν in problem 5, Exercise 7.3, point out that this sequence does not converge in length to the curve C_0. Let $\xi_\nu: [0,1] \to R^2$ be the reduced-length representation of C_ν, $\nu = 0,\ 1,\ 2,\ \ldots$ Establish directly from examination of the integral, $\int_0^1 |\dot\xi_\nu - \dot\xi_0|\, dt$, that this integral does not converge to zero.

2. Given that $J(C_\nu) = \int_0^1 F(\xi_\nu, \dot\xi_\nu)\, dt$, where ξ_ν is again the reduced-length representation of C_ν, $\nu = 0,1,2,\ldots$ consider the relation

$$J(C_\nu) - J(C_0) = \int_0^1 [F(\xi_\nu,\dot\xi_\nu) - F(\xi_0,\dot\xi_\nu)]\, dt + \int_0^1 [F(\xi_0,\dot\xi_\nu) - F(\xi_0,\dot\xi_0)]\, dt.$$

   Identify hypotheses on F under which the last integrand is dominated in absolute value by an expression of the form $k|\dot\xi_\nu - \dot\xi_0|$ and, granted this, prove that, if C_ν converges in length to C_0, than $J(C_\nu) \to J(C_0)$.

3. Given the nonparametric problem treated in our Chapters 2 and 3, suppose that it has been shown that $J(y_0) \le J(y)$ for all PWS functions y with the given end values. Let z be an AC function with these end values, let $\{p_\nu: \nu \in N\}$ be a sequence of piecewise linear functions converging in length to z, and interpret all integrals as Lebesgue. What conclusion is obtainable on the minimizing character of y_0?

4. If $W(C;F)$ in the statement of Theorem 7.16 is replaced by the corresponding Lebesgue integral, explain why the resulting statement is or is not valid, whichever is correct.

5. There is an extension [Reid (45e, p. 165)] of the du Bois Reymond Lemma which says that if m: $[t_0,t_1] \to R$ is a fixed measurable function on $[t_0,t_1]$ and if the integral $\int m\dot{\eta}\,dt$ taken over $[t_0,t_1]$ vanishes for every η: $[t_0,t_1] \to R$ that is lipschitzian on $[t_0,t_1]$ and vanishes at the endpoints, then $m(t)$ is constant on $[t_0,t_1]$ except possibly for a subset Z of measure zero of that interval, at points of which $m(t)$ remains undefined. With the proof of Theorem 2.2 as a guide and with the aid of Theorem 1.6, prove that, if y_0 minimizes $J(y)$ on the class $\mathscr{Y}$ of all AC functions y with fixed end values and if the integrand f has suitable properties, then equation (2.18) holds a.e. on $[t_0,t_1]$.

6. Given the integrand $f(t,y,r) = r^2$ for a nonparametric problem in the plane and the Euler necessary condition stated with the preceding problem, point out why the derivative $\dot{y}_0(t)$ of a function satisfying this condition is necessarily continuous a.e. on $[t_0,t_1]$. Extend this conclusion to a class of integrands f.

9.6 NORMED LINEAR SPACES

Although euclidean spaces E_n (defined in Section 1.10) play a major role in pure mathematical analysis and its applications, various other spaces are also important. We have made use of several metric spaces in this chapter and elsewhere. We turn now to a brief treatment of general normed linear spaces and then to special cases that appear frequently in modern variational theory.

In this section real numbers are denoted by lowercase Greek letters except that we use 0 and 1 with the customary meanings. Addition and multiplication of reals are respectively denoted by the symbol $\oplus$ and by juxtaposition.

Consider a nonempty set S with abstract elements denoted by Roman letters x, y, z, etc., except for a "zero element" θ. We suppose given a *binary operation*, alternatively stated a function $+$: $S \times S \to S$, called *addition*, together with a function from $R \times S$ to S, called *multiplication by a scalar*, that is, by a real number, a value of which is denoted by juxtaposition.

The set S with the structure implied by the following postulates is called a *real linear space* or a *real vector space*.

(i) $x + y = y + x$.
(ii) $x + (y + z) = (x + y) + z$.
(iii) $\exists$ *a unique* $\theta \in S$ *such that* $\theta + x = x, \forall x \in S$.
(iv) *Corresponding to each* $x \in S$, $\exists$ *a unique* $\bar{x} \in S$ *such that* $x + \bar{x} = \theta$.
(v) $1x = x$, $\quad \forall x \in S$ *and* $0x = \theta, \forall x \in S$.

(vi) $\alpha(\beta x) = (\alpha\beta)x, \quad \forall \alpha,\beta \in R, \quad \forall x \in S.$
(vii) $\alpha(x+y) = \alpha x + \alpha y, \quad \forall \alpha \in R, \quad \forall x,y \in S.$
(viii) $(\alpha \oplus \beta)x = \alpha x + \beta x, \quad \forall \alpha,\beta \in R, \quad \forall x \in S.$

One defines $-x \equiv (-1)x$ and proves that $-x = \bar{x}$. It can be proved from (iii) and certain of the other postulates that $x + \theta = x$.

A familiar example of a real linear space is the set $S = R^n$ of elements $x = (x^1, \ldots, x^n)$ with $x + y \equiv (x^1+y^1, \ldots, x^n+y^n)$, $\theta \equiv (0,0, \ldots, 0)$, $\bar{x} \equiv (-x^1, \ldots, -x^n)$ and $\alpha x \equiv (\alpha x^1, \ldots, \alpha x^n)$.

A function $\|\cdot\|\colon S \to R$ subject to the additional postulates that follow is called a *norm*.

(ix) $0 \leqslant \|x\| < \infty, \quad \forall x \in S.$
(x) $\|x\| = 0$ iff $x = \theta.$
(xi) $\|\alpha x\| = |\alpha|\,\|x\|, \quad \forall \alpha \in R, \quad \forall x \in S.$
(xii) $\|x+y\| \leqslant \|x\| + \|y\|, \quad \forall x,y \in S.$

We have already used the euclidean norm (1.30) with the linear space R^n to constitute the particular normed linear space E_n, the euclidean n-space.

As a consequence of (xii),

(9.23)
$$\left| \|x\| - \|y\| \right| \leqslant \|x+y\|,$$

in which $|\cdot|$ denotes ordinary absolute value of a real number. This is the norm of E_1.

To prove (9.23) use the inequalities

$$\|x\| = \|(x+y) - y\| \leqslant \|x+y\| + \|y\|$$

and

$$\|y\| = \|(x+y) - x\| \leqslant \|x+y\| + \|x\|.$$

Every normed linear space is a metric space in the sense that if we define

(9.24)
$$d(x,y) \equiv \|x - y\|,$$

then (S,d) is a metric space. A metric space is not in general either normed or linear.

Let $\{x_n\colon n \in N\}$ be a sequence in a normed linear space. The definitions of a limit x_0 of a sequence and of a Cauchy sequence in terms of the norm are, respectively, as follows.

(9.25) $\forall \epsilon > 0, \exists N_\epsilon$ such that $n > N_\epsilon \Rightarrow \|x_n - x_0\| < \epsilon,$

(9.26) $\forall \epsilon > 0, \exists N_\epsilon$ such that $m,n > N_\epsilon \Rightarrow \|x_m - x_n\| < \epsilon.$

A normed linear space is said to be *norm-complete* or simply *complete* if every Cauchy sequence (9.26) has a limit $x_0 \in S$. A complete normed linear space is called a *Banach space*.

A function $\cdot : S \times S \to R$ subject to the following postulates is called an *inner product* or a *scalar product*.

(xiii) $0 \leqslant x{\cdot}x < \infty, \quad \forall x \in S.$

(xiv) $x{\cdot}x = 0$ iff $x = \theta.$

(xv) $x{\cdot}y = y{\cdot}x, \quad \forall x,y \in S.$

(xvi) $x{\cdot}(y+z) = x{\cdot}y + x{\cdot}z, \quad \forall x,y,z \in S.$

(xvii) $(\alpha x){\cdot}y = \alpha(x{\cdot}y), \quad \forall \alpha \in R, \quad \forall x,y \in S.$

In the event that S is a real linear space with an inner product, we can define $\|x\| \equiv (x{\cdot}x)^{1/2}$ and verify that this norm has properties (ix) through (xii).

Theorem 9.9   (Cauchy–Buniakovski–Schwarz)

If x and y are elements of a real normed linear space with an inner product, then

(9.27) $$|x{\cdot}y| \leqslant \|x\| \, \|y\|.$$

PROOF

If $x = \theta$ or $y = \theta$, clearly (9.27) holds with the equality. Suppose next that neither x nor y is θ . Then

$$0 \leqslant (\alpha x - \beta y){\cdot}(\alpha x - \beta y) = \alpha^2 x{\cdot}x - 2\alpha\beta x{\cdot}y + \beta^2 y{\cdot}y.$$

Take $\beta = \|x\|$ and $\alpha = \|y\|$. It follows that

$$0 \leqslant 2\|x\|^2\|y\|^2 - 2\|x\| \, \|y\|x{\cdot}y = 2\|x\| \, \|y\|(\|x\| \, \|y\| - x{\cdot}y)$$

and hence that $x{\cdot}y \leqslant \|x\|\|y\|$. With $-x$ in place of x we find similarly that $-x{\cdot}y \leqslant \|x\| \, \|y\|$. These last two inequalities are equivalent to (9.27). We shall refer to (9.27) as the CBS inequality.

9.7   THE L_p-SPACES

Return for a moment to the setting of Chapter 8. Suppose given a universe X, a measure μ, and a measurable subset E of X. Let p be a positive real number and denote by $S_p(E)$, simply S_p, the set of all functions $x : E \to R^*$ each of which is measurable on E and such that $|x|^p$ is integrable over E.

Define a function $\|\cdot\|_p \colon S_p \to R$ called a *pseudo-norm* or *seminorm*;

$$(9.28) \qquad \|x\|_p \equiv \left(\int_E |x|^p \right)^{1/p}.$$

Theorem 9.10 (Hölder–Schwarz)

Given $x \in S_p(E)$ and $y \in S_q(E)$, where

$$p > 1, q > 1, \text{ and } \frac{1}{p} + \frac{1}{q} = 1,$$

then

    (i) *xy is integrable over E* and   (ii) *$\int_E |xy| \leq \|x\|_p \|y\|_q$.*

PROOF OF (i)

Given $t \in E$ such that $|y(t)| \leq |x(t)|^{p-1}$, then

$$(9.29) \qquad 0 \leq |x(t)y(t)| \leq |x(t)|^p,$$

while if $|y(t)| > |x(t)|^{p-1}$, then $|x(t)| < |y(t)|^{1/(p-1)} = |y(t)|^{q-1}$, whence

$$(9.30) \qquad 0 \leq |x(t)y(t)| < |y(t)|^q.$$

Since $x \in S_p$ and $y \in S_q$, x and y are understood to be measurable on E. Set

$$E_1 \equiv \{t \in E \colon |y(t)| \leq |x(t)|^{p-1}\}.$$

The set E_1 is then measurable by Theorem 8.11(v), Theorem 8.12(vi), and the proof used for Theorem 8.11(iii). Since xy is measurable on E, hence on E_1 by Theorem 8.10, and $|xy|$ is dominated on E_1 by an integrable function in (9.29), it follows by problem 7, Exercise 8.4, that xy is integrable over E_1. Similarly, xy is integrable over

$$E_2 \equiv \{t \in E \colon |x(t)| < |y(t)|^{q-1}\}.$$

By problem 8, Exercise 8.4, xy is integrable over E.

PROOF OF (ii)

If $x(t)y(t) = 0$ a.e. on E, the left member of conclusion (ii) is 0 and the inequality holds. Consider the contrary case in which $|x(t)y(t)| > 0$ on a subset of E of positive measure. Then E has positive measure as do the subsets of E on which $|x(t)| > 0$ and on which $|y(t)| > 0$. One verifies easily that

$$(9.31)$$

if $a > 0, b > 0, \alpha > 0, \beta > 0,$ and $\alpha + \beta = 1$, then $a^\alpha b^\beta \leq \alpha a + \beta b$

and that the last inequality is equivalent to the relation

$$(9.32) \qquad \alpha \ln a + \beta \ln b \leq \ln (\alpha a + \beta b).$$

Since $\alpha + \beta = 1$, this is clearly true if $a = b$. If $a < b$ and we grant that the logarithmic function is concave (that the function $-\ln$ is convex), then we have (9.32). This inequality under the stated restrictions (9.31) is simply a statement of the definition of concavity.

Next set

$$\alpha = 1/p, \quad \beta = 1/q, \quad a = |x|^p \Big/ \int_E |x|^p, \quad b = |y|^q \Big/ \int_E |y|^q.$$

From (9.31),

$$\frac{|xy|}{\left(\int_E |x|^p\right)^{1/p}\left(\int_E |y|^q\right)^{1/q}} \leq \frac{|x|^p}{p \int_E |x|^p} + \frac{|y|^q}{q \int_E |y|^q}.$$

After integrating each side over E, we have conclusion (ii).

Look ahead to definition (9.33). Since $|\int xy| \leq \int |xy|$, we see that conclusion (ii) of Theorem 9.10 is a sharper inequality for $S_2(E)$ than the CBS inequality (9.27).

Theorem 9.11 (Minkowski)

Given $x, y \in S_p(E)$ with $p \geq 1$, then

$$\text{(i)} \quad x + y \in S_p(E)$$

and

$$\text{(ii)} \quad \|x + y\|_p \leq \|x\|_p + \|y\|_p.$$

PROOF

If $p = 1$, both conclusions are obtained with the aid of problems 6(i) and 7, Exercise 8.4, and the triangle property $|x+y| \leq |x| + |y|$. If $x + y = 0$ a.e., then (i) and (ii) are clearly true.

Consider the case in which $p > 1$ and $|x+y| > 0$ on a set of positive measure. Define q by the equation $1/p + 1/q = 1$ and observe that then $(p-1)q = p$, hence that

$$|x+y|^{p-1} \in S_q(E).$$

We see that

$$\int_E |x+y|^p = \int_E |x+y||x+y|^{p-1} \leq \int_E |x||x+y|^{p-1} + \int_E |y||x+y|^{p-1}.$$

Apply the Hölder–Schwarz inequality to each integral on the right and then divide through by the positive quantity

$$\left(\int_E |x+y|^p \right)^{1/q}.$$

We see with reference to postulates (ix) through (xii) for a norm that the pseudo-norm $\|x\|_p$ has properties (ix) and (xi). It has property (xii) by Minkowski's inequality but fails to have property (x) because $\|x\|_p = 0$ if $x(t) = 0$ a.e. This suggests that we define $x, y \in S_p(E)$ to be *equivalent* if $x = y$ a.e. and set $\{x\} \equiv \{y \in S_p(E) : y(t) = x(t)$ a.e. *in* $E\}$, $\theta = \{y \in S_p(E) : y(t) = 0$ a.e. *in* $E\}$. We can then define a norm for the equivalence class $\{x\}$, namely

$$\|\{x\}\|_p \equiv \|y\|_p, \qquad y \text{ an arbitrary element of } \{x\},$$

and verify that this norm has all the properties (ix) through (xi). The set of all such equivalence classes so normed is a normed linear space called $L_p(E)$.

The notation used in introducing $L_p(E)$ is cumbersome. It is the common practice to use an arbitrary representative x of a class $\{x\}$ and to write $\|x\|_p$ for the norm. This causes no trouble if we simply remember that sets of measure zero are of no consequence and that any x can be replaced by any other element y of the class $\{x\}$.

Observe from Theorem 9.10(i) for the case $p = q = 2$ that, if x and y are in $S_2(E)$, so also is the product xy. In this case we can define

$$(9.33) \qquad x \cdot y \equiv \int_E xy$$

and verify that all postulates (xiii) through (xvii) for an inner product are satisfied except that $x \cdot x = 0$ does not imply that $x(t) \equiv 0$ but only that $x(t) = 0$ almost everywhere. However, if we reinterpret the left member of (9.33) in accord with the preceding paragraph as an abbreviation for $\{x\} \cdot \{y\}$, then $\{x\} \cdot \{x\} = 0$ if and only if $\{x\}$ is the θ defined above.

Exercise 9.4

1. Point out that each of the following classes of functions with suitable definitions of $+$, multiplication by a scalar, the element θ, and the additive inverse will constitute a real linear space:
   (i) the class $\mathcal{Y}$ of all AC functions $y : [a,b] \to R^n$ with the common domain $[a,b]$,
   (ii) the class $\mathcal{Y}_1$ of all PWS functions $y \in \mathcal{Y}$.
2. Given the class $\mathcal{Y}_1$ of problem 1, show that $\sup \{|y(t)| : t \in [a,b]\}$ is a norm.

3. Show that $\sup \{|y(t)|: t \in [a,b]\} + \sup \{|\dot{y}(t)|: t \in [a,b]^*\}$, where $[a,b]^*$ denotes the subset of $[a,b]$ on which $\dot{y}(t)$ exists, is also a norm on $\mathcal{Y}_1$.

9.8 SEPARABILITY OF THE SPACE $L_p([a,b])$

Any space Y in which limits of sequences have been defined is called *separable* if there exists a countable subset X of Y that is *dense in Y*. This means that $X = \{x_\nu \in Y : \nu \in N\}$ and that every $y \in Y$ is the limit of some subsequence of the sequentialized subset X. Thus separability abstracts a familiar property of the real numbers R. The set of rational reals or any of infinitely many other countable subsets of R will serve as the set X in the definition. The spaces R^n, $n > 1$, are similarly seen to be separable.

There is a classic approximation theorem of Weierstrass to the effect that every function $y: [a,b] \to R$ that is continuous on $[a,b]$ can be uniformly approximated by a polynomial, that is, given y and $\epsilon > 0$, there exists a polynomial p, of possibly high degree when ϵ is small, such that

$$|p(t) - y(t)| < \epsilon/2, \qquad \forall t \in [a,b].$$

Each of the real coefficients of p can be approximated as closely as desired by a rational real; consequently, there is a polynomial q with rational coefficients such that

$$|q(t) - p(t)| < \epsilon/2, \qquad \forall t \in [a,b].$$

The class of all such polynomials q is countable. It follows from these inequalities that

$$(9.34) \qquad |q(t) - y(t)| < \epsilon, \qquad \forall t \in [a,b].$$

The space $L_p([a,b])$ mentioned in the section heading is understood in the remainder of this chapter to be based on ordinary linear Lebesgue measure. Interval $[a,b]$ is fixed, and we shall usually suppress the symbol for it and write simply L_p. Functions x, x_ν, etc., are from $[a,b]$ to R^*.

Theorem 9.12

Given $x \in L_p$, there exists a sequence $\{x_\nu : \nu \in N\}$ of functions that are bounded and measurable on $[a,b]$ such that $\|x_\nu - x\|_p \to 0$ with $1/\nu$.

PROOF

Take x_ν: $[a,b] \to R$ to be the *two-sided truncation* of x, namely,

$$(9.35) \qquad x_\nu(t) \equiv \begin{cases} \nu & if \ x(t) > \nu, \\ x(t) & if \ |x(t)| \leq \nu, \\ -\nu & if \ x(t) < -\nu. \end{cases}$$

Clearly x_ν is bounded and $x_\nu(t) \to x(t)$. To see that x_ν is measurable, observe that x being in L_p is measurable by the definition of L_p and that $x_\nu = \sup [\inf (x,\nu), -\nu] = \inf [\sup (x,-\nu), \nu]$; consequently from either of these expressions and Theorem 8.12(i) and (ii), x_ν is measurable.

Now

$$|x_\nu - x| \leq |x_\nu| + |x| \leq 2|x|,$$

whence
$$(9.36) \qquad |x_\nu - x|^p \leq 2^p |x|^p.$$

It then follows from the integrability of $|x|^p$ and the Dominated Convergence Theorem 8.18 that $\|x_\nu - x\|_p \to 0$ as stated.

The next two approximation theorems are not only useful to the development of this section but in other places.

Theorem 9.13 (Egoroff)

If E is a Lebesgue measurable subset of the reals with $\mu(E) < \infty$ and $\{\phi_\nu: \nu \in N\}$ is a sequence of measurable functions $\phi_\nu: E \to R^*$ converging a.e. on E to a limit ϕ_0, then given $\epsilon > 0$, there exists a measurable subset E_ϵ of E with $\mu(E - E_\epsilon) < \epsilon$ such that ϕ_ν converges uniformly to ϕ_0 on E_ϵ.

PROOF

Given the positive integers k and ν, define

$$E_{k\nu} \equiv \bigcap_{i=\nu}^{\infty} \{t \in E: |\phi_i(t) - \phi_0(t)| < 1/k\}.$$

By the measurability of ϕ_i and theorems in Chapter 8, the set $E_{k\nu}$ is measurable. Define
$$L \equiv \{t \in E: \lim \phi_\nu(t) = \phi_0(t)\}.$$

Then, for each fixed k,

$$\bigcup_{\nu=1}^{\infty} E_{k\nu} \supset L.$$

With k fixed, the set $E_{k\nu}$ expands with ν, from which, with the preceding inclusion and (8.2_3),

$$\lim_\nu \mu(E_{k\nu}) = \mu\Big(\bigcup_1^{\infty} E_{k\nu}\Big) \geq \mu(L) = \mu(E).$$

It follows that $\mu(E - E_{k\nu}) \to 0$, hence that there is an integer K depending on k such that

$$\mu(E - E_{k\nu}) < \epsilon 2^{-k} \qquad if \quad \nu \geqslant K.$$

Set

$$E_\epsilon \equiv \bigcap_{k=1}^{\infty} E_{kK}.$$

E_ϵ is measurable by Theorem 8.9 and, since $E - \cap E_{kk} = \cup (E - E_{kK})$,

$$\mu(E - E_\epsilon) = \mu\left[\bigcup_{k=1}^{\infty} (E - E_{kK})\right] \leqslant \sum_{k=1}^{\infty} \mu(E - E_{kK}) < \epsilon \sum_{k=1}^{\infty} 2^{-k} = \epsilon.$$

It remains to show that $\phi_\nu \to \phi_0$ uniformly on E_ϵ.

By the definition of $E_{k\nu}$,

$$|\phi_i(t) - \phi_0(t)| < 1/k, \qquad \forall t \in E_{kK} \text{ and } \forall i \geqslant K,$$

hence, by the definition of E_ϵ, for all $t \in E_\epsilon$. This is the desired conclusion.

Theorem 9.14 (Lusin)

If $E \subset R$ with $\mu(E) < \infty$ and $\phi: E \to R^*$ is finite a.e. on E and measurable on E, then given $\epsilon > 0$, there exists a closed subset F_ϵ such that $\mu(E - F_\epsilon) < \epsilon$ and the restriction of ϕ to F_ϵ is continuous on F_ϵ.

PROOF

Suppose initially that ϕ is a simple function as defined following (8.32), hence that

$$\phi = \sum_{i=1}^{n} c_i \chi_{E_i},$$

where χ is the characteristic function defined in problem 3, Exercise 8.5.

Given $\epsilon > 0$ and the positive integer i, it follows from definitions (8.1) and (8.3) of Lebesgue outer measure μ^* and measure μ and from the fact that E_i as a measurable subset of E is of finite measure that there exists a closed subset F_i of E_i such that

$$\mu(E_i - F_i) < \epsilon/(n+1), \qquad i = 1, \ldots, n.$$

For the same reasons there exists a closed subset F_{n+1} of $E - \cup E_i$ such that

$$\mu\left[\left(E - \bigcup_1^n E_i\right) - F_{n+1}\right] < \epsilon/(n+1).$$

Define

$$F_\epsilon \equiv \bigcup_{i=1}^{n+1} F_i$$

and verify that $\mu(E-F_\epsilon) < \epsilon$. Since $\phi(t)$ is constant on each of the sets F_i, the restriction of ϕ to F_ϵ is clearly continuous on F_ϵ.

Consider next the case in which ϕ is a general nonnegative measurable function on E. Let $\{s_n : n \in N\}$ be a sequence of nonnegative simple functions tending monotonely from below to ϕ as in relation (8.34*). We know by the preceding case that, for each positive integer n, there is a closed subset F_n of E with

$$\mu(E-F_n) < \epsilon/2^{n+1}$$

and such that the restriction of s_n to F_n is continuous on F_n. Now define

$$F \equiv \bigcap_{n=1}^{\infty} F_n.$$

Then F is closed and

$$\mu(E-F) = \mu[\cup (E-F_n)] \le \sum \mu(E-F_n) < \sum \epsilon/2^{n+1} = \epsilon/2.$$

By Egoroff's Theorem with the present F in the role of E in that theorem, there is a measurable subset $E_{\epsilon/4}$ of F such that $\mu(F-E_{\epsilon/4}) < \epsilon/4$ and hence $\mu(E-E_{\epsilon/4}) < 3\epsilon/4$ such that s_n converges uniformly on $E_{\epsilon/4}$ to ϕ. The restriction of s_n to $E_{\epsilon/4}$, a set not dependent on n, is continuous on that set; hence, by the uniform convergence of s_n to ϕ on $E_{\epsilon/4}$, the restriction of ϕ to that set is continuous on that set.

Finally, there exists a closed subset F_ϵ of $E_{\epsilon/4}$ such that $\mu(E_{\epsilon/4}-F_\epsilon) < \epsilon/4$ and hence such that $\mu(E-F_\epsilon) < \epsilon$. Clearly the restriction of ϕ to F_ϵ is continuous on F_ϵ.

In the general case we can express ϕ in the form $\phi^+ - \phi^-$ and apply the preceding case to each of the nonnegative functions ϕ^+ and ϕ^-.

We now return to functions whose domains are the interval $[a,b]$.

Theorem 9.15

If $\phi: [a,b] \to R$ is bounded and measurable on $[a,b]$, there exists a sequence $\{\phi_\nu : \nu \in N\}$ of functions $\phi_\nu: [a,b] \to R$, each continuous on $[a,b]$ such that

    (i) $\phi_\nu(t) \to \phi(t)$ a.e. *on* $[a,b]$,

    (ii) $\sup |\phi_\nu(t)| \le \sup |\phi(t)|$ *on* $[a,b]$,

    (iii) $\|\phi_\nu - \phi\|_p \to 0$ *with* $1/\nu$.

PROOF

By Theorem 9.14, with the E of that theorem as the interval $[a,b]$, there exists a closed subset F_ν such that $\mu([a,b]-F_\nu) < 1/2^\nu$ and the restriction of ϕ to F_ν is continuous on F_ν. This means that if $\{t_i : i \in N\}$ is any sequence in F_ν having a limit t_0, necessarily in the closed set F_ν, then $\phi(t_i) \to \phi(t_0)$. Define ϕ_ν as follows:

$$(9.37) \quad \phi_\nu(t) \equiv \begin{cases} \phi(t) & \text{if } t = a \text{ or } b, \\ \phi(t) & \text{if } t \in F_\nu, \\ \phi(\alpha) + \dfrac{t-\alpha}{\beta-\alpha}[\phi(\beta) - \phi(\alpha)] & \text{if } t \in (\alpha,\beta), \\ & \text{any interval of the decomposition of the} \\ & \text{open set } (a,b) - F_\nu \text{ given by Theorem 8.5.} \end{cases}$$

Because of the simple form of ϕ_ν on the intervals (α,β) it requires only a careful but routine check to verify that $\phi_\nu(\tau) \to \phi_\nu(t)$ as $\tau \to t$, hence that ϕ_ν is continuous as required in the theorem. Property (ii) of ϕ_ν is immediate from definition (9.37). It remains to prove (i) and (iii).

Define

$$H_\nu \equiv \{t \in [a,b] : \phi_\nu(t) \neq \phi(t)\}.$$

It is clear from the second statement on the right in (9.37) and our choice of the set F_ν that $\mu(H_\nu) < (\frac{1}{2})^\nu$, therefore, using the subadditivity (8.2$_4$) of a measure that

$$(9.38) \qquad \mu\left(\bigcup_{\nu=m}^\infty H_\nu\right) \leq \sum_{\nu=m}^\infty \mu(H_\nu) < 2^{1-m}.$$

Define

$$(9.39) \qquad H \equiv \bigcap_{m=1}^\infty \bigcup_{\nu=m}^\infty H_\nu.$$

If $t \in [a,b] - H$, there exists an integer M depending on t such that $t \notin \bigcup_M^\infty H_\nu$, which by the definition of H_ν implies that

$$\phi_\nu(t) = \phi(t) \qquad \text{if } \nu \geq M.$$

The proof of conclusion (i) will be complete if $\mu(H) = 0$. To show this verify, from (9.38), (9.39), and the monotonicity (8.2$_3$) of a measure, that the inequality $\mu(H) \leq 2^{1-m}$ must hold for every positive integer m.

To prove the convergence property (iii), let K denote a bound for $|\phi(t)|$ on $[a,b]$. As a consequence of conclusion (ii),

$$|\phi_\nu - \phi| \leq |\phi_\nu| + |\phi| \leq 2K;$$

hence

$$|\phi_\nu - \phi|^p \leq 2^p K^p,$$

and the desired conclusion follows from the Dominated Convergence Theorem.

The next and final theorem of the section merely gathers together the conclusions provided by the Weierstrass Approximation Theorem, Theorem 9.12, and Theorem 9.15. The reader is asked to supply the details with appropriate use of $\epsilon/3$.

Theorem 9.16

If $x \in L_p$ and $\epsilon > 0$, there exists a polynomial q with rational coefficients such that $\|q - x\|_p < \epsilon$. Alternatively stated, $L_p([a,b])$ is separable and the countable set of polynomicals q on $[a,b]$ serves as one set X of the type described at the beginning of this section.

It is a further routine step to observe that a polynomial q can be uniformly approximated on an interval by means of step-functions with suitably short steps having only rational values and having discontinuities at rational values. The totality of such step-functions is another countable set X that is dense in L_p.

9.9   LINEAR FUNCTIONALS AND WEAK CONVERGENCE

Given a real linear space X, a function $f: X \to R$ is called a *linear functional* if

$$(9.40) \quad f(\alpha x + \beta y) = \alpha f(x) + \beta f(y), \quad \forall x,y \in X \text{ and } \forall \alpha, \beta \in R.$$

In the present book we are primarily interested in the case where X is a space $L_p([a,b])$.

A sequence $\{y_\nu \in L_p : \nu \in N\}$ is said to *converge weakly* to $y_0 \in L_p$ and we shall write $y_\nu \to y_0$ (wky) if

$$(9.41) \quad \int_a^b (y_\nu - y_0) \phi \to 0 \text{ with } 1/\nu, \quad \forall \phi \in L_q, 1/p + 1/q = 1.$$

Since the integral is required to converge to 0 with ϕ chosen arbitrarily in L_q, this must occur in particular if ϕ is the characteristic function of any measurable subset E of $[a,b]$, $\phi(t) = 1$ or 0 according as t is or is not an element of E. Thus if $y_\nu \to y_0$ (wky), then $\int_E (y_\nu - y_0) \to 0$ as $\nu \to \infty$ for every choice of E and, although $y_\nu(t)$ may differ widely from $y_0(t)$ on sets of measure zero, it appears that as ν increases $y_\nu(t) - y_0(t)$ must become close to zero on sets E of positive measure.

One's initial reaction to definition (9.41) may be to doubt the appropriateness of the term weak. However, weak convergence is always implied by convergence (9.25) in terms of the norm but not conversely, and convergence (9.25) is now called *strong convergence*. That strong convergence implies weak convergence follows from the Hölder inequality,

$$\left| \int (y_\nu - y_0)\phi \right| \leq \|y_\nu - y_0\|_p \|\phi\|_q.$$

To deny the converse consider the example with $p = q = 2$ and $y_\nu(t) =$ $\sin \nu t$, $\nu = 1, \ldots.$ Clearly $y_\nu \in L_2([0,\pi])$. Let ϕ be an arbitrary but fixed function in $L_2([0,\pi])$. The sequence of Fourier coefficients

$$b_\nu = (2/\pi) \int_0^\pi (\sin \nu t)\, \phi(t), \qquad \nu = 1, 2, \ldots,$$

is known to coverage to 0 as $\nu \to \infty$. It follows that $y_\nu \to \theta$ (wky) where θ is the identically zero function. However,

$$\|y_\nu - \theta\|^2 = \int_0^\pi \sin^2 \nu t = \pi/2,$$

and hence y_ν does not converge strongly to θ.

The gap between strong and weak convergence is partially filled by such theorems as the following.

Theorem 9.17. (F. Riesz)

If a sequence $\{y_\nu \in L_2(E) : \nu \in N\}$ converges weakly to $y_0 \in L_2(E)$ and if $\|y_\nu\|$ converges to $\|y_0\|$, then y_ν converges strongly to y_0.

PROOF

Consider the identity for real numbers,

$$a^2 = b^2 + 2b(a-b) + (a-b)^2.$$

If the last term is replaced by $c(a-b)^2$, $0 < c < 1$, then $=$ is replaced in the identity by $>$. Substitute y_ν and y_0 for a and b, respectively, and integrate over E to obtain the inequality

$$\int y_\nu^2 - \int y_0^2 > 2 \int y_0(y_\nu - y_0) + c \int (y_\nu - y_0)^2.$$

Under the stated hypotheses, the left member and the first term on the right both converge to zero. Since $c > 0$, it follows that $\|y_\nu - y_0\| \to 0$.

Theorem 9.18 (Banach–Saks)

If a sequence $\{y_\nu \in L_2(E) : \nu \in N\}$ converges weakly to $y_0 \in L_2(E)$, there exists a subsequence $\{z_\nu\}$ of $\{y_\nu\}$ such that the sequence $\{(1/k) \sum_1^k z_\nu : k \in N\}$ of arithmetic means converges strongly to y_0.

PROOF

That $y_\nu \to y_0$ (wky) is equivalent to the condition that $(y_\nu - y_0) \to \theta$ (wky); hence we can take the given limit y_0 to be θ with no real loss in generality.

The desired subsequence can be constructed as follows. Take $z_1 = y_1$. Let z_2 be a new label for the first y_ν beyond y_1 in the natural order $y_1, y_2, \ldots$ such that the inner product $z_1 \cdot y_\nu < 1$. That there is such a y_ν follows from the fact that $z_1 \cdot y_\nu = \int y_\nu y_1 \to 0$ with $1/\nu$. Proceeding inductively, suppose that $z_1, \ldots, z_n$ are terms appearing in that order in the given sequence $\{y_\nu\}$ with the property that

$$z_1 \cdot z_m < 1/(m-1), \ldots, z_{m-1} \cdot z_m < 1/(m-1), \qquad m = 2, \ldots, n.$$

Then select as z_{n+1} the first term y_ν in the given sequence beyond the one that has been denoted by z_n and which satisfies the inequalities

$$z_1 \cdot z_{n+1} < 1/n, \ldots, z_n \cdot z_{n+1} < 1/n.$$

We need the following lemma, a proof of which is given after the remaining steps in the proof of the theorem.

(α) *There exists $M \in R$ such that* $\|y_\nu\| \leqslant M$, $\quad \nu = 1, 2, \ldots$.

One verifies using the expansion of $(\Sigma_1^k z_\nu)^2$ that

$$\|(1/k) \sum_1^k z_\nu\|^2 \leqslant (1/k^2) [kM^2 + 2(1) + 4(1/2) + \cdots + 2(k-1)/(k-1)]$$

$$= (1/k^2) [kM^2 + 2(k-1)],$$

and the last expression clearly tends to zero with $1/k$.

The original Banach–Saks Theorem (Studia Math., Vol. 2 (1930), pp. 51–57) proves the stated conclusion for a given weakly convergent sequence in $L_p(E)$.

PROOF OF (α)

We shall prove this by showing that there is a real number M such that

(9.42) $\qquad |y_\nu \cdot x| \leqslant M\|x\|, \qquad \forall x \in L_2(E) \ and \ \forall \nu \in N.$

The hypothesis is that of Theorem 9.18, namely, that the sequence $\{y_\nu\}$ converges weakly to $y_0 \in L_2(E)$.

We know by the CBS inequality that

$$|y_\nu \cdot x| \leqslant \|y_\nu\| \, \|x\|, \qquad \forall x \in L_2(E).$$

Moreover, with y_ν fixed, the smallest M for which the inequality of statement (9.42) holds is $\|y_\nu\|$. This is clearly so if $y_\nu = \theta$. If $y_\nu \neq \theta$, if $\epsilon > 0$, and

$$|y_\nu \cdot x| \leqslant (\|y_\nu\| - \epsilon)\|x\|, \qquad \forall x \in L_2(E),$$

then this inequality must hold with $x = y_\nu$ and we find that $\|y_\nu\|^2 \leqslant \|y_\nu\|^2 - \epsilon\|y_\nu\|$, in contradiction with the choice of ϵ as positive. It follows that statement (9.42) implies statement (α).

A subset $\{x \in L_2(E) : \|x - a\| \leqslant r\}$ is called a *closed ball* in $L_2(E)$ of *radius* r and *center* $a \in L_2(E)$. Let $B(r,a)$ denote an arbitrary closed ball in $L_2(E)$. We show next that, if there is no number M for which (9.42) holds, then the set of real numbers

$$(9.43) \qquad \{|y_\nu \cdot x| \in R : \nu \in N, x \in B(r,a)\}$$

is unbounded. Proceeding contrapositively, suppose that for some real number K,

$$|y_\nu \cdot x| \leqslant K, \qquad \forall x \in B(r,a) \text{ and } \forall \nu \in N.$$

If $x \neq \theta$ but is an otherwise arbitrary element of $L_2(E)$, then

$$(r/\|x\|)x + a \in B(r,a).$$

It follows that

$$|(r/\|x\|)(y_\nu \cdot x) + (y_\nu \cdot a)| \leqslant K, \qquad \forall \nu \in N$$

and by the triangle inequality for absolute values that

$$(r/\|x\|)|y_\nu \cdot x| - |y_\nu \cdot a| \leqslant K,$$

whence by elementary algebra that

$$(9.44) \qquad |y_\nu \cdot x| \leqslant (1/r)[K + |y_\nu \cdot a|]\|x\|.$$

Under our hypothesis that y_ν converges weakly, the sequence $\{|y_\nu \cdot a|\}$ of real numbers is bounded. Therefore, the coefficient of $\|x\|$ in (9.44) is dominated by a real number M. Although $x = \theta$ was excluded in the preceding steps, (9.44) clearly holds for $x = \theta$ and hence for all x in $L_2(E)$.

The proof of (α) is completed by supposing that there is no number M for which (9.42) holds and obtaining a contradiction. Under this hypothesis every set (9.43) is unbounded.

Let $B(r_0, x_0)$ be an arbitrary closed ball of positive radius. There then exists $\nu_1 \in N$ and $x_1 \in B(r_0, x_0)$ such that $|y_{\nu_1} \cdot x_1| > 1$. It follows from the CBS inequality that, if $\|x - x_1\|$ is sufficiently small, then $|y_{\nu_1} \cdot x| > 1$.

Let r_1 be a positive real number with $r_1 \leqslant 1$ and also so small that

$$|y_{\nu_1} \cdot x| > 1, \qquad \forall x \in B(r_1, x_1).$$

The reader may find it suggestive to use a sketch in which circular discs stand for the two balls already introduced and for others with appropriate radii that follow.

Apply the same procedure to the ball $B(r_1, x_1)$, starting with the subsequence $\{y_\nu : \nu > \nu_1\}$ of $\{y_\nu\}$. There exists $\nu_2 > \nu_1$ and $x_2 \in B(r_1, x_1)$ such that $|y_{\nu_2} \cdot x_2| > 2$; consequently, by the continuity of the left member in terms of the L_2-norm, there is a ball $B(r_2, x_2)$ such that

$$|y_{\nu_2} \cdot x| > 2, \qquad \forall x \in B(r_2, x_2).$$

The positive radius r_2 is chosen so that this inequality holds, with $r_2 \leqslant \frac{1}{2}$, and also so small that $B(r_2, x_2) \subset B(r_1, x_1)$.

Continuing thus we obtain inductively a sequence of balls $\{B(r_k, x_k)\}$ each contained in its predecessor with $0 < r_k \leqslant 1/k$ together with a subsequence $\{z_k\}$ of $\{y_\nu\}$, $z_k \equiv y_{\nu_k}$, such that

(9.45) $|z_k \cdot x| > k, \qquad \forall x \in B(r_k, x_k).$

Since $B(r_{k+m}, x_{k+m}) \subset B(r_k, x_k)$, $k = 1, 2, \ldots$, it follows that

$$\|x_{k+m} - x_k\| \leqslant 1/k, \qquad k = 1, 2, \ldots \text{ and } m = 1, 2, \ldots,$$

hence that the sequence of centers of the balls is a Cauchy sequence (9.26) in $L_2(E)$. This space is known to be *complete*; that is, every Cauchy sequence (9.26) converges to some element of the space. Proofs are given in many books on real analysis or integration (for example, R. G. Bartle, *The Elements of Integration*, Wiley, New York, 1966, pp. 59–60). Therefore, there exists $\bar{x} \in L_2(E)$ such that $\|x_k - \bar{x}\| \to 0$ with $1/k$. It follows that

$$\bar{x} \in \bigcap_1^\infty B(r_k, x_k).$$

Since the given sequence $\{y_\nu\}$ converges weakly to y_0, so also does the subsequence $\{z_k\}$ satisfying (9.45); consequently, $z_k \cdot \bar{x}$ converges to the real number $y_0 \cdot \bar{x}$. But $\bar{x} \in B(r_k, x_k)$ for every k; hence we contradict (9.45) and must infer the truth of Lemma (α).

9.10 THE WEAK COMPACTNESS THEOREM

We consider next an adaptation to the space $L_2([a,b])$ of the selection process already used in the sequential compactness theorems of Sections 7.7, 7.8, and 7.9. Be reminded that, under the definition adopted in

Section 7.1, to say that a set K is sequentially compact means that every sequence in K has a subsequence converging to an element of K. Some writers would say that such a set K is sequentially compact in itself. The result we need is to be found in books on functional analysis, but to extract a complete proof one may have to refer to a succession of earlier results, including concepts related to but not essential to the present objectives. The author is indebted to M. Q. Jacobs for extracting and organizing the proof of the next theorem. The same conclusion for a general L_p, $p > 1$, was proved by F. Riesz in Math. Annalen, vol. 69 (1910), pp. 466–468, but we shall only use the case $p = 2$.

The subscript 2 on the symbol for the norm in L_2 will be omitted. The proof is slightly simplified by taking $[a,b] = [0,1]$ and hence understanding that L_2 means $L_2([0,1])$. There is really no loss in generality, since we can go from $[0,1]$ to $[a,b]$ by a linear transformation and this could be done at appropriate places in the proof. Any integral that appears without statement of the range of integration is to be understood as an integral over $[0,1]$.

Theorem 9.19

Given $\lambda > 0$, a closed ball B of radius λ in L_2 is sequentially compact in terms of weak convergence. Alternatively stated, given a sequence $\{y_\nu \in B: \nu \in N\}$, there exists a subsequence $\{z_\nu : \nu \in N\}$ of that sequence together with $z_0 \in B$ such that

$$(9.46) \qquad \lim \int z_\nu \phi = \int z_0 \phi, \qquad \forall \phi \in L_2.$$

PROOF

As a consequence of Theorem 9.16 there exists a sequence $\{r_\nu\}$ in L_2 that is dense in L_2. By the CBS inequality,

$$\left| \int y_\nu r_1 \right| \leq \|y_\nu\| \, \|r_1\|.$$

The left member is the general term in a sequence of real numbers in the compact interval $[0, \lambda \|r_1\|]$; hence there is a convergent subsequence

$$(9.47) \qquad \int y_{1\nu} r_1, \qquad \nu = 1, 2, \ldots.$$

With r_2 in place of r_1 in (9.47), we have a sequence that may not converge, but again there must exist a convergent subsequence,

$$\int y_{2\nu} r_2, \qquad \nu = 1, 2, \ldots.$$

Moreover, since $\{y_{2\nu}\}$ is a subsequence of $\{y_{1\nu}\}$, we can replace r_2 by r_1 and have a sequence with the same limit as that of (9.47).

Continuing thus we are led to a double sequence with the general term

$$(9.48) \qquad \int y_{m\nu} r_m, \qquad \nu = 1, 2, \ldots, \qquad m = 1, 2, \ldots.$$

Given $\phi \in L_2$ and $\epsilon > 0$, there exists r_m in the sequence dense in L_2 such that

$$(9.49) \qquad \|r_m - \phi\| < \epsilon/3\lambda.$$

If $\psi \in L_2$, we have by the CBS inequality that

$$(9.50) \qquad |\int \psi r_m - \int \psi \phi| = |\int \psi (r_m - \phi)| \leq \|\psi\| \, \|r_m - \phi\|.$$

Set $z_\nu \equiv y_{\nu\nu}$, the general diagonal term in the double sequence $\{y_{m\nu} \in L_2 : m, \nu \in N\}$. Since $\|y_{\nu\nu}\| \leq \lambda$ by hypothesis, it follows from (9.49) and (9.50) with $\psi = z_\nu$ that

$$(9.51) \qquad |\int z_\nu r_m - \int z_\nu \phi| < \epsilon/3.$$

With m fixed and μ and ν as positive integers, neither below m, diagonal terms z_μ and z_ν are new symbols for terms in the mth row of our double sequence $\{y_{m\nu}\}$. With m fixed in (9.48), we have a convergent simple sequence; therefore,

$$(9.52) \qquad \left|\int z_\mu r_m - \int z_\nu r_m\right| < \epsilon/3 \quad \textit{if } \mu, \nu \textit{ are sufficiently large.}$$

By the triangle inequality,

$$\left|\int z_\mu \phi - \int z_\nu \phi\right| \leq \left|\int z_\mu \phi - \int z_\mu r_m\right| + \left|\int z_\mu r_m - \int z_\nu r_m\right| + \left|\int z_\nu r_m - \int z_\nu \phi\right|.$$

It follows from (9.51) and (9.52) that $\int z_\nu \phi$, $\nu = 1, 2, \ldots$, is a Cauchy sequence of real numbers and therefore that the limit on the left in (9.46) exists. To complete the proof of the theorem we must show that there exists $z_0 \in L_2$ such that (9.46) holds.

Define a function $I : L_2 \to R$,

$$(9.53) \qquad I(\phi) \equiv \lim_{\nu \to \infty} \int z_\nu \phi.$$

One verifies that I satisfies definition (9.40) of a linear functional. It follows from (9.53), the CBS inequality, and the fact that $\|z_\nu\|$ is bounded that

$$(9.54) \qquad \textit{if } \|\phi_n - \phi\| \to 0 \textit{ with } 1/n, \textit{ then } |I(\phi_n) - I(\phi)| \to 0.$$

We need two lemmas, designated by (α) and (β), proofs of which are deferred to the end of the main proof.

(α)   *There exists $M > 0$ such that $|I(\phi)| \leqslant M \|\phi\|$, $\forall \phi \in L_2$.*

Define for each $t \in [0,1]$ a function $u(t,\cdot): [0,1] \to R$,

$$u(t,\tau) \equiv \begin{cases} 1 & \text{if } \tau \in [0,t), \\ 0 & \text{if } \tau \in [t,1], \end{cases}$$

and a function $g: [0,1] \to R$,

$$g(t) \equiv I[u(t,\cdot)] = \lim \int_0^1 z_\nu u(t,\cdot) = \lim \int_0^t z_\nu.$$

(β)   *g is AC on $[0,1]$.*

From (β), the Fundamental Theorem 8.38 of the Integral Calculus, and the fact that $g(0) = I[u(0,\cdot)] = 0$,

$$(9.55) \qquad g(t) = g(0) + \int_0^t \dot{g} = \int_0^1 u(t,\cdot)\dot{g}.$$

The remainder of the proof consists of showing that $\dot{g}$ will serve as the z_0 appearing in the theorem.

To show that $\dot{g}$ is in L_2 we first define

$$(9.56) \qquad x_n \equiv \sum_{k=1}^n c_{kn}\{u[k/n,\cdot] - u[(k-1)/n,\cdot]\}, \qquad n = 1, 2, \ldots.$$

With any choice of the real coefficients c_{kn}, the function $x_n: [0,1] \to R$ is bounded and measurable, hence in L_2. Since I is a linear functional,

$$I(x_n) = \sum_{k=1}^n (I\{u[k/n,\cdot]\} - I\{u[(k-1)/n,\cdot]\}).$$

By (9.55), (9.56), and the definition of g,

$$I(x_n) = \sum_{k=1}^n c_{kn}\left\{ \int u[k/n,\cdot]\dot{g} - \int u[(k-1)/n,\cdot]\dot{g} \right\} = \int x_n \dot{g}.$$

Now any bounded measurable function x is the almost-everywhere limit of a sequence (9.56) in view of Theorem 9.15 and the fact that a continuous function can be approximated by step-functions; hence by the Dominated Convergence Theorem

$$(9.57) \qquad \lim I(x_n) = \lim \int x_n \dot{g} = \int x\dot{g}.$$

It is easy to verify, for such an x and such a sequence $\{x_n\}$, that $\|x_n - x\| \to 0$, consequently from (9.54) that $I(x_n) \to I(x)$, which with (9.57) implies that

$$(9.58) \qquad I(x) = \int x\dot{g} \quad \text{if } x \text{ is bounded and measurable.}$$

Next define a sequence $\{\dot{g}_n : n \in N\}$ of functions bounded and measurable on the subset E of $[0,1]$ consisting of those t in $[0,1]$ at which $\dot{g}(t)$ exists,

$$
(9.59) \qquad \dot{g}_n(t) \equiv
\begin{cases}
n & \text{if } \dot{g}(t) > n, \\
\dot{g}(t) & \text{if } |\dot{g}(t)| \leqslant n, \\
-n & \text{if } \dot{g}(t) < -n, \qquad n = 1, 2, \ldots.
\end{cases}
$$

Clearly $|\dot{g}_n(t)| \leqslant n$. By Lemma ($\alpha$) and the definition of the L_2−norm,

$$
(9.60) \qquad |I(\dot{g}_n)| \leqslant M\|\dot{g}_n\| = M\left(\int \dot{g}_n^2\right)^{1/2}.
$$

On the other hand, by (9.58) with $x = \dot{g}_n$,

$$
(9.61) \qquad |I(\dot{g}_n)| \geqslant I(\dot{g}_n) = \int \dot{g}_n \dot{g}.
$$

One verifies from (9.59) that $\dot{g}_n \dot{g} \geqslant \dot{g}_n^2$ and hence that the last member of (9.61) dominates $\int \dot{g}_n^2$. It then follows from (9.60) and (9.61) that

$$
(9.62) \qquad \int \dot{g}_n^2 \leqslant |I(\dot{g}_n)| \leqslant M\left(\int \dot{g}_n^2\right)^{1/2},
$$

from which follows that

$$
(9.63) \qquad \|\dot{g}_n\| \leqslant M.
$$

We see further from definition (9.59) that $\dot{g}_n(t) \to \dot{g}(t)$ and that $|\dot{g}_n(t)| \leqslant |\dot{g}_{n+1}(t)|$, $n = 1, 2, \ldots$, consequently by the Dominated Convergence Theorem and relations (9.62) and (9.63) that

$$
\lim \int \dot{g}_n^2 = \int \dot{g}^2 \leqslant M^2
$$

and hence that $\dot{g}$ is in L_2.

Let ϕ be an arbitrary function in the class L_2. By Theorem 9.12 there is a sequence $\{\phi_n : n \in N\}$ of bounded measurable functions such that $\|\phi_n - \phi\| \to 0$ with $1/n$, whence by the CBS inequality,

$$
(9.64) \qquad |\int \dot{g}(\phi_n - \phi)| \leqslant \|\dot{g}\| \, \|\phi_n - \phi\|.
$$

Since $\|\dot{g}\|$ is a real constant, (9.64) implies that

$$
(9.65) \qquad \int \dot{g}\phi_n \to \int \dot{g}\phi \text{ as } n \to \infty.
$$

By (9.58) with its x as the present ϕ_n, we see that

$$
I(\phi_n) = \int \dot{g}\dot{\phi}_n, \qquad n = 1, 2, \ldots,
$$

while by the continuity (9.54) of I,

(9.67) $$I(\phi_n) \to I(\phi) \text{ as } n \to \infty.$$

As a consequence of (9.65), (9.66), and (9.67),

$$I(\phi) = \int \dot{g}\phi.$$

We see from definition (9.53) of $I(\phi)$ that this is the desired conclusion (9.46). We have constructed a function g whose derivative $\dot{g}$ serves as the z_0 mentioned in the theorem.

PROOF OF (α)

Proceeding contrapositively, suppose that (α) is false. Then

$$\forall M > 0, \exists \phi \in L_p \text{ such that } |I(\phi)| > M\|\phi\|;$$

hence there is a sequence $\{\phi_k\}$ in L_2 such that

(9.68) $$|I(\phi_k)| > k\|\phi_k\|, \qquad k = 1, 2, \ldots.$$

No term ϕ_k can be the zero function θ for, if so we would have 0 on each side of (9.68).

Set

$$\phi_k^* \equiv \phi_k/k\|\phi_k\|$$

and observe from property (xi) of a norm that

$$\|\phi_k^*\| = 1/k \to 0 \qquad \text{with } 1/k.$$

But by definition (9.53) of $I(\phi)$ and (9.68),

$$|I(\phi_k^*)| = |I(\phi_k)|/k\|\phi_k\| > 1,$$

and this with the preceding equation contradicts (9.54).
    We infer the truth of (α) as stated.

PROOF OF (β)

Let (a_i, b_i), $i = 1, \ldots, n$, denote an arbitrary finite set of disjoint open subintervals of $[a,b]$. Define

$$\sigma_i \equiv \text{sgn}[g(b_i) - g(a_i)],$$

where the function *sgn*, read "signum" or "sign," has the values sgn $t =$

1, 0, or -1 according as $t > 0$, $= 0$, or < 0, respectively. Then

$$(9.69) \qquad \sum_1^n |g(b_i) - g(a_i)| = \sum_1^n \sigma_i[g(b_i) - g(a_i)]$$

$$= \sum_1^n \sigma_i\{I[u(b_i,\cdot)] - I[u(a_i,\cdot)]\}$$

$$= I\Big\{\sum_1^n \sigma_i[u(b_i,\cdot) - u(a_i,\cdot)]\Big\}.$$

As a consequence of Lemma (α), the last term is dominated by

$$(9.70) \quad M\Big\|\sum_1^n \sigma_i[u(b_i,\cdot) - u(a_i,\cdot)]\Big\| = M\Big\{\int \sum_1^n \sigma_i[u(\ ) - u(\ )]^2\Big\}^{1/2}.$$

One verifies that the integrand on the right is the characteristic function of the set $\cup\ (a_i,b_i)$, namely, that its value is unity on that set and zero elsewhere. It then follows from (9.69) and (9.70) that

$$\sum_1^n |g(b_i) - g(a_i)| \leq M\Big(\sum_1^n |b_i - a_i|\Big)^{1/2},$$

from which we see with reference to definition (8.54) of absolute continuity that g is AC on $[a,b]$.

9.11   APPLICATIONS

The variety of functions $J: \mathcal{Y} \to R$ that occur as mathematical models for optimization questions is large and growing. For each type of function J and domain $\mathcal{Y}$ one wishes to know if there exists a minimizing or a maximizing element y_0 in $\mathcal{Y}$ and, if such a y_0 exists, to identify it.

EXAMPLE 9.1

Let $\mathcal{P}$ denote the class of all pairs (y,u) where $u \in L_2([0,1])$ and y is given by the equation

$$(9.71) \qquad y(t) \equiv \int_0^t u(\tau)\,d\tau, \qquad 0 \leq t \leq 1$$

subject to the terminal condition $y(1) = 1$. The initial condition $y(0) = 0$ is obtained by setting $t = 0$ in (9.71). The problem is to minimize

$$J(y,u) \equiv \int_0^1 u^2\,dt.$$

Discussion

For those u that are piecewise continuous in the sense of Section 1.8, it follows from (9.71) that $\dot{y}(t) = u(t)$ except at discontinuities of u and that, at a discontinuity, $\dot{y}^-(t) = u(t-)$ and $\dot{y}^+(t) = u(t+)$. With restriction to such functions u, we can eliminate u and reduce the problem to the classical fixed-endpoint problem of minimizing $\int_0^1 \dot{y}^2\,dt$ on the class of PWS functions y such that $y(0) = 0$ and $y(1) = 1$. If this were our problem, we could verify that Theorem 3.9 applies and that $y_0(t) = t$ furnishes a smaller value for $J(y)$ than does any other function y that is admitted to the competition in the classical problem.

Returning to the example as stated, we observe that the infimum γ of $J(y,u)$ is finite, in fact, nonnegative. Let $\{(y_\nu,u_\nu)\}$ be a sequence such that $J(y_\nu,u_\nu) < \gamma + 1/\nu$. Each u_ν is thus in the closed ball $B(r,\theta)$ of the given L_2-space with radius $r = (\gamma + 1)^{1/2}$ and center θ. In view of Theorem 9.19 we can suppose that the sequence $\{(y_\nu,u_\nu)\}$ has been so chosen that u_ν converges weakly to an element u_0 of this ball. Define y_0 by (9.71) with u_0 on the right. Then

$$y_\nu(t) - y_0(t) = \int_0^1 (u_\nu - u_0)\chi\,dt,$$

where χ denotes the characteristic function of the interval $[0,t]$, and $y_\nu(t) \to y_0(t)$ as a consequence of the weak convergence of u_ν to u_0. Since (y_ν,u_ν) is an admissible pair, $y_\nu(1) = 1$ for every ν and hence $y_0(1) = 1$. The properties of y_0 and u_0 that we have enumerated show that the pair (y_0,u_0) is in $\mathscr{P}$, the class within which we desire a minimizing element. It remains to establish that $J(y_0,u_0) = \gamma$. We notice that y as an integral of an integrable function u is AC (Theorem 8.33), hence that $\dot{y}(t) = u(t)$ a.e. (Theorem 8.37). One way in which to complete the proof would be to show that $\int \dot{y}^2(t)\,dt$ is lower semi-continuous and hence that lim inf $J(y_\nu) \geqslant J(y_0)$. We would then have the sequence of relations that

$$\gamma = \lim \int \dot{y}_\nu^2\,dt = \lim\inf \int \dot{y}_\nu^2\,dt \geqslant \int \dot{y}_0^2\,dt \geqslant \gamma$$

and hence the conclusion that $\int \dot{y}_0^2(t)\,dt = \gamma$. This is the method initiated by Tonelli and used by him and others in many places, including Section 7.11. We wish, however, to exhibit the use of L_2-space methods.

Let $\{w_k\}$ be the sequence of means $w_k = (1/k)\,\Sigma_1^k u_\nu$, which, by the Banach–Saks Theorem, converges strongly to u_0. If $\phi \colon R \to R$ is convex, we have, by the inequality of Jensen (p. 72 of Hardy, Littlewood, and Pólya, *Inequalities*, Cambridge Univ. Press, 1934, or McShane, *Jensen's inequality*, Bull. Amer. Math. Soc., vol. 43 (1937), pp. 521–527), that

$$(9.72) \qquad \phi(w_k) \leqslant (1/k)\,\sum_1^k \phi(u_\nu).$$

Our present integrand $\phi(u) = u^2$ is convex; therefore,

$$(9.73) \qquad \int_0^1 w_k^2\,dt \leqslant (1/k)\,\sum_1^k \int_0^1 u_\nu^2\,dt.$$

The right member is the first-order Cesáro mean of the first k terms of a sequence of real numbers converging to the infimum γ of $J(y,u)$. It is an early remark in every discussion of Cesáro means, which can be verified in a straightforward manner, that the sequence of such means converges to the original limit, in the present case to γ as $k \rightarrow \infty$. One verifies easily that if we define

$$z_k(t) \equiv \int_0^t w_k \, d\tau,$$

then $z_k(0) = 1$ as a consequence of the fact that $y_\nu(0) = 1, \nu = 1, \ldots, k$, and therefore from (9.73) that (z_k, w_k), $k = 1, 2, \ldots$, is a minimizing sequence of pairs in $\mathscr{P}$ for the integral J.

The equation

$$\int_0^1 (w_k^2 - u_0^2) \, dt = \int_0^1 2(w_k - u_0)u_0 \, dt + \int_0^1 (w_k - u_0)^2 \, dt$$

follows from the identical equality of the first integrand to the sum of the integrands on the right. The terms on the right converge to zero by the strong convergence and consequent weak convergence of w_k to u_0. Therefore, the left member tends to zero with $1/k$, and this yields the desired conclusion that

$$\int_0^1 u_0^2 \, dt = \gamma.$$

One wonders with reference to the first paragraph of this discussion whether the function y_0 corresponding to u_0 is or is not the linear function y_0, $y_0(t) = t$. Recall the standard procedure in Section 2.6 for deriving the Euler necessary condition under classical hypotheses.

Example 9.1 is equivalent to the problem of minimizing

$$J(y) = \int_0^1 \dot{y}^2 \, dt$$

on the class $\mathscr{Y}$ of all functions $y: [0,1] \rightarrow R$ that are AC on $[0,1]$ and have the end values $y(0) = 0$, $y(1) = 1$. Suppose that y_0 minimizes $J(y)$ and consider the equation

$$F(\epsilon) \equiv \int_0^1 (\dot{y}_0 + \epsilon \dot{\eta})^2 \, dt,$$

where $\eta: [0,1] \rightarrow R$ is chosen with reference to Reid (45e) as a function that is lipschitzian and hence AC on $[0,1]$ and such that $\eta(0) = \eta(1) = 0$. We find that

$$[F(\epsilon) - F(0)]/\epsilon = \int_0^1 2\dot{y}_0\dot{\eta} \, dt + \epsilon \int_0^1 \dot{\eta}^2 \, dt.$$

The second term on the right tends to zero with ϵ and we obtain as a necessary condition on y that

$$\int_0^1 \dot{y}_0 \dot{\eta} \, dt = 0$$

for every choice of η with the properties stated above. It follows from an extension of the du Bois Reymond lemma to the present setting (45e, p. 165) that $\dot{y}_0(t) = c = $ const. for almost all $t \in [0,1]$. Finally, after integration and determination of c so as to satisfy the condition $y_0(1) = 1$, we find that $y_0(t) = t$.

EXAMPLE 9.2

Let $\mathscr{T}$ denote the class of all triples $(y,v,u): [0,\pi] \to R^3$ such that

(i)   *y and v are AC on $[0,\pi]$,*

(ii) *u is Lebesgue measurable and $|u(t)| \leq 1$ a.e. on $[0,\pi]$,*

(iii) *$\dot{v} + y = u$ and $\dot{y} = v$ a.e. on $[0,\pi]$,*

and

(iv) *$y(0) = 0, v(0) = 0, y(\pi) = h, v(\pi) = m$.*

The problem is to investigate the existence of global extrema of

$$J(y,v,u) \equiv \int_0^\pi f(t,y,v) \, dt,$$

where $f: [0,\pi] \times R \times R$ is given to be continuous on its domain.

Discussion

By elimination of v between the constraints (iii) we obtain the familiar differential equation $\ddot{y} + y = u$ for forced oscillations. This is a Bolza problem but one that is not covered by the classical formulation of Chapter 5 in view of the presence of an inequality constraint and the fact that y, v, and u are not required to be PWS. In the language of control theory, we have two state variables y and v and one control u. Such a side-condition as $|u| \leq 1$, more generally $|u| \leq k$, is frequently encountered. It is not only a convenience in the mathematical analysis but, if the problem is a model for some material system, such a constraint is realistic. Our term u is dimensionally an acceleration, hence proportional to a force that is to be supplied by a control mechanism. Such a force would be bounded.

Such a pair of linear equations as those stated under (iii) have a unique solution (y,v) with preassigned initial values $y(0) = 0$, $v(0) = 0$. If u is an arbitrary bounded measurable control, the corresponding pair (y,v) obtained by variation of parameters is

(9.74)
$$y(t) = -(\cos t) \int_0^t u(\tau) \sin \tau \, d\tau + (\sin t) \int_0^t u(\tau) \cos \tau \, d\tau,$$
$$v(t) = (\sin t) \int_0^t u(\tau) \sin \tau \, d\tau + (\cos t) \int_0^t u(\tau) \cos \tau \, d\tau.$$

One sees easily that $y(t)$ and $v(t)$ are both bounded on $[0,\pi]$ independently of the choice of a function u with properties (ii). If the terminal values h and m given under condition (iv) are chosen arbitrarily, they may not be attainable. Clearly certain pairs (h,m) are attainable and we suppose in the remainder of this discussion that (h,m) is such a pair. It follows that both the infimum γ and the supremum Γ of $J(y,v,u)$ are finite.

Let $\{(y_\nu,v_\nu,u_\nu)\}$ be a sequence in the class $\mathcal{T}$ such that $J(y_\nu,v_\nu,u_\nu)$ converges to the infimum γ of all values of J on $\mathcal{T}$. Since each admissible u is measurable and $|u| \leq 1$, Theorem 9.19 applies. We can, therefore, suppose the minimizing sequence to have been chosen so that u_ν converges weakly to some function u_0 in the closed ball $B(\sqrt{\pi},\theta)$ in $L_2([0,\pi])$.

Define y_0 and v_0 by (9.74) with u_0 in the integrands in the right members. Then $v_\nu(t) - v_0(t)$ converges to zero boundedly as a consequence of the weak convergence of u_ν to u_0, and similarly for $y_\nu - y_0$. One verifies that y_0 and v_0 have properties (i), (iii), and (iv). To see that $|u_0(t)| \leq 1$ a.e., suppose on the contrary that $u_0(t) > 1$ on a subset E of $[0,\pi]$ of positive measure. By the weak convergence,

$$\int_E u_\nu \, d\tau \to \int_E u_0 \, d\tau > \mu(E),$$

from which follows that for large values of ν the integral mean of u_ν exceeds unity, in contradiction with condition (ii) in the example. A similar argument applies if $u_0(t) < 1$ on a subset E of positive measure. We infer that u_0 is admissible and hence that $(y_0,v_0,u_0) \in \mathcal{T}$.

That $J(y_0,v_0,u_0) = \gamma$ now follows from the Dominated Convergence Theorem.

Exercise 9.5

1. Establish for Example 9.2 that there also exists a triple (y_1,v_1,u_1) such that the value of J is a global maximum.
2. Modify Example 9.2 by taking an integrand of the form $f(t,y,v,u)$ and such that, for a given continuous $\phi(t,y,v)$,

$$|f(t,y,v,u_1) - f(t,y,v,u_2)| < \phi(t,y,v)|u_1 - u_2|.$$

   Then investigate the existence of global extrema of J.
3. Let $\mathcal{T}$ denote the class of all triples (y,v,u) such that $u \in L_2([0,1])$ and y and v satisfy the conditions

$$y(t) = \int_0^t v(\tau) \, d\tau, \qquad y(1) = 1$$

and

$$v(t) = \int_0^t u(\tau) \, d\tau, \qquad v(1) = 1.$$

   Prove that $J(y,v,u) \equiv \int_0^1 u^2(t) \, dt$ attains a global minimum.
4. In the event that a PWS triple (y_0,v_0,u_0) in $\mathcal{T}$ furnishes at least a weak extremum for problem 3 in comparison with PWS triples in $\mathcal{T}$,

replace the integral constraints by differential equations and find y_0, v_0, and u_0 by means of the Multiplier Rule. Show by means of Theorem 5.5 that this triple furnishes a smaller value for J than any other PWS triple in the class $\mathscr{T}$. Then investigate whether this triple (y_0, v_0, u_0) also provides a least value in competition with the entire class $\mathscr{T}$.

5. Replace the integrand u^2 in problem 3 by u^{2p}, p a positive integer, and let u be any function in $L_{2p}([0,1])$ with y and v determined as before. Again prove the existence of a global minimum.

Problem 5 is equivalent to that of minimizing the integral $\int_0^1 \dot{v}^{2p}\,dt$. If v is a velocity and hence $\dot{v}$ an acceleration and if p is large, then large values of $|\dot{v}(t)|$ tend to have a dominant effect on the value of the integral. The minimum problem is then a mathematical model that approximately represents an important type of question: If a moving object is to transfer from an initial state (position and velocity) to a terminal state, how can this be done in such a way that the maximum magnitude of acceleration is a minimum?

9.12   CONCLUDING REMARKS

Chapters 7 and 9 have introduced concepts and procedures that have been effective in proving the existence of global extrema. A wide variety of existence theorems is to be found in the research literature, for the most part more inclusive and with longer and more complex proofs than those included in this book. For examples see works of Tonelli, Graves, McShane, Cesari, and others in the bibliography.

To gain the sequential compactness needed in an existence proof it has been necessary to admit larger classes of functions y or curves C than those that are piecewise smooth. This move has entailed using other integrals than that of Riemann, and a number of different ones are to be found in work on optimization theory. The dissertation of J. K. Cole, University of Oklahoma, 1967 (a part of which is in the Jour. of Optimization Theory and Applications, vol. 2, 1968, pp. 199–204), employs an integral of S. Bochner.

And yet, having proved that there is an optimizing AC, BV, or L_2 function, one is pleased if the optimizing function can be shown to be piecewise smooth or perhaps even smooth, as in the case of Example 9.1. If this does not occur it can still be possible, as theorems in Section 9.8 will suggest, to approximate a minimizing function y_0 having undesirable features by y_1 with more acceptable continuity and differentiability and such that the value $J(y_1)$ is near the minimum value $J(y_0)$.

When the ultimate objective is optimal design of some device or system, y_0 might involve properties beyond those provided by available materials or techniques. In this event, such a y_1 is needed at the practical level.

Chapter 10

A MISCELLANY OF NONCLASSICAL PROBLEMS

10.1 INTRODUCTION

Preceding chapters have considered the existence and determination of local and global extrema for a variety of functionals J, some parametric, others nonparametric. Until about the middle of the twentieth century much of the work concerning such extrema consisted of refinements and extensions of problems dating back to Euler, Lagrange, Jacobi, and Weierstrass with impetus from the work of Hilbert, Lebesgue, Bolza, Bliss, Morse, and Tonelli and from developments in general topology and functional analysis.

Variational theory has always been motivated by questions from science and technology. Indeed the much-quoted brachistochrone problem proposed by John Bernoulli in 1696 is often regarded as its beginning. It is therefore not surprising that the question of optimal design or optimal performance of sophisticated modern systems would lead to mathematical models resembling Problems of Bolza but often with features not covered by the classical formulations. There may, for example, be game-theoretic or other stochastic ingredients, time-lags, infinite time intervals, or admissible functions having discontinuities.

A profusion of such nonclassical problems is to be found in the books of Bellman (IV, V, etc.), in the work of Pontryagin and his associates (XXXIII), in Isaacs (XXII), and in such articles as Leitmann (30a,b) and Miele (36b,c,d). It is the author's impression, at the date of this writing, that many of these problems have yielded only partial analyses. There is already a considerable body of existence theory and there are many papers which examine first-order necessary conditions but much less on second-order necessary conditions or sufficiency for local and global extrema.

This chapter injects the spirit of such problems while maintaining contact with what has gone before. That certain problems are selected for discussion is not to suggest that they are of special importance but merely that they are some with which the author has had experience that have novel features and potential applications and quite possibly interesting generalizations.

10.2 PROBLEMS MOTIVATED BY ROCKET PROPULSION

Some preparation for this topic is provided by Example 1.6, Section 5.12, and Exercise 5.5.

Supposing for the moment that all functions to be mentioned are PWS, consider the particle idealization of a rocket-propelled vehicle of variable mass m. Let $\mathbf{v}$ be the velocity of the vehicle and let $\mathbf{u}$ be a unit vector with the direction and sense of the thrust of the motor. We take for the magnitude of the latter a much used approximation, $-c\dot{m}$, in which the positive constant c is to be interpreted as the effective mean exit speed of products of combustion. Let $\mathbf{F}$ be the vector-sum of all forces other than the thrust.

The idealized vehicle then moves in accord with the two equations

(10.1) $m\dot{\mathbf{v}} = -c\dot{m}\mathbf{u} + \mathbf{F}$ *and* $\mathbf{v} = \dot{\mathbf{y}},$

subject to the constraint that

(10.2)   *m be nonincreasing in the time t.*

Each equation (10.1) is equivalent to one, two, or three equations in components of the respective vectors represented by symbols in boldface according as the motion is linear, planar, or in three-space. If, for example, the motion is on an upward-directed y axis subject to upward thrust and downward gravitational acceleration but to no other forces, then system (10.1) reduces to the scalar equations

(10.3) $m\dot{v} + c\dot{m} + mg = 0$ *and* $\dot{y} - v = 0.$

EXAMPLE 10.1

A vehicle starts from rest at the origin at $t = 0$ with initial mass M_0 and the preassigned burnout mass M_b and moves subject to (10.3). Is there a triple (y_0, v_0, m_0) of PWS functions satisfying (10.3) and the monotoneity constraint (10.2) such that the summit altitude Y_0 is a global maximum and, if so, what is this triple?

Discussion

It follows from (10.3) by integration that

$$(10.4) \qquad v = -c \ln (m/M_0) - gt$$

and

$$(10.5) \qquad y = -c \int_0^t [\ln(m/M_0)] \, d\tau - 1/2 g t^2.$$

Let $y(T)$ denote the maximum value Y of $y(t)$. Clearly $v(T)$ must be zero. There is reason to anticipate that $T \geq t_b = $ *burnout time*. With $t = T$ and $v(T) = 0$ in (10.4) we find that $T = (c/g) \ln(M_0/M_b)$, hence from (10.5) that

$$(10.6) \qquad Y(y,v,m) = y(T) = -\frac{c^2}{2g} \ln^2 \frac{M_0}{M_b} - c \int_0^T \left(\ln \frac{m}{M_0} \right) dt.$$

It is convenient to replace the integral by the expression

$$-c \int_0^T \left(\ln \frac{m}{M_b} \right) dt + cT \ln \frac{M_0}{M_b}.$$

To maximize $Y(y,v,m)$ on the class $\mathscr{T}$ of all PWS triples satisfying (10.3) and (10.2) together with the stated end-conditions is then the same as to minimize

$$(10.7) \qquad J(y,v,m) \equiv \int_0^T \left(\ln \frac{m}{M_b} \right) dt \quad on \, \mathscr{T}.$$

Upper limit T in (10.7) can be replaced by t_b since the integrand is $\ln(M_b/M_b) = 0$ for $t > t_b$.

Scrutiny of (10.7) reveals that if $m(t)$ decreases from M_0 to M_b during time t_b, then $J(y,v,m) \to 0$ with t_b. Clearly $J(y,v,m) \geq 0$; hence the infimum of all its values is zero. However, any PWS function $m:[0,t_b] \to R$ with the given initial and terminal values yields a positive value for the integral. Hence no triple in $\mathscr{T}$ for which $J(y,v,m,) = 0$ can exist.

When such a conclusion is reached, one then asks if it is possible to enlarge the class of admissible functions in such a way that the functional J extends to a larger domain that includes an extremizing element.

Let $\mathscr{T}^*$ denote the class of all triples (y,v,m) such that m is required only to be nonincreasing and to have the assigned initial and terminal values with v and y given in terms of m by (10.4) and (10.5). The infimum of integral (10.7) on $\mathscr{T}^*$ is again zero, and this value is realized if $m_0(t) = M_0$ or M_b according as $t = 0$ or $t > 0$, that is, if m_0 jumps from M_0 to M_b at $t = 0$.

Observe that the original differential equations (10.3) cannot be used in the formulation of the extremum problem on $\mathscr{T}^*$. Neither m nor v is AC when m is permitted to have discontinuities. In view of the Fundamental Theorem 8.38 of the Integral Calculus, differential equations (10.3) are not equivalent to (10.4) and (10.5); indeed, they are not even meaningful, since $\dot{m}(t)$ and $\dot{v}(t)$ can fail to exist or can have the respective values $-\infty$ and ∞. For example, the left member of the first equation (10.3) might reduce to the meaningless form $\infty - \infty$ at every rational t of an interval.

In Section 5.12 we analyzed a problem in which these difficulties were avoided by imposing a bound on the rate $\dot{m}$ of mass flow. Without such a restriction the mathematical models for problems in missile trajectory optimization continue to be problems of the Bolza type, but the classical formulation of Section 5.3 with differential side-conditions (5.5) does not suffice.

We have remarked in Section 6.16 and elsewhere that variational problems motivated by dynamical systems are generally nonparametric. We wish to show, however, that the nonclassical extremum problem on $\mathscr{T}^*$ analyzed above is equivalent to an essentially classical parametric problem of Mayer.

The parameter is an "artificial time" τ and the dot will denote differentiation with respect to τ. Set $l = c \ln m$ as in Section 5.12. Admissible functions are vector-valued with four components of the form (t,y,v,l): $[\tau_0,\tau_1] \to R^4$. More specifically, we admit the class $\mathscr{Q}$ of all such quadruples q that are AC on their respective intervals $[\tau_0,\tau_1]$ and satisfy on these intervals the side-conditions

$$(10.8) \qquad \begin{cases} \dot{v} + \dot{l} + g\dot{t} = 0 & \text{a.e.,} \\ \dot{y} - v\dot{t} = 0 & \text{a.e.,} \\ t \text{ is nondecreasing,} \\ l \text{ is nonincreasing,} \end{cases}$$

together with the end-conditions

$$(10.9) \qquad \begin{aligned} t(\tau_0) &= 0, \quad y(\tau_0) = 0, \quad v(\tau_0) = 0, \quad l(\tau_0) = L_0 \equiv c \ln M_0 \\ v(\tau_1) &= 0, \quad l(\tau_1) = L_1 \equiv c \ln M_b, \end{aligned}$$

The problem is

$$(10.10) \qquad\qquad y(\tau_1) = \textit{maximum on } \mathscr{Q}.$$

The problem clearly satisfies the homogeneity requirements in Section

6.10 and accordingly can be restated in terms of the class $\mathscr{C}$ of all Fréchet curves C each of which has representations $q \in \mathscr{Q}$.

Observe that, under the present formulation, it is possible for l to be strictly decreasing on a τ-interval on which t is constant. This coincidence corresponds to a discontinuity of the mass m with respect to "real time" t.

Let $\mathscr{Q}_1$ denote the subset of $\mathscr{Q}$ consisting of those quadruples $q \in \mathscr{Q}$ such that

(i) *the parameter interval* $[\tau_0,\tau_1]$ *is* $[0,1]$

and

(ii) $0 \leqslant t(\tau) \leqslant T$, $\tau \in [0,1]$, *where T is fixed and not less than the value mentioned preceding* (10.6).

Every admissible curve C has representations with property (i). We know from the analysis of the nonparametric version of the problem that an optimizing program exists. Accordingly, the maximum problem (10.10) is equivalent to the problem

$$(10.11) \qquad\qquad y(1) = maximum \; on \; \mathscr{Q}_1.$$

A quadruple q_0 furnishing the maximum (10.10) has property (ii) for a certain T and, since we are free to change the parameter, there will then exist a maximizing program with both of the properties (i) and (ii) that determine $\mathscr{Q}_1$.

We now prove the existence of a maximizing program in $\mathscr{Q}_1$. In the absence of previous knowledge of the problem this shift from (10.10) to (10.11) could be justified in other ways.

All first components t of quadruples in $\mathscr{Q}_1$ being nondecreasing are of total variation at most T. We understand that $L_1 < L_0$. All components l of quadruples q are nonincreasing and hence of total variation at most $L_0 - L_1$. It then follows from the first constraint (10.8) by integration that the total variation of every component v of a quadruple in $\mathscr{Q}_1$ is at most $L_0 - L_1 + gT$. The second equation (10.8) implies that for any choice of $q \in \mathscr{Q}_1$ and of $\tau,\tau' \in [0,1]$,

$$|y(\tau') - y(\tau)| = \left| \int_\tau^{\tau'} \dot{y}(\tau) \, d\tau \right| \leqslant \int_\tau^{\tau'} |v(\tau)| \dot{t}(\tau) \, d\tau$$

and hence that

$$(10.12) \qquad |y(\tau') - y(\tau)| \leqslant \max\{|v(\tau)| : \tau \in [\tau_0,\tau_1]\} \int_\tau^{\tau'} \dot{t}(\tau) \, d\tau.$$

One verifies that $|v(\tau)|$ is necessarily bounded on $[\tau_0,\tau_1]$. The value of the last integral is $t(\tau') - t(\tau)$. It follows from (10.12) that the total variation of y on $[0,1]$ is at most a constant multiple of T. As a consequence of these bounds and of Theorem 7.2, every curve C with a representation q in $\mathcal{Q}_1$ is of length at most some real number λ.

Let B denote a fixed closed ball in R^4 with the initial point $[t(0), y(0), v(0), l(0)] = (0,0,0,L_0)$ as center and with radius $r > \lambda$. No boundary point of B is *attainable*; that is, no terminal point $[t(1), y(1), v(1), l(1)] = [t(1), y(1), 0, L_1]$ can be at distance as great as r from the initial point.

Let $\{C_\nu : \nu \in N\}$ be a *maximizing sequence*, that is, a sequence of curves with representations in the class $\mathcal{Q}_1$ such that the corresponding sequence of summit heights $\{y_\nu(1) : \nu \in N\}$ converges to the supremum of all summit heights $y(1)$ for quadruples $q \in \mathcal{Q}_1$. In the remainder of this proof let $q_\nu \equiv (t_\nu, y_\nu, v_\nu, l_\nu) : [0,1] \to R^4$ be the reduced-length representation (Section 7.6) of C_ν.

In view of the compactness of the closed ball B, the bound λ on lengths of the curves C_ν and the Hilbert Compactness Theorem 7.10, we can suppose the sequence $(t_\nu, \ldots, l_\nu)$, $\nu = 1, 2, \ldots$, so chosen that it converges to a limit quadruple $q_0 = (t_0, y_0, v_0, l_0) : [0,1] \to R^4$. It follows that the summit height $y_\nu(1)$ converges to $y_0(1)$ and, since y_ν was chosen so that $y_\nu(1)$ converges to the supremum of summit heights, $y_0(1)$ must be this supremum. To establish that q_0 is the desired maximizing quadruple in $\mathcal{Q}_1$ it remains to verify that q_0 satisfies the side and end-conditions that determine the class $\mathcal{Q}$ and also the conditions (i) and (ii) that define the subclass $\mathcal{Q}_1$.

That the quadruple q_ν satisfying the end-conditions converges to q_0 implies that the latter must also satisfy these conditions. Clearly q_0 has the unit interval as its domain. Since the component t_ν of q_ν is nondecreasing and $t_\nu(\tau) \in [0,T]$ for all $\tau \in [0,1]$, its limit t_0 necessarily has these same properties. Since the component l_ν is nonincreasing, so also must be its limit l_0. We turn to the remaining side-conditions (10.8_1) and (10.8_2).

Verify that the expression $|\dot{y} - v\dot{t}|$ is convex in $\dot{t}$ and $\dot{y}$. With v held fixed, the graph of $u = |\dot{y} - v\dot{t}|$ in a $(\dot{t}, \dot{y}, u)$ rectangular coordinate system consists of two half-planes joined together in a line so as to form a dihedral angle with this line as the edge. It is intuitively clear that this surface is convex, but one can also find that definition (3.20) is satisfied. Since $|\dot{y} - v\dot{t}|$ is convex in $(\dot{t}, \dot{y})$ it is convex in $(\dot{t}, \dot{y}, \dot{v}, \dot{l})$. Clearly $|\dot{y} - v\dot{t}|$ has the continuity (i) and homogeneity (ii) required of a parametric integrand in Section 7.10; consequently, by Theorem 7.16 and the equality of Weirstrass and Lebesgue integrals, $\int |\dot{y} - v\dot{t}| \, d\tau$ is lower semi-continuous on the class of curves of length at most λ. Since $(t_\nu, y_\nu, u_\nu, l_\nu)$ satisfies (10.8_2), it now follows that

$$(10.13) \qquad 0 = \lim_{\nu \to \infty} \int_0^1 |\dot{y}_\nu - v_\nu \dot{l}_\nu| \, d\tau \geqslant \int_0^1 |\dot{y}_0 - v_0 \dot{l}_0| \, d\tau \geqslant 0$$

and therefore $\int_0^1 |\dot{y}_0 - v_0 \dot{l}_0| \, d\tau = 0$. The integrand is nonnegative and hence must vanish a.e. on $[0,1]$. Thus q_0 satisfies (10.8_2) and a similar argument shows that it satisfies (10.8₁). Proof of the existence of a maximizing quadruple in $\mathscr{Q}_1$ for Problem (10.10) and hence of such a quadruple in $\mathscr{Q}$ for Problem (10.9) is complete.

A complete analysis requires characterization of all maximizing quadruples. The last two constraints (10.8) translate into equations

$$\dot{i} - |\dot{w}| = 0 \qquad and \qquad \dot{l} + |\dot{z}| = 0$$

with the aid of two additional components w and z for our vector-valued functions. We now have a parametric Mayer Problem in six-space that resembles the pattern in Section 6.10 but differs from the problem of that section in two respects. Admissible sextuples are not now restricted to be PWS. Moreover, the functions ϕ_β in the side-conditions of Section 6.10 are understood, as in Section 5.3, to satisfy a blanket hypothesis. If we wish to use the Multiplier Rule, this includes the existence and continuity of first derivatives, but the absolute value function used above to provide the required homogeneity does not have a finite derivative at zero. The last defect is eliminated if we introduce $\dot{\omega} \equiv |\dot{w}|^{1/2}$ and $\dot{\zeta} \equiv |\dot{z}|^{1/2}$. The last two constraints (10.8) then become

$$(10.14) \qquad \dot{i} - \dot{\omega}^2 = 0 \qquad and \qquad \dot{l} + \dot{\zeta}^2 = 0.$$

Having used the advantages of parametric problems in our existence proof we now turn to a Mayer Problem that is seen to be nonparametric because left members in (10.14) lack the homogeneity (6.20).

We know that there is a maximizing AC quadruple (t_0, y_0, v_0, l_0) for the parametric problem in four-space. One sees readily that the sextuple $(t_0, y_0, v_0, l_0, \omega_0, \zeta_0)$, where

$$\omega_0(\tau) \equiv \int_0^\tau [\dot{l}_0(s)]^{1/2} ds \qquad and \qquad \zeta_0(\tau) \equiv \int_0^\tau [-\dot{l}_0(s)]^{1/2} ds,$$

will necessarily maximize the summit height for the nonparametric problem in the seven-space with points $(\tau, t, y, v, l, \omega, \zeta)$.

An extremizing AC vector-valued function often turns out to be PWS if the integrand (presently zero) and the side-conditions are of simple form. As a result of our choice of radius for the closed ball B in the existence proof and of the bound λ on the lengths of admissible curves for the parametric problem, the graph of the maximizing curve C_0 represented by q_0 consists of interior points of B. This circumstance assures that, if there happens to be a PWS maximizing sextuple for the nonparametric problem involving constraints (10.14), then the families of functions denoted by $y(\cdot, b)$ in the proof of the Multiplier Rule are present and equations (5.24) and (5.27) are necessary conditions on a PWS sextuple $(t_0, y_0, v_0, l_0, \omega_0, \zeta_0)$ that maximizes $y(1) = \int \dot{y} \, dt$.

The auxiliary functions (5.21) and (5.22) for the Lagrange Problem of maximizing this integral are

$$F(\ ) = \lambda_0 \dot{y} + \lambda_1(\tau)(\dot{v} + \dot{l} + g\dot{t})$$
$$+ \lambda_2(\tau)(\dot{y} - v\dot{t}) + \lambda_3(\tau)(\dot{l} - \dot{\omega}^2) + \lambda_4(\tau)(\dot{l} + \dot{\zeta}^2)$$

and

$$G(\ ) = e_1\tau_0 + e_2t_0 + e_3y_0 + e_4v_0$$
$$+ e_5(l_0 - L_0) + e_6(\tau_1 - 1) + e_7v_1 + e_8(l_1 - L_1).$$

The six Euler equations (5.24) are found to be

$$\lambda_1 g - \lambda_2 v + \lambda_3 = c_1,$$

$$\lambda_0 + \lambda_2 = c_2,$$

$$\lambda_1 = -\int_0^\tau \lambda_2(s)\dot{l}(s)\ ds + c_3,$$

$$\lambda_1 + \lambda_4 = c_4,$$

$$\lambda_3 \dot{\omega} = c_5,$$

$$\lambda_4 \dot{\zeta} = c_6.$$

Of the 14 transversality equations, there are 8 that serve only to determine the constants e_1 through e_8. The remaining 6, which can be used in identifying the desired sextuple, are

$$\lambda_3(\tau_0)\dot{\omega}(\tau_0) = 0, \qquad \lambda_4(\tau_0)\dot{\zeta}(\tau_0) = 0,$$

$$\lambda_1(\tau_1)g - \lambda_2(\tau_1)v(\tau_1) + \lambda_3(\tau_1) = 0, \qquad \lambda_0 + \lambda_2(\tau_1) = 0,$$

$$\lambda_3(\tau_1)\dot{\omega}(\tau_1) = 0,$$

and

$$\lambda_4(\tau_1)\dot{\zeta}(\tau_1) = 0.$$

Exercise 10.1

1. Given (10.4), (10.5), and (10.6), what is the maximum value of the summit height $Y(y,v,m)$ on the class $\mathcal{T}^*$ of triples (y,v,m) for Example 10.1?

2. Construct a heuristic argument for the nonparametric Bolza Problem discussed preceding these problems in support of the claim that one should be free to choose a burnout "time" $\tau_b \in (0,1)$ at pleasure and then to choose $l: [0,1] \to R$ as any nonincreasing function such that $l(0) = L_0$ and $l(\tau) = L_1$ on $[\tau_b, 1]$. In particular,

we can take $\tau_b = \frac{1}{2}$ and $l(\tau) = L_0 - 2(L_0 - L_1)\tau$ or L_1 according as $\tau \in [0,\frac{1}{2})$ or $[\frac{1}{2},1]$.

3. Using the end-conditions $t(0) = 0$, $y(0) = 0$, $v(0) = 0$, $l(0) = L_0 = c \ln M_0$, $v(1) = 0$, and $l(1) = L_1 = c \ln M_b$, verify that the sextuple $(t_0,y_0,v_0,l_0,\omega_0,\zeta_0)$ and multipliers $\lambda_0 = 1$, $\lambda_1(\tau),\ldots,\lambda_4(\tau)$ satisfy the Multiplier Rule for the nonparametric problem of this section.

If $\tau \in [0,\frac{1}{2}]$,

$$t_0(\tau) = 0, \qquad y_0(\tau) = 0, \qquad v_0(\tau) = 2(L_0 - L_1)\tau,$$

$$l_0(\tau) = L_0 - 2(L_0 - L_1)\tau, \qquad \omega_0(\tau) = 0, \qquad \zeta_0(\tau) = \tau[2(L_0 - L_1)]^{1/2},$$

$$\lambda_1(\tau) = -(L_0 - L_1)/g, \qquad \lambda_2(\tau) = -1, \qquad \lambda_3(\tau) = (L_0 - L_1)(1 - 2\tau),$$

$$\lambda_4(\tau) = 0.$$

If $\tau \in [\frac{1}{2},1]$,

$$t_0(\tau) = (L_0 - L_1)(2\tau - 1)/g, \qquad gy_0(\tau) = -2(L_0 - L_1)^2(\tau - \tfrac{1}{2})^2 + 2(L_0 - L_1)^2(\tau - \tfrac{1}{2}),$$

$$v_0(\tau) = 2(L_0 - L_1)(1 - \tau), \qquad l_0(\tau) = L_1,$$

$$\omega_0(\tau) = [2(L_0 - L_1)/g]^{1/2}(\tau - \tfrac{1}{2}), \qquad \zeta_0(\tau) = \tfrac{1}{2}[2(L_0 - L_1)]^{1/2},$$

$$\lambda_1(\tau) = 2(L_0 - L_1)(\tau - 1)/g, \qquad \lambda_2(\tau) = -1,$$

$$\lambda_3(\tau) = 0, \qquad \lambda_4(\tau) = -(L_0 - L_1)(2\tau - 1)/g.$$

4. Verify that the sextuple given in problem 3 describes the same program for Example 10.1 represented by the maximizing triple (y,v,m) in the class $\mathscr{T}^*$, hence that the sextuple furnishes a global maximum for the problem as initially formulated.

We remark that it is not possible, under restriction to the methods and results in this book, to establish that the above-described sextuple maximizes the summit height independently of the analysis of the maximum problem on $\mathscr{T}^*$. One may think of using Theorem 5.5, but it is not effective because of the term $v\dot{t}$ in the second condition (10.8).

10.3   A LEAST-SQUARES ESTIMATION

Suppose given that a certain function $y_0: [0,X] \to R$ is nonincreasing but that the only other information available is an approximation $\alpha: [0,X] \to R$ to y_0. Function α can be thought of as having been obtained by observation or experiment, so idealized that $\alpha(x)$ is known precisely for all $x \in [0,X]$. Function α may or may not be nonincreasing.

We assume that it is PWS and hence continuous under the definitions of Section 1.9.

By a *least-squares estimate* of y_0 we mean a function y_0 furnishing a global minimum for the integral

$$(10.15) \qquad J(y) \equiv \int_0^X [y - \alpha(x)]^2 \, dx$$

on the class $\mathcal{Y}$ of all PWS functions $y: [0,X] \to R$ satisfying the side-condition

$$(10.16) \qquad \dot{y} + \dot{z}^2 = 0,$$

and no end-conditions.

Thus far the problem is classical. We find with reference to Section 5.7 the two Euler equations (since $\lambda_0 = 1$ will do) that

$$(10.17) \qquad \begin{cases} \lambda(x) = 2 \int_0^x [y(\xi) - \alpha(\xi)] \, d\xi + c_1, \\ \lambda(x)\dot{z}(x) = c_2. \end{cases}$$

There are six transversality equations (5.27), of which two that yield useful information are

$$(10.18) \qquad \lambda(0) = 0 \quad and \quad \lambda(X) = 0.$$

It follows from (10.17) and (10.18) that $c_1 = c_2 = 0$ and, after multiplication of the respective equations (10.17) by $\dot{y}(x)$ and $\dot{z}(x)$, adding and using (10.16), that a necessary condition on an optimizing PWS y_0 is that

$$\dot{y}_0(x) \int_0^x [y_0(\xi) - \alpha(\xi)] \, d\xi = 0.$$

A PWS function y_0 satisfies this equation on $[0,X]$ if and only if that interval is a union of nonoverlapping subintervals on each of which either $\dot{y}_0(x)$ or the integral vanishes and hence on each of which either $y_0(x) = $ const. or $y_0(x) = \alpha(x)$.

If the given α is nonincreasing, then clearly $y_0(x) = \alpha(x)$ on $[0,X]$ furnishes the global minimum 0 for $J(y)$. If α is not monotone, one anticipates that an optimizing y_0 will consist of horizontal segments pieced together with nonincreasing segments of α .

EXAMPLE 10.2

$\alpha(x) = x + 1$ or $3 - x$ *according as* $x \in [0,1]$ *or* $[1,3]$.

Discussion

As a candidate for a function y_0 minimizing (10.15) among all PWS functions y satisfying (10.16) on the fixed interval $[0,X] = [0,3]$, consider

$y_0(x) = 3 - a =$ const. or $3 - x$ according as $x \in [0,a]$ or $[a,3]$. Integral (10.15) has the value

$$F(a) \equiv (a^3 - 6a + 6)/3, \quad a \in [0,3]$$

and $F(\sqrt{2}) = (6 - 4\sqrt{2})/3$ is the least value of $F(a)$.

Exercise 10.2

1. If a certain number of coincidences occur, the function y_0 in Example 10.2 corresponding to $a = \sqrt{2}$ will furnish the global minimum of $J(y)$. Identify a set of conditions under which this will be true.
2. Obtain a firm conclusion for Example 10.2 by applying Theorem 5.5.
3. Consider the nonclassical problem $J(y) \equiv \int_0^X [y - \alpha(x)]^2 \, dx =$ *minimum* on the class $\mathscr{Y}$ of all nondecreasing functions $y: [0,X] \to R$ with α now required only to be bounded and measurable on $[0,X]$. Prove that there exists $y_0 \in \mathscr{Y}$ such that $J(y_0) \leqslant J(y)$, $\forall y \in \mathscr{Y}$ with the aid of the Helly Theorem 7.12.
4. Modify problem 3 by fixing the end values of admissible functions and again prove an existence theorem.
5. Modify Example 10.2 by requiring that $y(0) = 2$ and $y(3) = 0$ and produce as complete an analysis as possible.
6. Investigate the problem $J(y) \equiv \int_0^X [y - \alpha(x)]^4 dx =$ *minimum* on the class $\mathscr{Y}$ of PWS nondecreasing functions $y: [0,X] \to R$.
7. Investigate the problem $J(y) \equiv \int_0^X |y - \alpha(x)| \, dx =$ *minimum* on the class $\mathscr{Y}$ of PWS nonincreasing functions $y: [0,X] \to R$.

Integral (10.15) is an example from a family of integrals, for which minimum values on classes of monotone functions are suggested by various applications. See papers of H. D. Brunk et al. (8a,b,c,d) and in particular Theorem 4.3 in (8d, p. 844), which characterizes optimizing functions for a number of cases and can be adapted to numerical methods. More recent related results are in a paper (45h) of W. T. Reid.

10.4 DESIGN OF A SOLENOID

It is shown in books on electricity and magnetism that a single circular conducting path of radius a in a plane normal to the x axis with center at position ξ on that axis and carrying electric current I produces an axial magnetic field at position x on the axis of magnitude $H = h(x - \xi)$, where

$$(10.19) \qquad h(x - \xi) = \frac{2\pi I a^2}{[a^2 + (x - \xi)^2]^{3/2}}.$$

Consider an *idealized solenoid* consisting of a distribution of turns along a circular cylinder of length $2L$ with $\phi(\xi)$ turns per unit length at $\xi \in [-L,L]$. We require that ϕ be Lebesgue integrable over the interval $[-L,L]$, hence (problem 5, Exercise 8.4) integrable over any subinterval $[-L,\xi]$ and set

$$(10.20) \qquad n(\xi) \equiv \int_{-L}^{\xi} \phi(s)\, ds = number\ of\ turns\ on\ [-L,\xi].$$

By Theorem 8.37, n has a derivative $\overset{\bullet}{n}(\xi)$ a.e. on $[-L,L]$ and $\overset{\bullet}{n}(\xi) = \phi(\xi)$ a.e. on that interval. We can thus replace $\phi(s)$ in (10.20) by $\overset{\bullet}{n}(s)$.

An idealized solenoid with $\overset{\bullet}{n}(\xi)$ coincident turns per unit length at $\xi \in [-L,L]$, each carrying current I, produces the same field at position x as does a single turn at position ξ carrying current $\overset{\bullet}{n}(\xi)I$. The axial field intensity $H(x;n)$ at position x that results from the choice of a particular function $n: [-L,L] \to R$ of the form (10.20) is then

$$(10.21) \qquad H(x;n) = \int_{-L}^{L} \overset{\bullet}{n}(\xi) h(x-\xi)\, d\xi \; .$$

It is desirable in certain measuring devices to produce a nearly constant axial field on a segment $[-b,b]$, $b \leqslant L$, of the axis of a solenoid. It is well known that a precisely constant axial field intensity is provided by an infinitely long uniformly wound $[\overset{\bullet}{n}(\xi) = \text{const.}]$ circular cylinder carrying current I. The result calculated by elementary techniques is

$$H(x;n) = \overset{\bullet}{n} \int_{-\infty}^{\infty} h(x-\xi)\, d\xi = 4\pi \overset{\bullet}{n} I.$$

We mention the less widely known result that a sphere of radius L with $\overset{\bullet}{n}(\xi) = \text{const.}$ turns per unit distance along a diameter to which all turns are normal has the constant axial field $\frac{8}{3}\pi \overset{\bullet}{n} I$ at points of the selected diameter, again found by elementary integration. Also known is the fact that a given total number of turns divided equally between two concentrated windings (Helmholtz coils) at suitable equal distances from the center of a circular cylinder will produce a better approximation to a constant axial field over an interval than will concentration of the same number of turns at the center.

Let $2N$ denote the total number of turns, again restricted to be on a circular cylinder:

$$(10.22) \qquad 2N = n(L) = \int_{-L}^{L} \overset{\bullet}{n}(\xi)\, d\xi \; .$$

With N and L fixed and both finite, one asks whether there exists in the class $\mathcal{N}$ of all functions $n: [-L,L] \to R$, each having an integrable extended real-valued derivative $\overset{\bullet}{n}$ a.e. on $[-L,L]$ and having end values $n(-L) = 0$, $n(L) = 2N$, a particular n_0 such that the axial field on the subinterval $[-b,b]$ of $[-L,L]$ is almost constant. If such a function n_0 exists, what is it?

298 CALCULUS OF VARIATIONS WITH APPLICATIONS

It seems desirable to translate such questions into the problem of existence and characterization of a function $n_0 \in \mathcal{N}$ furnishing a global minimum for some measure of deviations from constancy over $[-b,b]$ of $H(x;n)$. There is no unique measure of such deviations. Among those that suggest themselves are

$$(10.23) \qquad J(n) \equiv \int_{-b}^{b} [H(x;n) - H(0;n)]^2 \, dx,$$

$$(10.24) \qquad J(n) \equiv \int_{-b}^{b} [H(x;n) - H_0]^2 \, dx, \quad H_0 = \text{const.},$$

and

$$(10.25) \qquad J(n) \equiv \sup \{|H(x;n) - H(0;n)| : x \in [-b,b]\}.$$

Although a least value for the last of these would appear to be especially effective, the first two are of more familiar form and we elect to consider (10.23). Using (10.21) we see that

$$(10.26) \qquad J(n) = \int_{-b}^{b} \left\{ \int_{-L}^{L} \dot{n}(\xi) [h(\xi) - h(x-\xi)] \, d\xi \right\}^2 dx.$$

There is reason to believe that no choice of $n \in \mathcal{N}$ will reduce the inner integral to zero and, although the infimum of $J(n)$ on $\mathcal{N}$ conceivably may be a zero, it seems likely that it is positive.

If we restrict n to the subclass $\mathcal{N}_1$ of $\mathcal{N}$ consisting of all $n \in \mathcal{N}$ such that $\dot{n}^2$ is integrable over $[-L,L]$, an application of the Cauchy–Buniakovski–Schwarz inequality to the inner integral with unity as the first factor of the integrand yields the relation

$$(10.27) \qquad \left\{ \int_{-L}^{L} \dot{n} [\quad] \, d\xi \right\}^2 \le 2L \int_{-L}^{L} \dot{n}^2 [h(\xi) - h(x-\xi)]^2 \, d\xi;$$

consequently from (10.26),

$$(10.28) \qquad J(n) \le J_1(n) \equiv 2L \int_{-b}^{b} \int_{-L}^{L} \dot{n}^2 [h(\xi) - h(x-\xi)]^2 \, d\xi \, dx.$$

Although the value of an iterated integral depends in general upon the order in which the integrations are performed, a theorem of Fubini, various versions of which are to be found in books on integration, justifies reversing the order in (10.28) and writing

$$(10.29) \qquad J_1(n) = \int_{-L}^{L} \dot{n}^2 I(\xi,b) \, d\xi,$$

where

$$I(\xi,b) \equiv 2L \int_{-b}^{b} [h(\xi) - h(x-\xi)]^2 \, dx.$$

If we restrict n further to the subclass $\mathcal{N}_2$ of $\mathcal{N}_1$ consisting of all PWS functions $n \in \mathcal{N}_1$, the auxiliary problem $J_1(n) = $ *global minimum* on $\mathcal{N}_2$ falls under Theorem 3.9 in view of the absence of n from the integrand. The least value $J_1(n_0)$ of $J_1(n)$ dominates the infimum on $\mathcal{N}_2$ of the original integral $J(n)$ and will be a better approximation to the latter the smaller the ratio b/L. However, such considerations leave the minimum problem of $J(n)$ on $\mathcal{N}$ untouched.

Our class $\mathcal{N}$ was chosen somewhat arbitrarily. It may be that a different mathematical model for the raw problem of achieving a "most nearly constant" axial field will be more tractable. Changes in sign of $\phi(s)$ in (10.20) can be interpreted as reversals in the direction in which turns are wound and hence in the direction of current flow in a continuously wound coil carrying current I. Such sign reversals can occur for functions $n \in \mathcal{N}$. However, in view of the fact that, if $\dot{n}(\xi)$ $[\dot{n}(\xi) = \phi(\xi) \text{ a.e}]$ in (10.21) is constant, then $H(x;n)$ decreases as $|x|$ increases, intuition suggests that deviations in $H(x;n)$ will be most efficiently reduced by increasing the turn density $\phi(\xi)$ with $|\xi|$ with no reversals in sign. If so, then functions ϕ of the form $\phi = \phi_0 + \phi_1$, where $\phi_0(\xi)$ is a positive constant and $\phi_1(\xi)$ increases with $|\xi|$, are important. Symmetry of the given cylinder with respect to a plane normal to its axis at the center ensures that, whatever class of functions n may be admitted, we can immediately restrict attention to those that are even. Such observations suggest various functionals J in addition to those already mentioned, the infima and possible minima of which would be germane to the original physical problem.

For example, with ϕ nonnegative let $\mathcal{N}_3$ denote the class of all nondecreasing functions n: $[-L,L] \to R$ with end values $n(-L) = 0$, $n(L) = 2N$. Since such an n is not in general AC nor even continuous, it is clear from Theorems 8.35 and 8.38 that integrals involving $\dot{n}$ must be avoided. We can, however, replace (10.21) by the relation

$$(10.30) \qquad H(x;n) = \int_{-L}^{L} h(x - \xi) \, dn.$$

The right member can be regarded as a Lebesgue–Stieltjes integral (not discussed in this book), as a Riemann–Stieltjes integral, or as the Weierstrass integral with the integrand F in (7.33) of the special form $F(x,r) = r\phi(x)$. With $H(x;n)$ now given by (10.30), one can ask for a function $n_0 \in \mathcal{N}_3$ minimizing such expressions as (10.23) through (10.25).

Given any one of these nonnegative expressions and a class $\mathcal{N}_j$ of functions n for which it is meaningful, there necessarily exists a sequence n_ν, $\nu = 1, 2, \ldots$ in $\mathcal{N}_j$ such that $J(n_\nu)$ converges to the nonnegative infimum of all values of $J(n)$ on $\mathcal{N}_j$. If further the class $\mathcal{N}_j$ is sequentially compact and if $J(n)$ is continuous or even lower semi-continuous in n, then we have all ingredients needed for an existence theorem following the plan of Theorem 7.1.

Exercise 10.3

1. Establish that $J_1(n)$ has a global minimum on the class $\mathcal{N}_2$ of PWS functions n: $[-L,L] \to R$ with end values $n(-L) = 0$, $n(L) = 2N$ with the aid of Theorem 3.9. Exhibit the unique minimizing function n_0 for $J_1(n)$.

2. Let Φ be the class of all functions ϕ: $[-L,L] \to R$ each of which is even and nondecreasing on $[0,L]$ and hence nonincreasing on $[-L,0]$ and each of which has the fixed positive value $\phi(0)$ at zero. Given

$$H(x;\phi) \equiv \int_{-L}^{L} \phi(\xi)h(x-\xi)\,d\xi$$

and

$$J(\phi) = \int_{-b}^{b} |H(x;\phi) - H(0;\phi)|\,dx$$

set $\gamma \equiv \inf\{J(\phi): \phi \in \Phi\}$. Let ϕ_ν, $\nu = 1, 2, \ldots$, be a sequence in Φ such that $J(\phi_\nu) \to \gamma$ as $\nu \to \infty$. Show, using the conclusion of problem 2, Exercise 7.4, and Fatou's Lemma (Theorem 8.17), that ϕ_ν, can be chosen to have a limit ϕ_0 and that

$$\gamma = \lim_{\nu\to\infty} J(\phi_\nu) \geq J(\phi_0) \geq \gamma.$$

3. With $H(x;n)$ defined by (10.30) and $J(n)$ by (10.23), let n_ν, $\nu = 1, 2, \ldots$, be a sequence in the class $\mathcal{N}_3$ defined preceding (10.30) so chosen that $J(n_\nu)$ converges to the infimum γ of $J(n)$ on $\mathcal{N}_3$. Point out that the Helly Theorem 7.12 is applicable and hence that we can suppose n_ν so chosen, $\nu = 1, 2, \ldots$, as to have a limit n_0. It is known that such a sequence as $H(x;n_\nu) = \int_{-L}^{L} h(x-\xi)\,dn_\nu$, $\nu = 1, 2, \ldots$, then converges to $H(x;n_0)$. (See, for example, L. M. Graves, *Theory of Functions of Real Variables*, McGraw-Hill, New York, 1954, 2nd ed. 1956, Theorem 22, p. 282 in both editions.) Granted this result show that $J(n_0) \leq J(n)$ for all $n \in \mathcal{N}_3$.

4. Let Φ denote the class of all functions ϕ each of which is measurable on $[-L,L]$ and satisfies the relation $|\phi(\xi)| \leq M$, where M is independent of the choice of $\phi \in \Phi$. Point out that the class $\mathcal{N}$ of functions n given by (10.20) and corresponding to $\phi \in \Phi$ is equilipschitzian and therefore sequentially compact. Show next that, for each fixed $n \in \mathcal{N}$ there exists $x_n \in [-b,b]$ such that $|H(x_n;n) - H(0;n)| = \sup\{|H(x;n) - H(0;n)|: x \in [-b,b]\}$. Then let n_ν, $\nu = 1, 2, \ldots$ be a sequence in $\mathcal{N}$ such that $J(n_\nu)$ of the form (10.25) converges to the infimum γ of $J(n)$ on $\mathcal{N}$. Establish finally with the aid of appropriate theorems that there is an $n_0 \in \mathcal{N}$ furnishing a global minimum for $J(n)$.

The author met the physical problem behind this section in connection with the design of a piece of testing equipment. An "engineering solution" that provides a field intensity $H(x;n)$ with small ripples is obtainable by generalizing the two Helmholtz coils into a system of discrete windings with the numbers of turns and the spacing chosen so that $H(x;n)$ has the same value at each of a suitable set of positions x_i on the axis of the cylinder. It is not difficult to prove the existence of minimum values for various functionals $J(n)$ of the types that have been introduced but the author knows of no characterization of a minimizing function nor of any literature on such problems.

10.5   CONFLICT ANALYSIS, GAMES

Given two abstract nonempty sets $\mathscr{X}$ and $\mathscr{Y}$ and a function $J: \mathscr{X} \times \mathscr{Y} \to R$, think of a *maximizing player* $P_{\max}$ and a *minimizing player* $P_{\min}$ who select $x \in \mathscr{X}$ and $y \in \mathscr{Y}$, respectively, and whose respective objectives are to maximize $J(x,y)$ and to minimize $J(x,y)$. Each player knows the function J and is aware of the other player and his objective.

A cogent argument is given in books on game theory to the effect that the players should select $x^* \in \mathscr{X}$ and $y_* \in \mathscr{Y}$ such that

$$(10.31) \qquad \forall x \in \mathscr{X}, \qquad J(x,y_*) \leqslant J(x^*,y_*) \leqslant J(x^*,y), \qquad \forall y \in \mathscr{Y}.$$

In the present brief discussion we simply define a *game* as the problem of establishing existence of at least one pair (x^*,y_*) satisfying inequalities (10.31) together with a characterization of all such pairs (x^*,y_*).

This definition includes the simple two-person zero-sum game with a payoff matrix, for which $\mathscr{X}$ and $\mathscr{Y}$ are the rows and columns of the matrix and $J(x,y)$ is an element in a matrix of real numbers. We are interested here in cases where $\mathscr{X}$ and $\mathscr{Y}$ are function spaces.

Observe that, if (x^*,y_*) satisfies (10.31), then $J(x^*,y_*)$ is simultaneously the global maximum of a function $J(\cdot,y_*)$ on $\mathscr{X}$ and the global minimum of a function $J(x^*,\cdot)$ on $\mathscr{Y}$. A pair (x^*,y_*) satisfying (10.31) is a *maximinimizing pair* for the problem $J(x,y) = maximinimum$ on $\mathscr{X} \times \mathscr{Y}$.

Theorem 10.1

Given nonempty sets $\mathscr{X}$ and $\mathscr{Y}$ and a function $J: \mathscr{X} \times \mathscr{Y} \to R^*$,

$$(10.32) \qquad \sup_x \inf_y J(x,y) \leqslant \inf_y \sup_x J(x,y).$$

PROOF

For an arbitrary pair $(x,y) \in \mathscr{X} \times \mathscr{Y}$,

$$J(x,y) \leqslant \sup_x J(x,y),$$

hence

$$\inf_y J(x,y) \leqslant \inf_y \sup_x J(x,y).$$

The left member is an extended real value $F(x)$ depending only on x. The right member depends on neither x nor y; consequently, the supremum of the left member must satisfy the same inequality and this relation is (10.32).

Theorem 10.2

If the nonempty sets $\mathscr{X}$ and $\mathscr{Y}$ are sequentially compact and $J: \mathscr{X} \times \mathscr{Y} \to R$ is continuous on the stated domain, then

(10.33) $$\max_{x \in \mathscr{X}} \min_{y \in \mathscr{Y}} J(x,y) \quad and \quad \min_{y \in \mathscr{Y}} \max_{x \in \mathscr{X}} J(x,y)$$

both exist.

PROOF

The real values $J(x,y)$ have a finite supremum G, for otherwise there would exist a sequence $\{(x_\nu, y_\nu) : \nu \in N\}$ in $\mathscr{X} \times \mathscr{Y}$ converging to a limit $(x_0, y_0) \in \mathscr{X} \times \mathscr{Y}$ with $J(x_\nu, y_\nu)$ tending to ∞ contrary to the continuity of J at (x_0, y_0).

With x a fixed element of $\mathscr{X}$, set

$$\gamma(x) \equiv \inf\{J(x,y) : y \in \mathscr{Y}\}.$$

There necessarily exists a sequence $\{y_{x\nu} : \nu \in N\}$ in $\mathscr{Y}$ such that $J(x, y_{x\nu}) \to \gamma(x)$ and, since $\mathscr{Y}$ is sequentially compact, we can suppose $y_{x\nu}$ so chosen that it converges to an element y_x of $\mathscr{Y}$. Moreover, by the continuity of J, $J(x, y_x) = \gamma(x)$.

Similar consideration of a sequence $\{x_n : n \in N\}$ in $\mathscr{X}$ so chosen that $J(x_n, y_n^*)$, in which y_n^* means y_x with $x = x_n$, converges to the supremum of $J(x, y_x)$ on $\mathscr{X}$, yields a pair $(x_1, y_1) \in \mathscr{X} \times \mathscr{Y}$ such that $J(x_1, y_1)$ is the first expression (10.33).

The existence of $(x_2, y_2) \in \mathscr{X} \times \mathscr{Y}$ such that $J(x_2, y_2)$ is the second expression (10.33) is similarly established.

By Theorem 10.1, the first quantity (10.33) is dominated by the second. Additional hypotheses sufficient for the equality of these quantities and conditions under which equality occurs for a single pair (x,y) will not be considered here. This was done for a special case by von Neumann, the originator of game theory and various extensions of this *minimax theorem* are to be found in the literature. See, for example, Nikaidô (40a), Kakutani in the Duke Math. Jour., vol. 8 (1941), pp. 457–459, and papers on this topic in several volumes of the *Contributions to the Theory of Games Annals of Mathematics Studies*, Princeton Univ. Press, Princeton, N.J., 1950–1959.

EXAMPLE 10.3

$J(x,y) \equiv y^2 - x^2$ and $\mathscr{X} = \mathscr{Y} = [0,1]$.

One verifies easily that $(x^*, y_*) = (0,0)$ satisfies (10.31). If J is replaced by $-J$, then $(1,1)$ satisfies (10.31) but $(0,0)$ does not. Variables x and y do not play symmetric roles, as is already indicated by Theorem 10.1.

EXAMPLE 10.4

$J(x,y) \equiv \int_0^1 y[1 - q^x]\, dt$, where q is a given constant, $0 < q < 1$, and $\mathscr{X}$ and $\mathscr{Y}$ are, respectively, the classes of all functions x: $[0,1] \to R$ and y: $[0,1] \to R$ that are integrable over $[0,1]$ and satisfy the side conditions

$$x(t) \geq 0 \qquad and \qquad y(t) \geq 0,$$

$$\int_0^1 x\, dt = M > 0 \qquad and \qquad \int^1 y\, dt = N > 0.$$

One verifies that the pair (x^*, y_*) such that $x^*(t) \equiv M$ and $y_*(t) \equiv N$ satisfies (10.31).

Exercise 10.4

1. With hints from Chapters 2 and 3, consider an integral $J(x,y) \equiv \int f(t,x,y,\dot{x},\dot{y})\, dt$, where x and y are required to be PWS on $[t_0, t_1]$ and to have fixed end values and the integrand f is suitably differentiable. Obtain an Euler necessary condition on a pair (x^*, y_*) satisfying (10.31) and explore the question of a theory of necessary conditions and of sufficient conditions for a game-theoretic problem based on such an integral.

2. Given $J(x,y) \equiv \int (\dot{y}^2 + 2\dot{x}\dot{y} - \dot{x}^2)\, dt$ and that $y(0) = 0$, $y(1) = 1$, $x(0) = 1$, and $x(1) = 0$, find a pair (x^*, y_*) of PWS functions with these end values satisfying (10.31).

3. Explore with hints from Chapter 5 the possibility of a pair (x^*, y_*) of PWS functions satisfying (10.31) for the integral, $\int (\dot{x}^2 + \dot{y}^2)\, dt$, subject to side-conditions $\dot{x} = 1 - \sin \dot{w}$ and $\dot{y} = \dot{z}^2$ and to end-conditions $t_0 = 0$, $t_1 = 1$, $x(t_0) = 2$, and $y(t_0) = 0$.

See Isaacs (XXII), Fleming (14a,b), and Berkovitz (4a) for a sample of the work on so-called differential games and for further references.

10.6 PROBLEMS WITH STOCHASTIC INGREDIENTS

This section presupposes familiarity with such concepts from probability theory as a *random variable* $\mathbf{x}$, the *expected value* $\bar{x}$ of a random variable, and a cumulative distribution function $F: R \to [0,1]$,

$$F(x) = Pr\{\mathbf{x} \leq x\}, \qquad \textit{the probability that } \mathbf{x} \leq x.$$

If and only if F is AC, there exists in accord with an extension of Theorem 8.38 a function $f: R \to R$ called a *probability density* such that

$$(10.34) \qquad F(x) = \int_{-\infty}^{x} f(\xi) \, d\xi, \qquad \forall x \in R.$$

Given an integrable function $g: R \to R$ and that $\mathbf{x}$ has a density function f, the expected value $\bar{g}(x)$ is given by the integral

$$(10.35) \qquad\qquad \int_{-\infty}^{\infty} g(x) f(x) \, dx.$$

If in particular $g(x) = x$, then

$$\bar{x} = \int_{-\infty}^{\infty} x f(x) \, dx.$$

Probability theory is so constructed that all probabilities are real numbers on the closed unit interval; consequently, a density $f(x)$ is nonnegative and

$$\int_{-\infty}^{\infty} f(x) \, dx = 1.$$

We are motivated by circumstances of the following sort. A given system has certain inputs and responses with some of the former and hence of the latter affected by random disturbances. One feels intuitively both that realistic mathematical models for many physical systems should include random inputs and that these models ought to generate useful optimization problems.

We shall ignore the important task of formulating a stochastic model for a given system. The problems considered will be abstract and overly simple but will serve to suggest some of the possibilities.

Suppose given a suitably differentiable integrand ϕ with values $\phi(t, \mathbf{x}, y, \dot{y})$, where $\mathbf{x}$ is a random variable with a known probability density function $f: R \to R$ and hence a known distribution function F given by (10.34). Corresponding to each function y in a given class $\mathscr{Y}$, define a random variable $J(y, \mathbf{x})$ by the equation

$$J(y, \mathbf{x}) \equiv \int_{t_0}^{t_1} \phi(t, \mathbf{x}, y, \dot{y}) \, dt$$

and suppose that the desideratum is a small value of $J(y, \mathbf{x})$ in some sense.

Among possible optimization questions, each of which is an approach to the stated objective and has uses depending upon the nature of the problem and the context within which it has arisen, are the following:

(i) Does there exist $y_0 \in R$ such that the expected value $\bar{J}(y_0)$ of $J(y_0, \mathbf{x})$ is a global minimum?

(ii) Does there exist $y_0 \in R$ such that the probability $\Pr\{J(y_0,\mathbf{x})$ $\leqslant k\}$ is a global maximum?

If the answer to any such question is affirmative, one then wishes to identify a function y_0 with the extremizing property.

EXAMPLE 10.5

$J(y,\mathbf{x}) \equiv \int (\dot{y}^2 + 4\mathbf{x}y) \, dt$, $\mathcal{Y}$ consists of all PWS functions y: $[0,1] \to R$ with end values $y(0) = 0$ and $y(1) = 1$ and $\mathbf{x}$ has a given density function f.

Discussion

The expected (or average) value of $J(y,\mathbf{x})$, for each fixed $y \in \mathcal{Y}$, is, by (10.35),

$$\int_{-\infty}^{\infty} J(y,\mathbf{x}) f(x) \, dx = \int_{-\infty}^{\infty} \int_{0}^{1} (\dot{y}^2 + 4xy) f(x) \, dt \, dx$$
$$= \int_{0}^{1} \int_{-\infty}^{\infty} (\dot{y}^2 + 4xy) f(x) \, dx \, dt = \int_{0}^{1} (\dot{y}^2 + 4\bar{x}y) \, dt.$$

It is then left as an exercise to show by the methods of Chapters 2 and 3 that

$$y_0(t) = \bar{x}t^2 + (1 - \bar{x})t$$

furnishes a global minimum for the last integral.

EXAMPLE 10.6

$J(y,\mathbf{x}) \equiv \int (\dot{y}^2 + \mathbf{x}y^2) \, dt$, $\mathcal{Y}$ is the same class as in Example 10.5 and $\mathbf{x}$ has a density function f that vanishes for all negative values of x; hence $\Pr\{x < 0\} = 0$.

Discussion

Turning to question (ii), consider

(10.36) $\Pr\{J(y,\mathbf{x}) \leqslant k\} = \Pr\left\{\int_{0}^{1} \dot{y}^2 \, dt + \mathbf{x} \int_{0}^{1} y^2 \, dt \leqslant k\right\}.$

The inequality on the right is equivalent to

(10.37) $$\mathbf{x} \leqslant \frac{k - \int_{0}^{1} \dot{y}^2 \, dt}{\int_{0}^{1} y^2 \, dt}$$

provided that the divisor is not zero. Therefore, the probability (10.36) is of the form $F_J[\rho(y)]$, where F_J is the cumulative distribution function for $J(y,\mathbf{x})$ and $\rho(y)$ denotes the ratio on the right in (10.37).

A distribution function is necessarily nondecreasing, hence (10.36) has a greatest value if and only if ρ has a greatest value. The problem is then to maximize the functional $\rho(y)$ on $\mathcal{Y}$. We have developed no theory of extrema for examples of this form and must decide whether to seek necessary conditions on a maximizing function, to try to prove an existence theorem, or to start with some other move.

Exercise 10.5

1. Let the number k in (10.36) be positive and consider the admissible functions y_n with values $y_n(t) = [2n^{1/4}/(n+2)]t^{(n+2)/2}$ for arbitrarily large values of n. What conclusions do you reach for the problem considered under Example 10.6?

2. Investigate the problem of minimizing the expected value of the integral $J(y,\mathbf{x})$ of Example 10.6.

3. $J(y,\mathbf{x}) = \int (\dot{y} - \mathbf{x})^2 \, dt$ and $\mathcal{Y}$ is the class of all PWS functions y: $[0,2] \rightarrow R$ with end values $y(0) = y(2) = 0$. Show that the probability $\Pr\{J(y,\mathbf{x}) \leqslant k\}$ has a global maximum that is independent of the choice of $k \in [0,1]$.

This section points to a field, which cannot even be described in a few paragraphs. Elementary random variables do not suffice. Among the prerequisites is a knowledge of what are called stochastic processes. Back of this lies parts of real and complex analysis not presupposed nor included in the present book. One should make contact with the developments initiated by Norbert Wiener. To obtain an impression of work in this sector and additional references, see Part Two of reference (VI), Chapter VII of Pontryagin et al. (XXXIII), Stratonovich (52a), and various works of Bellman.

10.7 PROBLEMS WITH LAGS

Questions concerning the optimal design or optimal performance of modern systems have brought to attention phenomena in which the response to a given input occurs not instantaneously but after the elapse of a certain time. In contrast with the equations of classical dynamics in which position $y(t)$, velocity $\dot{y}(t)$, and acceleration $\ddot{y}(t)$ are all evaluated at the same time t, the dynamical equations of a system with lags will include some values at t and others at $t - \tau, \tau > 0$. In more complicated examples, several different lags may appear or, in place of a constant lag τ, there may be a variable lag of the form $\tau(t)$.

Among the alternative names of such equations are *lag differential equations, delay differential equations, differential-difference equations,* or *functional differential equations.* There is a scattered literature on these equations going back a number of years with wide interest and many publications since about 1950. For information see Oğuztöreli (XXX, Part I) and his extensive bibliography; El'sgol'ts, *Introduction to the Theory of Differential Equations with Deviating Arguments*, Holden-Day, San Francisco, 1966; or Bellman and Cooke, *Differential-Difference Equations*, Academic Press, New York, 1963.

EXAMPLE 10.7

$$\dot{y}(t) = y(t-1).$$

Discussion

Experience with equations having no lags will suggest the possibility that the given equation has a family of solutions and that, moreover, a suitable initial condition should determine a unique member of this family. What type of initial condition will achieve this is not immediately obvious.

Since the example is linear, one may simply try $y(t) = Ae^{rt}$ and find that this is indeed a solution for an arbitrary A if and only if r is the unique real root of the equation $r = e^{-r}$. An assigned initial value $y(0)$ yields a unique value $A = y(0)$. That there are many other solutions having this same initial value will be pointed out. To obtain a unique solution for $t \geqslant 0$, we need not merely an initial value $y(0)$ but must require that

$$(10.38) \qquad y(t) = \alpha(t), \qquad \forall t \in [-1,0],$$

where $\alpha(t)$ is the value at t of a given *initial function* α: $[-1,0] \to R$. If, for instance, $\alpha(t) \equiv 1$, we can calculate $y(t)$ on $[0,1]$ by direct integration from the given differential equation as follows.

$$y(t) = \int_0^t y(s-1)\, ds = \int_{-1}^t \alpha(s)\, ds = 1+t, \qquad t \in [0,1].$$

We now know $y(t)$ on $[0,1]$ and, by repeating the process with $y(t-1) = 1 + (t-1)$, we find for $t \in [1,2]$ that

$$y(t) = \int_1^t y(s-1)\, ds = \int_1^t [1 + (s-1)]\, ds = 1 + t + \frac{(t-1)^2}{2}, \qquad t \in [1,2].$$

Continuing in this stepwise manner, an interval of length $\tau = 1$ at each step, the solution y can be extended as far as may be desired. For a more complicated or nonlinear example it would likely not be possible to replace the sequence of expressions for $y(t)$ by a single elementary expression but, because of the simplicity of this example, one finds that

$$(10.39) \qquad y(t) = \sum_0^n \frac{(t-k+1)^k}{k!}, \qquad t \in [n-1, n], n = 1, 2, \ldots.$$

The step by step process, illustrated by Example 10.7, applies generally to a first-order equation $\dot{y} = g[t, y(t-\tau), y(t)]$ provided that g is lipschitzian in its last two arguments [actually somewhat less suffices (XXX, p. 29)] to yield a unique solution y on some interval $0 \leqslant t \leqslant a$ and such that $\dot{y}(t)$ coincides in value on $[0,\tau] \cap [0,a]$ with $g[t, \alpha(t-\tau), y(t)]$, where α is a given continuous initial function from $[-\tau, 0]$ to R.

If and only if $\alpha(t) = e^{rt}$, with $r = e^{-r}$, will we obtain an exponential solution $y(t) = e^{rt}$ of Example 10.7 having the initial value $y(0) = 1$. Any function α having a different indefinite integral from the first one, even one with the same value $\alpha(0) = 1$ at zero, will yield a different solution y of $\dot{y}(t) = y(t-1)$. For instance, (10.39) is not of the form e^{rt}, even though each of them is a solution of Example 10.7 and has the same value 1 at $t = 0$.

For a system of first-order equations with delay, again written $\dot{y} = g[t, y(t-\tau), y(t)]$, but with y, $\dot{y}$, and g now each having n components, the initial function α must now have n components. A single higher-order equation with a lag is equivalent to a system of first-order equations, as in the lag-free case described in Section 1.12. For example, if $\ddot{y} = g[t, y(t-\tau), y(t), \dot{y}(t-\tau), \dot{y}(t)]$, we can set $\dot{y} = z$ and obtain the pair of equations

$$\dot{y} = z \quad and \quad \dot{z} = g[t, y(t-\tau), y(t), z(t-\tau), z(t)].$$

Any physical system idealized by a system of lag differential equations generates variational problems with lags in the same manner as do systems without lags in Chapter 5. Existence theory for such variational problems has progressed more rapidly than the development of criteria for characterizing functions y that furnish extreme values. See Oğuztöreli (42a,b,c) and (XXX, Part 2 and bibliography); also M. Q. Jacobs (24a, Chap. V).

Euler necessary conditions for certain problems are mentioned without details by El'sgol'ts in (11a) and (XVIII, pp. 215–229). Pontryagin et al. (XXXIII, pp. 213–226) discuss their Maximum Principle (see Section 11.6) for a control problem with a lag.

An analogue of the classical fixed-endpoint nonparametric variational problem in $(n+1)$-space is the following. Let $\mathcal{Y}$ denote the class of all PWS functions $y \equiv (y^1, \ldots, y^n): [a-\tau, b] \to R^n$ with fixed values $y(t) = \alpha(t)$ on $[a-\tau, b]$ and $y(b) = \beta$, the given initial function $\alpha: [a-\tau, b] \to R^n$ being PWS. Given the functional $J: \mathcal{Y} \to R$,

$$(10.40) \qquad J(y) \equiv \int_a^b f[t, x(t), y(t), \dot{x}(t), \dot{y}(t)]\, dt,$$

in which $x(t) \equiv y(t-\tau)$ and $\dot{x}(t) \equiv \dot{y}(t-\tau)$, we desire necessary conditions and sufficient conditions on an extremizing function $y_0 \in \mathcal{Y}$.

Among the results of D. K. Hughes in (21a,b) are analogues, for this problem, of the necessary conditions of Euler, Weierstrass, Legendre, and Jacobi together with a sufficient condition for a global minimum such as that in Section 3.12. The following is a statement of his first theorem.

Theorem 10.3

If y_0 minimizes $J(y)$ on $\mathcal{Y}$, then there exists a constant n-vector c such that y must satisfy the following integrodifferential-difference equations subject to convention (2.17):

$$(10.41) \qquad f_r(t,x,y,\dot{x},\dot{y}) + f_q(t+\tau,y,z,\dot{y},\dot{z})$$

$$= \int_{b-\tau}^{t} [f_y(s,x,y,\dot{x},\dot{y}) + f_x(s+\tau,y,z,\dot{y},\dot{z})] \, ds + c, \qquad \forall t \in [a,b-\tau]$$

and

$$(10.42) \qquad f_r(t,x,y,\dot{x},\dot{y}) = \int_{b-\tau}^{t} f_y(s,x,y,\dot{x},\dot{y}) \, ds + c, \qquad \forall t \in [b-\tau,b].$$

Each vector-equation (10.41) and (10.42) yields n equations in the respective components. Symbol x is defined above and $z(t) \equiv y(t+\tau)$. The integrand $f: [a,b] \times R^{4n}$ is subject to a blanket continuity and differentiability hypothesis and has values denoted by $f(t,x,y,q,r)$; hence respective subscripts in (10.41), (10.42) denote the vectors whose components are partial derivatives of f with respect to components of the indicated arguments.

The conclusions (10.41) and (10.42) are obtained by an adaptation of the proof of Theorem 2.2 using an η vanishing on $[a-\tau, a]$ and at b and the classical du Bois Reymond Lemma (our Theorem 2.1). An Erdmann corner condition at $b-\tau$ and also one at other possible corners are obtained as corollaries.

Since the Euler equations involve both lead and lag terms, particular examples quickly outstrip the author's ability to obtain elementary solutions. Investigating all details for the following examples is suggested as an exercise.

EXAMPLE 10.8.   (Hughes)

$$J(y) = \int_0^3 (\dot{y}^2 - \ddot{x}^2)^2 \, dt,$$

with lag $\tau = 1$, $\alpha(t) \equiv -t$, and $y(3) = 1$.

EXAMPLE 10.9.   (Hughes)

$$J(y) = \int_0^3 (\dot{y} + \ddot{x})^2 \, dt,$$

with $\tau = 1$, $\alpha(t) \equiv -t$, and $y(3) = 2$.

The Euler equations are

$$2\dot{y}(t) + \dot{y}(t+1) + \dot{y}(t-1) = c, \qquad 0 \leqslant t \leqslant 2,$$
$$\dot{y}(t) + \dot{y}(t-1) = c, \qquad 2 \leqslant t \leqslant 3.$$

A solution of these equations with $c = \frac{1}{2}$ is $y(t) = -t$, $3t/2$, $3 - 3t/2$, or $2t - 4$ according as $t \in [-1,0]$, $(0,1]$, $(1,2]$, or $(2,3]$, respectively. This function y has a corner at $t = 1$.

In view of the nonnegative integrand we suspect that there is no maximizing function y and that we have found a function y that furnishes some kind of a minimum, weak local, strong local, or global. By an application of a sufficiency theorem of Hughes that has been mentioned above, it can be shown that $J(y)$ is a global minimum.

10.8   CONCLUDING REMARKS

The purpose of this chapter has been to illustrate by a sample the diversity of problems that can occur. Our selection is far from inclusive either of features that may be encountered or of concepts and devices that may be an aid to analysis.

There are a number of active fronts for which the theories are still in process of development at the time this is written and in which relatively few particular examples of any complexity have been completely analyzed.

We have simplified the discussion by isolating nonclassical features into separate sections, but a realistic mathematical model for a given system may include combinations of such features. For example, an optimal maneuver of an aircraft might involve stochastic and game-theoretic elements as well as time-delays and inequality constraints.

Chapter 11

HAMILTON–JACOBI THEORY

11.1 INTRODUCTION

The ideas to be presented come from work of Hamilton and Jacobi in the early nineteenth century. Each of them was motivated by dynamical systems in rational mechanics, but Carathéodory has called attention (XII, p. 251) to a relationship with results of Huygens in optics published in 1690. Thus the Hamilton–Jacobi approach to variational theory has certain roots of about the same age as other parts of the theory.

In principle, all of the calculus of variations could be developed from this viewpoint, with little dependence upon the methods surveyed in preceding chapters. In practice the discussion of Hamilton–Jacobi methods often assumes previous knowledge of necessary conditions and of fields.

This chapter is a brief introduction under classical hypotheses that permit us to draw from Chapters 2 and 3. Our limited objective is to open the subject, not to survey its present state or even to treat in detail all aspects of the simplest problems. For more extensive developments including Bolza Problems see Hestenes (XXI). Carathéodory continues to provide an important source in his book (XII, pp. 203–388) and in a number of his papers (XIII). Hamilton–Jacobi methods are discussed in many books on the calculus of variations, including Gelfand and Fomin

(XX), Bliss (IX), Bolza (XI), and Osgood (XXXI). A number of advanced books on physics describe parts of the theory.

Two recent developments, each in the Hamilton–Jacobi tradition, are the Maximum Principle of L. S. Pontryagin and the Dynamic Programming of Richard Bellman. Since both have received extensive coverage in books and articles, our discussion here is confined to pointing out relationships among classical Hamilton–Jacobi theory and the approaches of Bellman and of Pontryagin.

For detailed treatment of the Maximum Principle in a number of settings with PWC controls u and with state variables x that are PWS in the sense of Section 1.9, see Hestenes (XXI). One learns from Pontryagin and his associates (XXXIII) that the Maximum Principle also extends to the case of Lebesgue measurable controls u and absolutely continuous state variables x. An important feature of the problems of Bolza type treated by these and other authors is the manner in which side-inequalities [in place of or in addition to the equality side-conditions (5.5)] fit into the theory without the introduction of new variables as in Section 5.12.

Dynamic Programming is also applicable to problems with inequality constraints and to discrete problems, that is, to optimization problems in which one seeks a function x whose domain is not an interval but a finite set. The works of Bellman, including (IV), together with those of Bellman and Dreyfus (V) and Dreyfus (XVII) discuss the methods and scope of Dynamic Programming.

11.2 THE CANONICAL FORM OF THE EULER CONDITION

Consider the fixed-endpoint nonparametric problem formulated in Section 2.2 with the integrand f again subject to the blanket differentiability hypothesis of that section. We continue to denote a derivative of f with respect to its third argument by a subscript r, even though various symbols will occupy the third position.

Suppose that f, f_r, etc., have as common domain the set $A \equiv [t_0, t_1] \times R \times R$. Cases in which the domain is a suitable proper subset of A would require only minor revision of what follows. Denote the range of f_r by R_1, set

(11.1) $$v = f_r(t, y, r),$$

and restrict attention to those integrands f such that

(11.2)   *equation (11.1) defines a one-one correspondence between triples*
         $(t, y, r) \in A$ *and triples* $(t, y, v) \in B \equiv [t_0, t_1] \times R \times R_1, R_1 \subset R.$

Given $(t,y,r) \in A$, we have a unique $v \in R_1$ from (11.1) and hence a unique triple $(t,y,v) \in B$. Consequently, (11.2) is realized if there exists a function $P: B \to R$ such that, if we substitute $P(t,y,v)$ for r in (11.1), that equation becomes an identity on B. A sufficient condition for such a function to be determined implicitly by (11.1) is that f_r be strictly increasing in r for each fixed $(t,y) \in [t_0,t_1] \times R$ and a sufficient condition for this is that $f_{rr}(t,y,r)$ be positive on A. This is the condition III_R stated in Section 3.1.

Define a *hamiltonian function* $H: B \to R$,

$$(11.3) \qquad H(t,y,v) \equiv vP(t,y,v) - f[t,y,P(t,y,v)].$$

As a result of the blanket hypothesis, f, H and P have finite first-order partial derivatives and, by differentiation of (11.3),

$$(11.4) \qquad \begin{aligned} H_t(t,y,v) &= -f_t[t,y,P(t,y,v)], \\ H_y(t,y,v) &= -f_y[\qquad\qquad], \\ H_v(t,y,v) &= P(t,y,v). \end{aligned}$$

Theorem 11.1

If $y: [t_0,t_1] \to R$ is a smooth solution on $[t_0,t_1]$ of the Euler equation for f, then y and v satisfy Hamilton's so-called canonical equations

$$(11.5) \qquad \dot{y} = H_v(t,y,v) \qquad and \qquad \dot{v} = -H_y(t,y,v).$$

Conversely, if the pair (y,v) satisfies this system on $[t_0,t_1]$, then y is a smooth solution of the Euler equation (2.24).

PROOF

Suppose first that y is a smooth solution of the equation (2.24), that is, that

$$(11.6) \qquad f_y[t,y(t),\dot{y}(t)] = \frac{d}{dt} f_r[t,y(t),\dot{y}(t)] \quad on \quad [t_0,t_1].$$

By (11.1), (11.4), and the relation $\dot{y} = P(t,y,v)$, with $\dot{y}$ now replacing the r in (11.1), it follows from (11.4$_3$) that

$$\dot{y}(t) = P[t,y(t),v(t)] = H_v[t,y(t),v(t)]$$

and then from (11.1), (11.6), and (11.4$_2$) that

$$\dot{v}(t) = \frac{d}{dt} f_r[t,y(t),\dot{y}(t)] = f_y[t,y(t),\dot{y}(t)]$$

$$= f_y\{t,y(t),P[t,y(t),v(t)]\} = -H_y[t,y(t),v(t)].$$

If, conversely, the pair (y,v) satisfies (11.5) on $[t_0,t_1]$, observe from the form (11.3) of H that H_v and H_y are continuous in (t,y,v) under the blanket hypothesis on f. Moreover, from (11.5), the integrals

$$y(t) = \int_{t_0}^{t} H_v[\tau,y(\tau),v(\tau)]\, d\tau + y(t_0)$$

and

$$v(t) = -\int_{t_0}^{t} H_y[\qquad\qquad]\, d\tau + v(t_0)$$

are both continuous in the upper limit t on $[t_0,t_1]$ so that $H_v[t,y(t),v(t)]$ is continuous in t on $[t_0,t_1]$ and, by (11.5$_1$), $\dot{y}(t)$ is continuous on that interval. That y satisfies the Euler equation (11.6) on $[t_0,t_1]$ then follows from (11.5$_2$), (11.4$_2$), and definition (11.1) of v.

Theorem 11.1 says that, under the hypotheses of this section, the canonical system (11.5) and the Euler equation (11.6) are equivalent.

Exercise 11.1

1. Given that $f(t,y,r) = (1+r^2)^{1/2}$, what are $P(t,y,v)$ and $H(t,y,v)$? Solve the canonical equations $\dot{y} = v/(1-v^2)^{1/2}$ and $\dot{v} = 0$ and point out that the results are consistent with Example 2.1 of Section 2.7 and Theorem 11.1.

2. Extend the discussion of this section to the fixed-endpoint nonparametric problem in $(n+1)$-space, showing that we again obtain equations (11.5) provided that all members are properly interpreted as vectors.

The canonical equations provide another viewpoint from which to approach the theory, but like the Euler equation they seldom have solutions expressible in elementary form. For discussion of certain special cases see Osgood (XXXI, pp. 410–446).

11.3   TRANSVERSALS TO A FIELD

Let (S,p) be an arbitrary but fixed field in the sense of Section 3.2 and define

$$(11.7)\quad W(t,y) \equiv \int_{(a,b)}^{(t,y)} \{f[\tau,\eta,p(\tau,\eta)] - p(\tau,\eta)f_r[\quad]\}\, d\tau + f_r[\quad]\, d\eta.$$

We see from definition (3.6) and Theorem 3.2 that $W(t,y)$ is the value of the Hilbert invariant integral $J^*(y)$ from an arbitrary fixed point (a,b) of S to a variable point (t,y) of that set.

As a consequence of the continuity of f and f_r under our blanket hypothesis and continuity of the slope-function p of the field under restrictions of Section 3.2, which are retained in this section, one can verify that $W_t(t,y)$ and $W_y(t,y)$ are continuous on S. Moreover, the expression to be integrated in (11.7) is the differential dW, hence

(11.8)
$$W_t(t,y) = f[t,y,p(t,y)] - p(t,y)f_r[t,y,p(t,y)],$$
$$W_y(t,y) = f_r[t,y,p(t,y)].$$

Hypothesis (11.2) applied to (11.8$_2$) assures that the slope function p of the given field and the function P defined implicitly by (11.1) satisfy the equation

(11.9)
$$p(t,y) = P[t,y,W_y(t,y)], \qquad \forall (t,y) \in S.$$

As a consequence of (11.1), (11.3), and (11.8) we also verify that W is a solution of the *Hamilton–Jacobi partial differential equation*

(11.10)
$$\frac{\partial W}{\partial t} + H\left(t,y,\frac{\partial W}{\partial y}\right) = 0,$$

or, alternatively stated, that

$$W_t(t,y) + H[t,y,W_y(t,y)] = 0, \qquad \forall (t,y) \in S.$$

Suppose that the equation $W(t,y) = c \equiv$ const. determines implicitly a function Y on some t-interval I, that is, that there exists $Y: I \to R$ such that

$$W[t,Y(t)] = c, \qquad \forall t \in I.$$

Then, by differentiation,

$$W_t[t,Y(t)] + W_y[t,Y(t)]\dot{Y}(t) = 0,$$

and, after substitution from (11.8),

(11.11) $$f\{t,Y(t),p[t,Y(t)]\} + \{\dot{Y}(t) - p[t,Y(t)]\}f_r[t,Y(t),p[t,Y(t)]] = 0.$$

This shows that the transversality condition (2.57) (see problem 11, Exercise 2.6) holds at each intersection where $\dot{Y}(t)$ exists and is finite of a function $\phi(\cdot,\alpha)$ in the family that generates the given field (S,p) with the graph of $W(t,y) = c$. In the special case where $W_y(t,y)$ vanishes at such a point and the classical implicit function theorem (Theorem 1.1) does not apply, we have from (11.8$_2$) that

(11.12)
$$f_r[t,y,p(t,y)] = 0.$$

This is the special form of the transversality condition found in problem 10, Exercise 2.6.

11.4 THE FORMALISM OF DYNAMIC PROGRAMMING

Given the classical nonparametric problem of Section 2.2, suppose that $y_0: [t_0,t_1] \to R$ is a smooth function with the given end values and that $J(y_0)$ is a minimum at least in the weak local sense. Suppose further that y_0 is embedded in a field. Sufficient conditions for the existence of a field are given by Theorem 3.1, but we shall not use this theorem explicitly.

One verifies from definition (11.7) of W and the additivity of a curvilinear integral over two paths with a common endpoint that, if t and $t+\Delta$ are both in the interval $[t_0,t_1]$, then

$$(11.13) \qquad W[t+\Delta,y_0(t+\Delta)] - W[t,y_0(t)]$$
$$= \int \{f[\tau,\eta,p(\tau,\eta)] - p(\tau,\eta)f_r[\quad]\}\, d\tau + f_r[\quad]\, d\eta.$$

The integral is from $[t,y_0(t)]$ to $[t+\Delta,y_0(t+\Delta)]$.

The present τ and η, respectively, replace symbols t and y of our discussion in Sections 3.3 and 3.4 of the Hilbert integral J^* and the E-function. Recall with reference to the proof of Theorem 3.3 that in the present notation.

$$(11.14) \qquad\qquad J(y_0) = J^*(y_0) = J^*(\eta),$$

where $J^*(\eta)$ is the right member of (11.13). It follows from (11.13) and (11.14) that

$$(11.15) \qquad -W[t,y_0(t)] = \int_t^{t+\Delta} f[\tau,y_0(\tau),\dot{y}_0(\tau)]\, d\tau - W[t+\Delta,\, y_0(t+\Delta)].$$

Let $\mathscr{Y}(y_0,t,\Delta)$ denote the class of all PWS functions $y: [t,t+\Delta] \to R$ that are coterminal with y_0 on the interval $[t,t+\Delta]$, $\Delta > 0$. This means that

$$(11.16) \qquad\qquad y(t) = y_0(t) \qquad and \qquad y(t+\Delta) = y_0(t+\Delta).$$

Since y_0 is a minimizing function for the problem of Section 2.2 by hypothesis, then the restriction of y_0 to any subinterval $[t, t+\Delta]$ of $[t_0,t_1]$ necessarily furnishes the same type of minimum on the class $\mathscr{Y}(y_0,t,\Delta)$. It follows from (11.15), with $y_0(t+\Delta)$ in the last term replaced by $y(t+\Delta)$ so as not to lose the effect of $\dot{y}(t)$, when we expand the last term, that

$$(11.17) \qquad -W[t,y_0(t)] = \min_y \left\{ \int_t^{t+\Delta} f(\tau,y,\dot{y})\, d\tau - W[t+\Delta,\, y(t+\Delta)] \right\}.$$

By the Mean Value Theorem of the Differential Calculus,

$$(11.18) \quad W[t+\Delta, y(t+\Delta)] = W[t,y(t)]$$

$$+ W_t[t+\theta\Delta, y(t+\theta\Delta)]\Delta + W_y[\quad]\dot{y}(t+\theta\Delta)\Delta, \quad \theta \in (0,1).$$

By the first mean value theorem for integrals, the integral in (11.17) has the value

$$(11.19) \qquad f[t+\theta'\Delta, y(t+\theta'\Delta), \dot{y}(t+\theta'\Delta)]\Delta, \quad \theta' \in [0,1].$$

We substitute (11.18) and (11.19) into (11.17), observing from (11.16₁) that the left member of (11.17) and the first term on the right in (11.18) are equal. After deleting these terms and dividing out the positive factor Δ we have that

$$0 = \min_{y} \{f[t+\theta'\Delta, y(t+\theta'\Delta), \dot{y}(t+\theta'\Delta)]$$

$$- W_t[t+\theta\Delta, y(t+\theta\Delta)] - W_y[t+\theta\Delta, y(t+\theta\Delta)]\dot{y}(t+\theta\Delta)\}.$$

If it were possible to make a direct determination of this minimum we might then be led to a useful necessary condition on the minimizing function y_0 by letting $\Delta \to 0$. There is no apparent way in which to carry out this procedure and, although we know that such operators as *min* and *lim* generally do not commute, we proceed at this juncture to let $\Delta \to 0$ first and hence to make the questionable assertion that

$$(11.20) \qquad 0 = \min_{y} \{f[t,y(t),\dot{y}(t)] - W_t[t,y(t)] - W_y[t,y(t)]\dot{y}(t)\}.$$

The class $\mathscr{Y}(y_0,t,\Delta)$ of functions y on $[t, t+\Delta]$ has now disappeared except for the initial values $y(t)$ and $\dot{y}(t)$ but $y(t) = y_0(t)$ is fixed by condition (11.16₁). It seems plausible to regard the minimum indicated in (11.20) as being with respect to the value $\dot{y}(t)$. Our intention, although tenuously based, is clarified by shifting notation to a new symbol q in place of $\dot{y}(t)$, moving the term that is free of q to the left member and expressing (11.20) in the form

$$(11.21) \qquad W_t[t,y_0(t)] = \min_{q} \{f[t,y_0(t),q] - W_y[t,y_0(t)]q\},$$

known as the *Bellman functional differential equation*.

Proceeding in the same exploratory manner, let us suppose that the minimum in (11.21) exists and that the minimizing value of q corresponds to vanishing of the derivative with respect to q. If so, then

$$(11.22) \qquad\qquad f_r[t,y_0(t),q] - W_y[t,y_0(t)] = 0.$$

At this juncture return to the hypotheses stated in the opening paragraph of this section and strengthen them by requiring not only that y_0 be a minimizing function but that each function $y: [t_0, t_1] \to R$ in the family that generates the field shall be a minimizing function, at least in the weak local sense, in competition with other PWS functions having the same end values as y. Then, if all steps leading to (11.22) are valid for such functions y, we can replace $y_0(t)$ by $y(t)$ in (11.22) and $[t, y(t)]$ is now any point (t, y) in the set S associated with the given field (S, p). In place of (11.22), we now have

(11.22*) $f_r(t, y, q) - W_y(t, y) = 0.$

If this equation determines $q(t, y)$ implicitly on the set S, a condition essentially equivalent to hypothesis (11.2), we then have for $(t, y) \in S$ that

$$W_y(t, y) = f_r[t, y, q(t, y)],$$

which with the Bellman equation (11.21) yields the further result that

$$W_t(t, y) = f[t, y, q(t, y)] - q(t, y) W_y(t, y).$$

Thus, granted conditions which justify all steps that precede, the Bellman approach leads to the classic equations (11.8) of Hamilton–Jacobi theory. In Section 11.6 we use Hamilton–Jacobi methods to obtain a special case of the Pontryagin Maximum Principle and from it we recover the Bellman equation (11.21). Under suitable restrictions that we shall not identify here, (11.21) is a necessary condition on a smooth minimizing function y_0 for the variational problem of Section 2.2.

For discussion of some of the difficulties see H. Osborn. *On the foundations of dynamic programming*, Jour. of Math. and Mechanics, vol. 8 (1959), pp. 867–872; also Pontryagin et al. (XXXIII, pp. 69–73) and Boltyanskii (6a); also Sagan (XL, Sec. 5.2).

11.5   EXAMPLES

EXAMPLE 11.1

$J(y) = \int \dot{y}^2 \, dt$ with fixed endpoints $(0,0)$ and $(1,0)$. Clearly $y_0 = 0$ minimizes, but we wish to illustrate methods of this chapter.

Discussion

Equation (11.21) is

(11.23) $W_t(t, y) = \min_q \left[q^2 - q W_y(t, y) \right].$

One finds by elementary calculus [see conditions (2.7)] that for each fixed (t,y) the bracketed expression is a global minimum and hence a local minimum provided that $2q - W_y(t,y) = 0$. Using this in (11.23) we find that

$$(11.24) \qquad 4\frac{\partial W}{\partial t} = -\left(\frac{\partial W}{\partial y}\right)^2,$$

which is the Hamilton–Jacobi equation (11.10) for the present example.

We wish to avoid discussing ways and means for solving partial differential equations and hence shall work indirectly. By Chapters 2 and 3 the linear function joining (a,c) to $(1+e, 0)$ furnishes a global minimum on the class $\mathscr{Y}(a,c)$ of PWS functions y with the stated endpoints. The slope of this linear function is

$$\overset{\bullet}{y}(t) = -c/(1+e-a).$$

With e fixed, the family of all linear functions through $(1+e, 0)$ generates a field of the type presupposed at the beginning of Section 11.4.

We verify that $W(a,c)$ is given by (11.7) with $(1+e, 0)$ and (a,c), respectively, in place of (a,b) and (t,y). Therefore, in view of (11.14) we can replace the Hilbert integral J^* in (11.7) by J and have that

$$(11.25) \qquad W(a,c) = \int_{1+e}^{a} \overset{\bullet}{y}{}^2(t)\, dt = c^2/(1+e-a).$$

We can now replace (a,c) by (t,y) and verify by differentiation of $W(t,y)$ that this is a solution of (11.24) on the entire (t,y) plane, except for the line $t = 1+e$. Hence our W is a solution on the strip bounded by $t = 0$ and $t = 1$ in which we are interested.

The minimum value of the given integral is $W(0,0) = 0$, but we have not yet identified the minimizing function y_0. The field is an important part of the theory. Clearly the slope $p[t,y_0(t)] = \overset{\bullet}{y}_0(t)$; hence from equation (11.8$_2$) as applied to our example

$$\overset{\bullet}{y}_0(t) = y_0(t)/[t-(1+e)],$$

and we find by elementary integration and the initial condition $y_0(0) = 0$ that $y_0(t) \equiv 0$.

In principle we should be able to find (11.25) directly from (11.24) and could then have obtained y_0 as has been done above. The function y_0 is only a candidate for a minimizing function and we must show that it satisfies some sufficiency theorem before arriving at any firm conclusion. If we deny ourselves the use of anything not in the present chapter it will require considerable work to establish sufficiency, and we would likely succeed only if we retraced some of the work of Jacobi and Weierstrass.

The term Dynamic Programming has several shades of meaning. It is sometimes a descriptive term for computational techniques associated with equation (11.17). A first approximation to the integral in that relation is $f[t,y(t),\overset{\bullet}{y}(t)]\Delta$. Therefore after a shift to a minimum with respect to q, as in (11.21), we have that

$$(11.26) \quad W[t,y(t)] = \min_{q} \{f[t,y(t),q]\Delta - W[t+\Delta, y(t+\Delta)] + \rho\},$$

where ρ denotes a remainder term that goes to zero with Δ. For our Example 11.1, (11.26) becomes

$$(11.27) \qquad W[t,y(t)] = \min_{q} \{q^2\Delta - W[t+\Delta, y(t+\Delta)] + \rho\}.$$

To generate an approximation y to a candidate for minimizing function, ignore the remainder term ρ, start by taking $[t+\Delta, y(t+\Delta)]$ to be the given right endpoint $(1,0)$ and work backward step by step with repeated use of (11.27). By (11.25), $W(1,0) = 0$. Moreover, $q^2\Delta$ assumes its global minimum if $q = 0$. Since q really means $\overset{\bullet}{y}(t)$, we obtain the approximations

$$y(1-\Delta) = 0 \qquad and \qquad W[1-\Delta, y(1-\Delta)] = 0.$$

Repeating the calculation with $t+\Delta$ in (11.27) now as $1-\Delta$, we find similarly that

$$y(1-2\Delta) = 0 \qquad and \qquad W[1-2\Delta, y(1-2\Delta)] = 0.$$

If we take $\Delta = 1/n$ and proceed through n such steps we arrive at the values

$$y(1-n\Delta) = y(0) = 0 \qquad and \qquad W[0,y(0)] = 0.$$

This is an example of a direct computational method, one of those mentioned in the second paragraph of Section 7.1.

For this overly simple example the successive values $y(1-n\Delta)$ are independent of the choice of a value for Δ; hence we may hope that the function y: $[0,1] \to R$, $y(t) \equiv 0$ minimizes the given integral. We can show that it furnishes a global minimum with the aid of Theorem 3.9.

Given a complex example or one of unfamiliar form, one must devise means for verifying that the computational procedure converges to an admissible function together with some way of proving that this limit function, of which we would generally have only some approximations, actually furnishes some type of minimum. Without such results we have only some computed values for what is hoped to be a satisfactorily close approximation to what is hoped to be a minimizing function.

To avoid such uncertainties insofar as possible, one should make the fullest use of variational theory that he can in conjunction with computing. We have seen repeatedly in Chapter 2, 3, 5, and 7 and elsewhere that convexity of integrands is involved in the necessary conditions of Legendre and Weierstrass, in sufficiency theorems, and in existence theorems. Illustrative examples are frequently chosen with quadratic integrands and linear side-conditions, and everything then seems to work beautifully.

It is easy, however, to find simple-looking examples for which the procedure applied to Example 11.1 breaks down in one way or another or for which it converges to a function y_0 that fails to satisfy the Jacobi condition and hence cannot minimize.

11.6 THE PONTRYAGIN MAXIMUM PRINCIPLE

Consider again the fixed-endpoint nonparametric problem in the plane but recast in the form

$$(11.28) \qquad J(y,u) \equiv \int_a^b f(t,y,u)\, dt = minimum$$

on the class $\mathscr{P}$ of all pairs (y,u): $[a,b] \to R^2$ such that y is PWS and satisfies the side-condition

$$(11.29) \qquad \overset{\bullet}{y} = u$$

and the end-conditions

$$(11.30) \qquad y(a) = h_1, \qquad y(b) = h_2.$$

It is clear from (11.29) that u is piecewise continuous. If u were eliminated from (11.28) by means of (11.29) we would recover the formulation of Section 2.2, in which y is PWS and hence $\overset{\bullet}{y}$ is PWC.

Define with Hestenes (XXI) a hamiltonian function $H:[a,b] \times R^3 \to R$.

$$(11.31) \qquad H(t,y,u,v) \equiv uv - f(t,y,u).$$

If we now denote the function (11.3) of three arguments by H^*, one verifies easily that

$$H[t,y,P(t,y,v),v] = H^*(t,y,v),$$

where P is the function introduced by means of condition (11.2) on equation (11.1).

It follows from (11.31) by differentiation that

$$H_t(t,y,u,v) = -f_t(t,y,u)$$

and

$$H_y(t,y,u,v) = -f_y(t,y,u),$$

for all $(t,y,u,v) \in S \equiv [a,b] \times R^2 \times R_1$, where R_1 is the set of reals introduced in (11.2). Observe further of (11.31) that

$$(11.32) \qquad H_u(t,y,u,v) = v - f_r(t,y,u).$$

Suppose that the admissible pair (y_0,u_0) furnishes a local minimum for $J(y,u)$. Define

$$(11.33) \qquad v_0(t) \equiv f_r[t,y_0(t),u_0(t)].$$

It then follows from (11.32) that $H_u[t,y_0(t),u_0(t),v_0(t)] = 0$. From (11.31) and (11.33),

$$(11.34) \qquad H[t,y_0(t),u_0(t),v_0(t)]$$
$$= u_0(t)f_r[t,y_0(t),u_0(t)] - f[t,y_0(t),u_0(t)]$$

and also

$$(11.35) \qquad H[t,y_0(t),q,v_0(t)] = qf_r[t,y_0(t),u_0(t)] - f[t,y_0(t),q],$$

where q is an arbitrary real number. By subtraction of (11.35) from (11.34) we see with the aid of (11.29) that

$$(11.36) \qquad H[t,y_0(t),u_0(t),v_0(t)] - H[t,y_0(t),q,v_0(t)] = E[t,y_0(t),u_0(t),q],$$

in which E denotes the Weierstrass excess-function (2.29).

By hypothesis, $J(y_0,u_0)$ is a minimum; consequently, (y_0,u_0) is an admissible pair and satisfies (11.32). It is then a consequence of Theorem 2.5 that the right member of (11.36) is necessarily nonnegative for all $(t,q) \in [a,b] \times R$ in the event that $J(y_0,u_0)$ is a strong local minimum. If $J(y_0,u_0)$ is only a weak local minimum, we can get a similar conclusion with the aid of problem 4, Exercise 2.4. These considerations provide a proof for the following theorem.

Theorem 11.2.(Pontryagin Maximum Principle)

If $J(y_0,u_0)$ is a strong (weak) local minimum for (11.28) subject to conditions (11.29) and (11.30), then for each $t \in [a,b]$,

$$(11.37) \qquad H[t,y_0(t),u_0(t),v_0(t)] = \max_q H[t,y_0(t),q,v_0(t)],$$

the maximum being with respect to all real values of q or with respect to all q such that, for every $t \in [a,b]$, $|q - u_0(t)| < h$, a positive number independent of t, respectively, according as $J(y_0,u_0)$ is a strong or weak minimum.

One sees from (11.36) and (11.37) that, for the present problem, the maximum principle is simply a rephrasing of the Weierstrass necessary condition.

The Maximum Principle can be stated as a Minimum Principle with H replaced by $-H$ in (11.37). If this is done and if we also eliminate H by means of (11.34) and (11.35), we obtain the condition that

$$(11.38) \qquad f[t,y_0(t),u_0(t)] - u_0(t)f_r[t,y_0(t),u_0(t)] \\ = \min_q \{f[t,y_0(t),q] - qf_r[t,y_0(t),q]\}.$$

We find with reference to Section 11.4 that (11.38) is equivalent to the Bellman equation (11.21).

11.7 CONCLUDING REMARKS

Results presented in Sections 11.2 through 11.6 represent special cases. Corresponding results for higher-dimensional and more complex problems are known. Hamilton–Jacobi theory for both parametric and nonparametric Problems of Bolza was discussed by M. K. Landers (XV, years 1938–1941). For extensions of Sections 11.4 and 11.6, see the titles cited in Section 11.1 and at the end of Section 11.4.

The Maximum Principle is not in general equivalent to the classical necessary condition of Weierstrass, as it was for the problem discussed in Section 11.6, but is rather an extension of that condition to types of problems not covered by the classical theory.

Chapter 12

CONCLUSION AND ENVOY

12.1 COMMENTS AND SUGGESTIONS

Few people begin the study of variational theory with really adequate preparation. In contrast with certain portions of mathematical thought which admit almost completely self-contained treatments, the calculus of variations draws motivation and methods from a diversity of sources that cut across traditional boundaries between courses and disciplines. This feature is both an indication of broad relevance and an impediment to progress.

A reader who has substantially mastered the content of preceding chapters will have identified gaps in the treatment as well as in his own preparation. He should be able to ask questions that he cannot answer or that are not answered in this text or perhaps anywhere else. On the positive side he has established a position from which he can begin to investigate such gaps and questions or aided by books and by papers in the journals to identify and attack some of the untreated or partially treated problems.

The published literature covers a wide spectrum of difficulty, style, and sophistication reflecting an enthusiasm for optimization problems from various points of view and at various levels. At one extreme is mature and modern mathematical exposition consisting of precisely formulated mathematical problems together with theorems and proofs. At the other extreme is work that goes from a loose description of some optimization question of a practical nature directly to the manipulation of Euler equations, to the techniques of Dynamic Programming, or to calculations based on Pontryagin's Maximum Principle or some direct method. Although material of the latter type seldom goes beyond an examination of necessary conditions, it can be suggestive and may serve to introduce the reader to problems with new features.

We have not attempted to cover everything that the title of this book may include. Brief descriptions of some of the important omitted topics constitute the closing sections.

12.2   GENERALIZED CURVES

These mathematical objects introduced by L. C. Young in (57a) and other papers in proving existence of the global minimum for a problem without side-conditions were adapted by McShane (33g,h,i) to obtain further existence theorems for such problems and also for the Problem of Bolza. More recently this approach has been applied to optimal control formulations by Warga (55a,b,e), Gambill (15a), Cesari (9h, II), and McShane (33m).

One avoids the effect of rapid and violent changes in the direction of an admissible curve by an averaging process. By this means, to each curve C in the sense of Chapter 6, now called an *ordinary curve*, corresponds a generalized curve C^*. The class of generalized curves represents an enlargement of the class of admissible curves in somewhat the same sense that the set of real numbers is an enlargment of the set of rationals. There is a one-one correspondence between ordinary curves C and a proper subclass of the generalized curves C^*.

Given a curve-function $J: \mathscr{C} \to R$ to be minimized, let the same symbol J denote the extension to $\mathscr{C}^*$, a class of generalized curves, of the original function. The convergence questions encountered in proving an existence theorem are more easily handled for the minimum problem on $\mathscr{C}^*$ than for the original minimum problem on $\mathscr{C}$. Under suitable hypotheses there not only exists a minimizing generalized curve C_0^* but it turns out to be one of those that corresponds to an ordinary curve C_0 in $\mathscr{C}$, and hence $J(C_0)$ properly interpreted is also the global minimum for the original minimum problem. There remain functionals J for which the infimum of values $J(C^*)$ on $\mathscr{C}^*$ is provided by a generalized curve C_0^* not corresponding to an ordinary curve. Thus, by enlarging the class of admissible entities, the class of solvable minimum problems also is enlarged.

12.3   THE CALCULUS OF VARIATIONS IN THE LARGE

This intriguing title describes an important part of the larger subject, analysis in the large, with Marston Morse as the leading creator and contributor over many years beginning in the 1920s. For a first description see Morse (38f) or the Introduction of H. Seifert and W. Threlfall, *Variationsrechnung im Grossen*, Teubner, Berlin, 1938.

Given a suitably differentiable function $\phi \colon M \to R$ whose domain M is a suitable subset of the plane, suppose that the first-order partials of ϕ vanish at an isolated interior point of M, which we take to be the origin. Then second-order terms in a Taylor expansion of ϕ about the origin will, after reduction to a sum of squares by rotation, be of one of the types

$$(12.1) \qquad x^2 + y^2, \quad x^2 - y^2, \quad or \quad -x^2 - y^2.$$

These forms correspond to the respective cases in which $[0, 0, \phi(0,0)]$ is a minimum point, a saddle point, or a maximum point. If the domain of ϕ is a subset of R^n, the number of minus signs in the blocks of squared terms can be any integer from 0 to n inclusive, and there are $n - 1$ possible kinds of saddle points.

The Morse theory of critical values of a point-function ϕ is concerned with relationships between topological characteristics (Betti or connectivity numbers) of the graph of ϕ and the numbers of local minima, local maxima, and different kinds of saddle points. In contrast with the strictly local considerations in the usual discussion of critical points based on Taylor expansions, we are concerned here with the whole function ϕ. This is the connotation of the phrase "in the large." Various questions from mathematical physics concerning stable points in a force field with an associated potential are related to this part of the Morse theory.

Variational theory in the large is an extension of these considerations to real-valued functionals J. We have made a first contact with properties in the large in sections of this book dealing with global extrema. Certain admissible functions y in a function-space $\mathcal{Y}$ correspond to extreme values, others to saddle values $J(y)$. For more information see Morse (38f,h), (XXIX), and his other many publications.

The work of Morse stands in a noble tradition going back to that of H. Poincaré and G. D. Birkhoff on dynamical systems and in particular on the problem of three bodies. It has implications for both pure and applied mathematics including questions on stability and control.

The calculus of variations in the large depends upon previous knowledge of other parts of variational theory, together with some experience in general and algebraic topology and differential geometry. The coincidence of a professor with an active interest together with a group of students who have adequate preparation seems to occur infrequently, and there are regrettably few universities where courses or seminars in this field are available.

12.4   THE THEORY OF AREA

This is an extension of the theory of length discussed in Sections 6.2 through 6.4 and 7.3 through 7.7 but not an easy or routine extension. In view of definition (7.2) of the length $\mathcal{L}(x)$ of a function as the supremum of lengths of inscribed piecewise linear functions, it was thought at one time that a similar definition for the area of a function x: $D \to R^3$, $D \subset R^2$, would be satisfactory. Triangulate the domain D of x and consider the sum of the elementary areas of the triangles in R^3 whose vertices are the respective triples of image points under x of the vertices of the curvilinear triangles in D. The supremum of all such sums is analogous to the supremum in definition (7.2).

Unfortunately this is not an acceptable definition of area. It was pointed out by H. A. Schwarz about 1890 that, if the graph of x is a circular cylinder of radius r and height h, the inscribed polyhedrons with triangular faces can be arbitrarily crinkly in such a manner that the sum of areas of faces has no finite upper bound. Rather than the familiar expression $2\pi rh$, the proposed supremum is ∞, which must be rejected for several reasons.

Various competing definitions for area have been devised to avoid this defect. That of Lebesgue, which has led to a large literature and a rather complete theory, is the most widely preferred. Fréchet distance $\rho(x,y)$ between mappings $x: D_1 \to R^3$ and $y: D_2 \to R^3$ can be defined in the pattern of (6.9). If D is a polygonal subset of R^2 and $p: D \to R^3$ is continuous on D and linear on each of a finite set of triangular subsets of D whose union is D, then p is called *piecewise linear*. The Lebesgue area $\mathcal{L}(x)$ is defined by the statement that

$$(12.2) \qquad \mathcal{L}(x) \equiv \lim \inf E(p) \text{ as } \rho(p,x) \to 0,$$

where $E(p)$, the *elementary area* of the piecewise linear mapping p, is the sum of the areas of the triangular faces (some of which may be degenerate) of the graph of p in R^3. Definition (12.2) does not require that these faces be inscribed in the graph of x. If we were to add this restriction, then (12.2) would yield an area $\mathcal{L}_1(x) \geqslant \mathcal{L}(x)$. Whether the relation $>$ can ever occur has been frequently asked but remains unanswered insofar as the author is aware.

For an introduction to the theory see Cesari (9i; XIV, Chaps. 1 and 2) and Cesari and Radó (44d). For detailed development see Cesari (XIV) and Radó (XXXIV). Among the many contributions are papers of C. Goffman, R. G. Helsel, K. Iseki, J. H. Michael, E. J. Mickle, P. V. Reichelderfer, E. Silverman, L. C. Young, and J. W. T. Youngs, some of which are listed in our bibliography.

A much used result of Tonelli in the theory of length is that given by our Theorem 9.7. It was natural to seek an analogous theorem concerning the area integral, which in the classic notation of differential geometry is written

$$(12.3) \qquad \int \int [EG - F^2]^{1/2} \, du \, dv.$$

Radó identified (44a) in 1942 a class of representations of surfaces for which the Lebesgue integral of form (12.3) is the Lebesgue area but did not prove that every surface of finite Lebesgue area necessarily has such a representation. That such is indeed the case was affirmed by Cesari, *On the representation of surfaces*, Amer. Jour. of Math., vol. 72 (1950), pp. 335–346. Leaning on this result, the present author defined a Weierstrass integral (12d) for a restricted class of representations, which turned out to be similar to the class used by Radó in (44a). For any of these representations of a surface S, the Lebesgue area is given by a supremum with features analogous to that in definition (7.2).

If one's major interest is in the calculus of variations he tends to regard the theory of area as introductory to that of multiple integral variational problems. However, area theory is so massive in its own right that it also has the status of a separate field.

12.5 MULTIPLE INTEGRAL PROBLEMS

As a natural analogue of the fixed-endpoint problem of Section 2.2, consider the problem

$$(12.4) \qquad J(z) \equiv \int \int f(x,y,z,z_x,z_y) \, dy \, dz = extremum$$

on the class $\mathscr{Z}$ of all continuous functions $z: D \to R$ having partial derivatives z_x and z_y that are continuous except on the boundaries of a finite number of subsets of D whose union is D and such that

$$(12.5) \qquad z(x,y) = \phi(x,y) \text{ on the boundary of } D.$$

Extrema of double integrals received some attention as early as the Bernoullis and by Lagrange in 1760. Many results analogous to those for single integral variational problems are now available. For an introduction see Bolza (XI, Chap. 13), Pars (XXXII, Chap. 10), or Bliss (VIII). Further results are in the Chicago dissertations of M. Coral, B. Cosby, II, J. H. Levin, E. A. Nordhaus, J. E. Powell, and A. W. Raab in the several volumes of the *Contributions* (XV) and that of J. E. Wilkins, Jr. (56a). A bibliography on multiple integrals going back to the 1890's is included with Raab's paper. See also Klötzler (XXXIX).

Existence theory for double integrals has been distinctly more difficult than for single integral problems. Tonelli proved such theorems for nonparametric double integrals in the early 1930's. For references to the work of McShane, Morrey, and others see Morrey (XXVII) and his extensive bibliography in (37a).

Existence theorems for parametric double integrals with a special integrand were published by McShane in the 1930s. Comparable results for more general integrands are by Sigalov (49a), Cesari (9a), and Danskin (10a).

Although one hopes that results for double integrals extend to general m-fold integrals, treatment of such matters tends to be cumbersome and there seems to be relatively little literature.

12.6   TRENDS

Just as there are intimate relations between single integral variational problems and ordinary differential equations, so also are partial differential equations related to multiple integral problems. For instance, such equations appear as side-conditions in multiple integral Bolza Problems. The Euler equations analogous to (2.24) or (2.25) are now partial differential equations. Optimal design or optimal control of dynamical systems described by partial differential equations leads to such Bolza Problems.

At the time this is written there seem to be relatively few published results on such control problems but there is an ample supply of systems that can generate such problems, and we can expect to hear more about them as time goes on. See the collection of papers, *Functional Analysis and Optimization*, E. R. Caianello (editor), Academic Press, New York, 1966; also Balakrishnan (3b), Russell (47a), Neustadt (39g), Halkin and Neustadt (19c), and Cesari (9j,k). The methods employed include Banach spaces and other tools of functional analysis, together with rather general topological spaces.

The potential usefulness of the calculus of variations for the very kinds of applications that motivate the theory has always been limited by the difficulty of computing or otherwise identifying at the practical level a suitable approximation to an optimizing admissible function y_0.

The author has not worked in this important sector and can only make an occasional comment based on second-hand information. It is now feasible with the aid of modern computers to approximate certain functions y satisfying Euler equations or functions W satisfying the equations of Hamilton–Jacobi theory. Direct numerical methods and bibliographies thereto are available in such books as Balakrishnan and Neudstadt (III) and Leitmann (XXV, Chaps. 2, 6, and 9). There is a continuing need for attention to such questions by persons skilled both in numerical analysis and in the calculus of variations.

Variational theory after a long history remains active and growing and can be expected to continue as new classes of problems appear and methods for dealing with them must be borrowed or created.

Bibliography

This bibliography is a small sample from a vast and far-flung literature and there have no doubt been important omissions. Many of the books and some of the articles supplement material in the present text, as indicated by numerous text citations. Others are included as information on trends. That a work is listed or cited does not necessarily imply its priority but rather that it is relevant and in most cases is widely available. The list purposely covers a wide range of subject matter and of style from some that is formal yet suggestive to very meticulous mathematical exposition. In many instances the titles by a particular author comprise only a fraction of his published work in this area.

BOOKS

I. Akhiezer, N. I., *The Calculus of Variations*, Ginn (Blaisdell), Boston, 1962.

II. Athans, M., and Falb, P. L., *Optimal Control*, McGraw-Hill, New York, 1966.

III. Balakrishnan, A. V., and Neustadt, L. W. (editors), *Computing Methods in Optimization Problems*, Academic Press, New York, 1964.

IV. Bellman, R. E., *Dynamic Programming*, Princeton Univ. Press, Princeton, N.J., 1957.

V. Bellman, R. E., and Dreyfus, S. E., *Applied Dynamic Programming*, Princeton Univ. Press, Princeton, N.J., 1962.

VI. Bellman, R. E. (editor), *Mathematical Optimization Techniques*, Univ. California Press, Berkeley, 1963.

VII. Bliss, G. A., *Calculus of Variations* (Carus Monograph No. 1), Open Court, Chicago, 1925, and subsequent reprints.

VIII. Bliss, G. A., *The Calculus of Variations Multiple Integrals*, University of Chicago lectures, summer quarter (mimeographed), 1933.

IX. Bliss, G. A., *Lectures on the Calculus of Variations*, Univ. Chicago Press, Chicago, 1946, and subsequent reprints in paperback as Phoenix Science Series No. 504 by the same press.

X. Bolza, O., *Lectures on the Calculus of Variations*, Univ. Chicago Press, Chicago, 1904, and subsequent reprints by Stechert, New York, 1931, and Chelsea, New York, 1961.

XI. Bolza, O., *Vorlesungen über Variationsrechnung*, Teubner, Berlin, 1909, and reprints by Koehler's Antiquarium, Leipzig, 1933, and Chelsea, New York, 1963.

XII. Carathéodory, C., *Variationsrechnung und partielle Differentialgleichungen erster Ordnung*, Teubner, Berlin, 1935, and an English translation, Holden-Day, San Francisco, 1965 and 1967.

XIII. Carathéodory, C., *Gesammelte mathematische Schriften*, C. H. Beck Verlagsbuchhandlung, Munich, 1954–1955.

XIV. Cesari, L., *Surface Area* (Annals of Mathematics Studies, No. 35), Princeton Univ. Press, Princeton, N. J., 1956.

XV. Chicago, University of, *Contributions to the Calculus of Variations*, four volumes designated by respective years 1930, 1931–1932, 1933–1937, and 1938–1941, Univ. Chicago Press, Chicago.

XVI. Courant, R., *Calculus of Variations and Supplementary Notes and Exercises*, 1945–1946, revised and amended by J. Moser, supplementary notes by M. Kruskal and H. Rubin, New York University, Courant Institute of Mathematical Sciences (multigraphed), 1962.

XVII. Dreyfus, S. E., *Dynamic Programming and the Calculus of Variations* (R-441-PR), the RAND Corp., Santa Monica, Calif., 1965, and Academic Press, New York, 1966.

XVIII. El'sgol'ts, L. (Elsgolc, L.), *Qualitative Methods in Mathematical Analysis*, American Mathematical Society, Providence, R.I., 1964.

XIX. Funk, P., *Variationsrechnung und ihre Anwendung in Physik und Technik* (Die Grundlehren der mathematischen Wissenschaften, Vol. 94), Springer, Berlin, 1962.

XX. Gelfand, I. M., and Fomin, S. V., *Calculus of Variations*, Prentice-Hall, Englewood Cliffs, N.J., 1963.

XXI. Hestenes, M. R., *Calculus of Variations and Optimal Control Theory* Wiley, New York, 1966.

XXII. Isaacs, R., *Differential Games*, Wiley, New York, 1965.

XXIII. Lanczos, C., *The Variational Principles in Mechanics*, Univ. Toronto Press, Toronto, 1949. 2nd ed., 1962; 3d ed., 1966.

XXIV. Lawden, D. F., *Optimal Trajectories and Space Navigation*, Butterworth, London, 1963.

XXV. Leitmann, G. (editor), *Optimization Techniques with Applications to Aerospace Systems*, Academic Press, New York, 1962.

XXVI. Mikhlin, S. G., *The Problem of the Minimum of a Quadratic Functional*, Holden-Day, San Francisco, 1965.

XXVII. Morrey, C. B., Jr., *Multiple Integrals in the Calculus of Variations* (Die Grundlehren der mathematischen Wissenschaften, Vol. 130), Springer, Berlin, 1966.

XXVIII. Morse, M., *The Calculus of Variations in the Large*, American Mathematical Society Colloquium Publications, Vol. 18, American Mathematical Society, Providence, R.I., 1934.

XXIX. Morse, M., *Introduction to Analysis in the Large*, lectures at the Institute for Advanced Study, Princeton, N.J., 2nd ed. (mimeographed), 1951, reprinted 1956.

XXX. Oğuztöreli, M. N., *Time-Lag Control Systems*, Academic Press, New York, 1966.

XXXI. Osgood, W. F., *Mechanics*, Macmillan, New York, 1937.

XXXII. Pars, L. A., *An Introduction to the Calculus of Variations*, Wiley, New York, 1962.

XXXIII. Pontryagin, L. S., et al., *The Mathematical Theory of Optimal Processes*, Interscience, New York, 1962, and Macmillan, New York, 1964.

XXXIV. Radó, T., *Length and Area*, American Mathematical Society Colloquium Publications, Vol. 30, American Mathematical Society, Providence, R.I., 1948.

XXXV. Tonelli, L., *Fondamenti di Calcolo delle Variazioni*, Vols. 1 and 2, Zanichelli, Bologna, 1921 and 1923.

XXXVI. Tonelli, L., *Opere Scelte*, a cura dell'Un. Mat. Ital., Edizioni Cremonese, Rome, 1961, 1962.

XXXVII. Weinstock, R., *Calculus of Variations with Applications to Physics and Engineering*, McGraw-Hill, New York, 1952.

ARTICLES

1. Aronszajn, N.
   (a) *Quelques recherches sur l'intégrale de Weierstrass*, Revue Scientifique, vol. 77 (1939), 490–493; II, ibid., vol. 78 (1940), 165–167; III, ibid., vol. 78 (1940), 233–239.

2. Ayer, M. C.
(a) *On convergence in length*, Proc. Nat. Acad. of Science, vol. 31 (1945), 261–266. See also Chapter III, 3, Arc Length, in Radó's *Length and Area* (our ref. XXXIV) and the author's remarks ll. 1 and 2, p. 265.
(b) [With T. Radó], *A note on convergence in length*, Bull. Amer. Math. Soc., vol. 54 (1948), 533–539. Also see (8a).

3. Balakrishnan, A. V.
(a) *An operator theoretic formulation of a class of control problems and a steepest descent method of solution*, SIAM Jour. on Control, vol. 1 (1963), 109–127.
(b) *Optimal control problems in Banach spaces*, SIAM Jour. on Control, vol. 3 (1965), 152–180.

4. Berkovitz, L. D.
(a) *A variational approach to differential games*, in *Advances in Game Theory*, Princeton Univ. Press, Princeton, N.J., 1964.
(b) [With S. E. Dreyfus], *A dynamic programming approach to the nonparametric problem in the calculus of variations* (RM-4329-PR), the RAND Corp., Santa Monica, Calif. Dec. 1964.

5. Bliss, G. A.
(a) *The problem of Lagrange in the calculus of variations*, Amer. Jour. of Math., vol. 52 (1930), 673–744.
(b) *The calculus of variations for multiple integrals*, Amer. Math. Monthly, vol. 49 (1942), 77–89.

6. Boltyanskii, V. G.
(a) *Sufficient conditions for optimality and the justification of the dynamic programming method* SIAM Jour. on Control, vol. 4 (1966), 326–361.

7. Bolza, O.
(a) *Über den anormalen Fall beim Lagrangeschen und Mayerschen Problem mit gemischten Bedingungen und Variabeln Endpunkten*, Math. Annalen, vol. 74 (1913), 430–436.

8. Brunk, H. D.
(a) [With M. C. Ayer, G. M. Ewing, W. T. Reid, and E. Silverman], *An empirical distribution function for sampling with incomplete information*, Annals of Math. Stat., vol. 26 (1955), 641–647.
(b) [With G. M. Ewing and W. T. Reid], *The minimum of a certain definite integral suggested by the maximum likelihood estimate of a distribution function*, Bull. Amer. Math. Soc., vol. 60 (1954), 535.
(c) [With G. M. Ewing and W. R. Utz], *Some Helly theorems for monotone functions*, Proc. Amer. Math. Soc., vol. 7 (1956), 776–783.
(d) [With G. M. Ewing and W. R. Utz], *Minimizing integrals in certain classes of monotone functions*, Pacific Jour. of Math., vol. 7 (1957), 833–847.

9. Cesari, L.
(a) *An existence theorem of the calculus of variations for integrals on parametric surfaces*, Amer. Jour. of Math., vol. 74 (1952), 265–295.
(b) *Rectifiable curves and the Weierstrass integral*, Amer. Math. Monthly, vol. 65 (1958), 485–500.
(c) *Recent results in surface area theory*, Amer. Math. Monthly, vol. 66 (1959), 173–192.
(d) *Semicontinuità e convessità nel calcolo delle variazioni*, Annali della Scuola Norm. Sup. di Pisa, Ser. III, vol. 18 (1964), 389–423.
(e) *Un teorema di esistenza in problemi di controlli*, Annali della Scuola Norm. Sup. di Pisa, Ser. III, vol. 19 (1965), 35–78.
(f) *An existence theorem in problems of optimal control*, SIAM Jour. on Control, vol. 3 (1965), 7–22.
(g) *Existence theorems for optimal solutions in Pontryagin and Lagrange problems*, SIAM Jour. on Control, vol. 3 (1965), 475–498.

334   BIBLIOGRAPHY

(h) *Existence theorems for weak and usual optimal solutions in Lagrange problems with unilateral constraints*. I, Trans. Amer. Math. Soc., vol. 124 (1966), 369–412, and same title, II, *Existence theorems for weak solutions*, ibid., vol. 124 (1966), 413–430.

(i) *Surface area*, in *Studies* in *Mathematics*, vol. 4, 123–146, S. S. Chern (editor), Mathematical Association of America, distributed by Prentice-Hall, Englewood Cliffs, N.J., 1967.

(j) *Existence theorems for multidimensional problems of optimal control*, in *Differential Equations and Dynamical Systems*, pp. 115–132, Academic Press, New York, 1967.

(k) *Existence theorems for multidimensional Lagrange problems*, Jour. of Optimization Theory and Applications, vol. 1 (1967), 87–112.

Also see (44d).

10. Danskin, J. M.

(a) *On the existence of minimizing surfaces in parametric double integral problems in the calculus of variations*, Riv. di Mat. della Univ. di Parma, vol. 3 (1952), 43–63.

(b) *The theory of max-min, with applications*, SIAM Jour. on Appl. Math., vol. 14 (1966), 641–664.

11. El'sgol'ts, L. (Elsgolc, L.)

(a) *Variational problems with a delayed argument*, Amer. Math. Soc. Translations, Ser. 2, vol. 16 (1960), 468–469.

12. Ewing, G. M.

(a) *Sufficient conditions for a non-regular problem in the calculus of variations*, Bull. Amer. Math. Soc., vol. 43 (1937), 371–376.

(b) *Sufficient conditions for an ordinary problem in the calculus of variations*, Boletin Matematico, vol. 12 (1938), 13–15.

(c) *Variation problems formulated in terms of the Weierstrass intergral*, Duke Math. Jour., vol. 14 (1947), 675–687.

(d) *Surface integrals of the Weierstrass type*, Duke Math. Jour., vol. 18 (1951), 275–286.

(e) *Lipschitzian parameterizations and existence of minima in the calculus of variations*, Proc. Amer. Math. Soc., vol. 11 (1960), 87–89.

(f) *A fundamental problem of navigation in free space*, Quart. of Appl. Math., vol. 18 (1961), 355–362.

(g) [With W. R. Haseltine], *Optimal programs for an ascending missile*, SIAM Jour. on Control, vol. 2 (1964), 66–88.

(h) *Sufficient conditions for global extrema in the calculus of variations*, Jour. of the Astronautical Sciences, 1965, 102–105.

(i) *Thrust direction programs for maximal range*, Jour. of Math. Anal. and Appl., vol. 16 (1966), 347–354.

Also see (8a,b,c,d) and (38g).

13. Filippov, A. F.

(a) *On certain questions in the theory of optimal control*, SIAM Jour. on Control, vol. 1 (1962), 76–89.

14. Fleming, W. H.

(a) *On a class of games over function space and related variational problems*, Annals of Math., vol. 60 (1954), 578–599.

(b) *A note on differential games of prescribed duration*, in *Contributions to the Theory of Games* (Annals of Mathematics Studies, No. 39), vol. 3, 407–412, Princeton Univ. Press, Princeton, N.J., 1957.

(c) *An example in the problem of least area*, Proc. Amer. Math. Soc., vol. 7 (1956), 1063–1074.

(d) *On the oriented plateau problem*, Rend. Circolo Mat. di Palermo, Ser. 2, vol. 11 (1962), 69–90.

15. Gambill, R. A.

(a) *Generalized curves and the existence of optimal controls*, SIAM Jour. on Control, vol. 1 (1963), 246–260.

16. Goffman, C.
   (a) *Lower semi-continuity and area functionals*, Rend. Circolo Mat. di Palermo, Ser. 2, vol. 2 (1953), 203–235.
17. Goldstein, A. A.
   (a) *Minimizing functionals on normed linear spaces*, SIAM Jour. on Control, vol. 4 (1966), 81–89.
18. Graves, L. M.
   (a) *On the existence of the absolute minimum in space problems of the calculus of variations*, Annals of Math., vol. 28 (1927), 153–170.
   (b) *Discontinuous solutions in the calculus of variations*, Bull. Amer. Math. Soc., vol. 36 (1930), 831–846.
   (c) *On the Weierstrass condition for the problem of Bolza in the calculus of variations*, Annals of Math., vol. 33 (1932), 747–752.
   (d) *A transformation of the problem of Lagrange in the calculus of variations*. Trans. Amer. Math. Soc., vol. 35 (1933), 675–682.
19. Halkin, H.
   (a) *On the necessary conditions for optimal control of non-linear systems*, Applied Mathematics and Statistics Laboratory (Report 116), Stanford Univ., Stanford, Calif., 1963.
   (b) *The principle of optimal evolution*, in *International Symposium on Nonlinear Differential Equations and Nonlinear Mechanics*, J. P. LaSalle and S. Lefschetz (editors), Academic Press, New York, 1963.
   (c) [With L. W. Neustadt], *General necessary conditions for optimization problems*, (Report USCEE 173), Univ. Southern California, School of Engineering, Los Angeles, Calif., June 1966.
   (d) *A maximum principle of the Pontryagin type for systems described by nonlinear difference equations*, SIAM Jour. on Control, vol. 4 (1966), 90–111.
20. Hestenes, M. R.
   (a) *Sufficient conditions for the problem of Bolza in the calculus of variations*, Trans. Amer. Math. Soc., vol. 36 (1934), 793–818.
   (b) *Sufficient conditions for the isoperimetric problem of Bolza in the calculus of variations*, Trans. Amer. Math. Soc., vol. 60 (1946), 93–118.
   (c) *An alternate sufficiency proof for the normal problem of Bolza*, Trans. Amer. Math. Soc., vol. 61 (1947), 256–264.
   (d) *An indirect sufficiency proof for the problem of Bolza in nonparametric form*, Trans. Amer. Math. Soc., vol. 62 (1947), 509–535.
   (e) *Sufficient conditions for multiple integral problems in the calculus of variations*, Amer. Jour. of Math., vol. 70 (1948), 239–276.
   (f) *On variational theory and optimal control theory*, SIAM Jour. on Control, vol. 3 (1965), 23–48.
21. Hughes, D. K.
   (a) *Contributions to the theory of variational and optimal control problems with delayed argument*, doctoral dissertation, Univ. Oklahoma, Norman, Okla., 1967.
   (b) *Variational and optimal control problems with delayed argument*, Jour. of Optimization Theory and Applications, vol. 2 (1968), 1–14.
22. Hughs, R. E.
   (a) *Length for discontinuous curves*, Arch. for Rat. Mech. and Anal., vol. 12 (1963), 213–222.
23. Iseki, K.
   (a) *On certain properties of parametric curves*, Jour. Math. Soc. of Japan, vol. 12 (1960), 129–173.
24. Jacobs, M. Q.
   (a) *The existence of optimal controls*, doctoral dissertation, Univ. of Oklahoma, Norman, Okla., 1966.

(b) *Some existence theorems for linear optimal control problems*, SIAM Jour. on Control, vol. 5 (1967), 418–437.

(c) *Attainable sets in linear systems with unbounded controls*, in *Proceedings of the Conference on the Mathematical Theory of Control at the University of Southern California*, Academic Press, New York, 1967, 46–63.

(d) [With P. L. Falb], *Differentials in locally convex spaces*, Jour. of Differential Equations, vol. 4 (1968), 444–489.

(e) *Attainable sets in systems with unbounded controls*, Jour. of Differential Equations, vol. 4 (1968), 408–423.

(f) *Remarks on some recent extensions of Filippov's implicit functions lemma*, SIAM Jour. on Control, vol. 5 (1967), 622–627.

(g) *Measurable multivalued mappings and Lusin's Theorem*, Trans. Amer. Math. Soc., vol. 134 (1968), 471–481.

25. Kalman, R. E.

(a) *Contributions to the theory of optimal control*, Bol. de la Soc. Mat. Mexicana, vol. 5 (1960), 102–119.

(b) *The theory of optimal control and the calculus of variations*, Chap. 16 in ref. VI.

26. LaSalle, J. P.

(a) *The time-optimal control problem*, in *Contributions to the Theory of Non-Linear Oscillations*, vol. V, 1–24, Princeton Univ. Press, Princeton, N.J., 1960.

(b) *Stability and Control*, SIAM Jour. on Control, vol. 1 (1962), 3–15.

27. Lawden, D. F.

(a) *Discontinuous solutions of variational problems*, Jour. Australian Math. Soc., vol. 1 (1959), 27–37.

28. Lee, E. B., and Markus, L.

(a) *Optimal control for non-linear processes*, Arch. for Rat. Mech. and Anal., vol. 8 (1961), 36–58.

29. Lefkowitz, L., and Eckman, D.P.

(a) *Application and analysis of a computer control system*, Trans. of the ASME, Ser. D, Jour. of Basic Engineering, vol. 81 (Dec. 1959), 569–577.

30. Leitmann, G.

(a) *On a class of variational problems in rocket flight*, Jour. of the Aero/Space Sciences, vol. 26 (1959), 586–591.

(b) *Comment on a class of variational problems in rocket flight*, Jour. of the Aero/Space Sciences, vol. 27 (Feb. 1960).

31. Mancill, J. D.

(a) *Identically non-regular problems in the calculus of variations*, Revista, Ser. A, Mat. y Fisica Teorica, Univ. Nac. del Tucumán, vol. 7 (1950), 131–139.

32. Markus, L.

(a) *Controllability of nonlinear processes*, SIAM Jour. on Control, vol. 3 (1965), 78–90.

33. McShane, E. J.

(a) *Semicontinuity in the calculus of variations and absolute minima for isoperimetric problems*, in *Contributions to the Calculus of Variations*, 1930 (see ref. XV).

(b) *Existence theorems for ordinary problems in the calculus of variations*, Parts I and II and an addendum, Annali della R. Scuola Norm. Sup. di Pisa, Ser. II, vol. 3 (1934), 183–210, 239–241, 287–315.

(c) *Semi-continuity of integrals in the calculus of variations*, Duke Math. Jour., vol. 2 (1936), 597–616.

(d) *A navigation problem of the calculus of variations*, Amer. Jour. of Math., vol. 59 (1937). 327–334.

(e) *On multipliers for Lagrange problems*, Amer. Jour. of Math., vol. 61 (1939), 809–819.

(f) *Curve-space topologies associated with variational problems*, Annali della R. Scuola Norm. Sup. di Pisa, Ser. II, vol. 9 (1940), 45–60.

(g) *Generalized curves*, Duke Math. Jour., vol. 6 (1940), 513–536.

(h) *Necessary conditions in generalized-curve problems of the calculus of variations*, Duke Math. Jour., vol. 7 (1940), 1–27.

(i) *Existence theorems for Bolza Problems in the calculus of variations*, Duke Math. Jour., vol. 7 (1940), 28–61.

(j) *On the second variation in certain abnormal problems of the calculus of variations*, Amer. Jour. of Math., vol. 63 (1941), 516–530.

(k) *Sufficient conditions for a weak relative minimum in the problem of Bolza*, Trans. Amer. Math. Soc., vol. 52 (1942), 344–379.

(l) [With R. B. Warfield, Jr.], *On Filippov's implicit function lemma*, Proc. Amer. Math. Soc., vol. 18 (1967), 41–47.

(m) *Relaxed controls and variational problems*, SIAM Jour. on Control, vol. 5 (1967), 438–485.

34. Menger, K.

(a) *Die metrische Methode in der Variationsrechnung*, Ergebnisse eines mathematischen Kolloquiums, Wien, Heft 8 (1937), 1–37.

(b) *A theory of length and its applications to the calculus of variations*, Proc. Nat. Acad. of Science, vol. 25 (1939), 474–478.

35. Michael, J. H.

(a) *Approximation of functions by means of Lipschitz functions*, Jour. Australian Math. Soc., vol. 3 (1963), 134–150.

(b) *The equivalence of two areas for nonparametric discontinuous surfaces*, Ill. Jour. of Math., vol. 7 (1963), 59–78.

36. Miele, A.

(a) *Generalized variational approach to the optimum thrust programming for the vertical flight of a rocket*, Zeits. für Flugwissenschaften, vol. 6 (1958), 69–77, and part II [with C. R. Cavoti], ibid., 102–109.

(b) *Flight mechanics and variational problems of linear type*, Jour. of the Aero/Space Sciences, vol. 25 (1958), 581–590.

(c) *A survey of the problem of optimizing flight paths of aircraft and missiles*, Chap. 1 in ref. VI.

(d) *Extremal problems in aerodynamics*, SIAM Jour. on Control, vol. 3 (1965), 129–141.

37. Morrey, C. B., Jr.

(a) *Multiple integral problems in the calculus of variations and related topics*, Annali della Scuola Norm. Sup. di Pisa, Ser. III, vol. 14 (1960), 1–61.

38. Morse, M.

(a) *Sufficient conditions in the problem of Lagrange with fixed endpoints*, Annals of Math., vol. 32 (1931), 567–577.

(b) *Sufficient conditions in the problem of Lagrange with variable end conditions*, Amer. Jour. of Math., vol. 53 (1931), 517–546.

(c) *Sufficient conditions in the problem of Lagrange without assumptions of normality*, Trans. Amer. Math. Soc., vol. 37 (1935), 147–160.

(d) *A special parameterization of curves*, Bull. Amer. Math. Soc., vol. 42 (1936), 915–922.

(e) *Functional topology and abstract variational theory*, in *Mémorial des Sciences Mathématiques*, vol. 93, Gauthier-Villar, Paris, 1939.

(f) *What is analysis in the large?*, Amer. Math. Monthyl, vol. 49 (1942), 358–364. Reprinted with an added section in *Studies in Mathematics*, vol. 4, 5–15, S. S. Chern (editor), Mathematical Association of America, distributed by Prentice-Hall, Englewood Cliffs, N.J., 1967.

(g) [With G. M. Ewing], *The variational theory in the large including the non-regular case—first paper*, Annals of Math., vol. 44 (1943), 339–353, and *second paper*, ibid., vol. 44 (1943), 354–374.

(h) *Recent advances in variational theory in the large*, in *Proceedings of the International Congress of Mathematicians, Cambridge, Mass., 1950*, vol. 2, 143–56, American Mathematical Society, Providence, R.I., 1952.

39. Neustadt, L. W.
   (a) *Synthesizing time optimal control systems*, Jour. Math. Anal. and Appl., vol. 1 (1960), 484–493.
   (b) *The moment problem and weak convergence in L_2*, Pacific Jour. of Math., vol. 11 (1961), 715–721.
   (c) *Minimum effort control systems*, SIAM Jour. on Control, vol. 1 (1962), 16–31.
   (d) *Time optimal control systems with position and integral limits*, Jour. Math. Anal. and Appl., vol. 3 (1961), 406–427.
   (e) *The existence of optimal controls in the absence of convexity conditions*, Jour. Math. Anal. and Appl., vol. 7 (1963), 110–117.
   (f) *A general theory of minimum fuel space trajectories*, SIAM Jour. on Control, vol. 3 (1965), 317–356.
   (g) *An abstract variational theory with applications to a broad class of optimization problems I, General theory*, SIAM Jour. on Control vol. 4 (1966), 505–527, and *II, Applications*, ibid., vol. 5 (1967), 90–137.
   Also see (19c).
40. Nikaidô, H.
   (a) *On von Neumann's Minimax theorem*, Pacific Jour. of Math., vol. 4 (1954), 65–72.
41. Nikodym, O. M.
   (a) *A theorem on infinite sequences of finitely additive real-valued measures*, Rend. del Seminario Mat. della Univ. di Padova, vol. 24 (1955), 265–286.
42. Oğuztöreli, M. N.
   (a) *Relay type control problems with retardation and switching delay*, SIAM Jour. on Control, vol. 1 (1963), 275–289.
   (b) *A time-optimal control problem for systems described by differential difference equations*, SIAM Jour. on Control, vol. 1 (1963), 290–310.
   (c) *Optimal pursuit strategy processes with retarded control systems*, SIAM Jour. on Control, vol. 2 (1964), 89–105.
43. Pauc, C. Y.
   (a) *Les méthodes directes en calcul des variations et en géométrie différentielle*, Thèses présentées a la Faculte des Sciences de l'Université de Paris, Hermann, Paris, 1941.
   (b) *La méthode métrique en calcul des variations* (Actualités Sci. et Ind., No. 885), Hermann, Paris, 1941.
44. Radó, T.
   (a) *On the semi-continuity of double integrals in parametric form*, Trans. Amer. Math. Soc., vol. 51 (1942), 336–361.
   (b) *On semi-continuity*, Amer. Math. Monthly, vol. 49 (1942), 446–450.
   (c) *Convergence in area*, Duke Math. Jour., vol. 16 (1949), 61–71.
   (d) [With L. Cesari], *Applications of area theory in analysis*, in *Proceedings of the International Congress of Mathematicians, Cambridge, Mass., 1950*, vol. 2, 174–179, American Mathematical Society, Providence, R.I., 1952.
   Also see (2b).
45. Reid, W. T.
   (a) *Analogues of the Jacobi condition for the problem of Mayer in the calculus of variations*, Annals of Math., vol. 35 (1934), 836–848.
   (b) *Sufficient conditions by expansion methods for the problem of Bolza in the calculus of variations*, Annals of Math., vol. 38 (1937), 662–678.
   (c) *A direct expansion proof of sufficient conditions for the non-parametric problem of Bolza*, Trans. Amer. Math. Soc., vol. 42 (1937), 183–190.
   (d) *Isoperimetric problem of Bolza in non-parametric form*, Duke Math. Jour., vol. 5 (1939), 675–691.
   (e) *A note on the du Bois Reymond equations in the calculus of variations*, Bull. Amer. Math. Soc., vol. 52 (1946), 158–166.
   (f) *Expansion methods for the isoperimetric problem of Bolza in non-parametric form*, Amer.

Jour. of Math., vol. 71 (1949), 946–975.

(g) *Ordinary linear differential operators of minimum norm*, Duke Math. Jour., vol. 29 (1962), 591–606.

(h) *A simple optimal control problem involving approximation by monotone functions*, Jour. of Optimization Theory and Applications, vol. 2 (1968), 365–377.

Also see (8a,b).

46. Roxin, E. O.

(a) *The existence of optimal controls*, Mich. Math. Jour., vol. 9 (1962), 109–119.

(b) *A geometric interpretation of Pontryagin's Maximum principle*, in *International Symposium on Nonlinear Differential Equations and Nonlinear Mechanics*, J. P. LaSalle and S. Lefschetz (editors), Academic Press, New York, 1963.

47. Russell, D. L.

(a) *Optimal regulation of linear symmetric hyperbolic systems with finite dimensional controls*, SIAM Jour. on Control, vol. 4 (1966), 276–294.

48. Serrin, J.

(a) *On a fundamental theorem of the calculus of variations*, Acta Math., vol. 102 (1959), 1–22.

(b) *On the definition and properties of certain variational integrals*, Trans. Amer. Math. Soc., vol. 101 (1961), 139–167.

49. Sigalov, A. G.

(a) *The existence of an absolute minimum for double integrals of the calculus of variations in parametric form*, Doklady Akad. Nauk, vol. 70 (1950), 769–772.

50. Silverman, E.

(a) *Definitions of Lebesgue area for surfaces in metric spaces*, Riv. di Mat. della Univ. di Parma vol. 2 (1951), 47–76.

(b) *An intrinsic property of Lebesgue area*, Riv. di Mat. della Univ. di Parma, vol. 2 (1951), 195–201.

(c) *A miniature theory of Lebesgue area*, Amer. Math. Monthyl, vol. 67 (1960), 424–430.

(d) *A problem of least area*, Pacific Jour. of Math., vol. 14.(1964), 309–331.

Also see (8a).

51. Smiley, M. F.

(a) *The Jacobi condition for extremaloids*, Duke Math. Jour., vol. 6 (1940), 425–427.

52. Stratonovich, R. L.

(a) *A new representation for stochastic integrals and equations*, SIAM Jour. on Control, vol. 4 (1966), 362–371.

53. Utz, W. R.

See (8c, d).

54. Vachino, R. F.

(a) *Steepest descent with inequality constraints on the control variables*, SIAM Jour. on Control, vol. 4 (1966), 245–261.

55. Warga, J.

(a) *Relaxed variational problems*, Jour. of Math. Anal. and Appl., vol. 4 (1962), 111–128.

(b) *Necessary conditions for minimum in relaxed variational problems*, Jour. of Math. Analysis and Appl., vol. 4 (1962), 129–145.

(c) *Unilateral variational problems with several inequalities*, Mich. Math. Jour., vol. 12 (1965), 449–480.

(d) *Minimax problems and unilateral curves in the calculus of variations*, SIAM Jour. on Control, vol. 3 (1965), 91–105.

(e) *Variational problems with unbounded controls*, SIAM Jour. on Control, vol. 3 (1965), 424–438.

56. Wilkens, J. E., Jr.

(a) *Multiple integral problems in parametric form in the calculus of variations*, Annals of Math., vol. 54 (1944), 312–324.

57. Young, L. C.

(a) *Generalized curves and the existence of an attained absolute minimum in the calculus of variations*, Comptes Rend. de la Soc. des Sciences et des Lettres de Varsovie, class III, vol. 30 (1937), 212–234.

(b) *Necessary conditions in the calculus of variations*, Acta Math., vol. 69 (1938), 229–258.

(c) *Generalized surfaces in the calculus of variations*, Annals of Math., vol. 43 (1942), 530–554.

58. Youngs, J. W. T.

(a) *The additivity of Lebesgue area*, Bull. Amer. Math. Soc., vol. 49 (1943), 779–784.

(b) *Curves and surfaces*, Amer. Math. Monthly, vol. 51 (1944), 1–11.

(c) *On surfaces of class K_1*, Bull. Amer. Math. Soc., vol. 51 (1945), 669–673.

(d) *The representation problem for Fréchet surfaces*, (Memoirs Amer. Math. Soc., No. 8), American Mathematical Society, Providence, R.I., 1951.

Supplementary Bibliography

BOOKS

XXXVIII.   Berkovitz, L.D., *Optimal Control Theory* (Applied Mathematical Sciences, vol. 12), Springer-Verlag, New York, Heidelberg, Berlin, 1974.

XXXIX.   Klötzler, R., *Mehrdimensionale Variationsrechnung*, Birkhäuser Verlag, Basel und Stuttgart, 1970.

XL.   Sagan, H., *Introduction to the Calculus of Variations*, McGraw-Hill, New York, 1969.

XLI.   Warga, J., *Optimal Control of Differential and Functional Equations*, Academic Press, New York and London, 1972.

XLII.   Young, L. C., *Lectures on the Calculus of Variations and Optimal Control Theory*, W. B. Saunders, Philadelphia, 1969.

ARTICLES

12. Ewing, G. M.

(j) *Sufficient conditions for global minima of suitably convex functionals from variational and control theory*, SIAM Review, vol. 19 (1977), 202–220.

(k) *Sufficient conditions for asymptotic optimal control*, Jour. of Optimization Theory and Appl., vol. 32 (1980), 307–325.

These papers contain refinements and extensions of material treated in Sections 3.11, 3.12, and 5.11.

Index

A CATALOG OF SELECTED
DOVER BOOKS
IN ALL FIELDS OF INTEREST

A CATALOG OF SELECTED DOVER BOOKS IN ALL FIELDS OF INTEREST

CONCERNING THE SPIRITUAL IN ART, Wassily Kandinsky. Pioneering work by father of abstract art. Thoughts on color theory, nature of art. Analysis of earlier masters. 12 illustrations. 80pp. of text. 5⅜ × 8½. 23411-8 Pa. $2.95

LEONARDO ON THE HUMAN BODY, Leonardo da Vinci. More than 1200 of Leonardo's anatomical drawings on 215 plates. Leonardo's text, which accompanies the drawings, has been translated into English. 506pp. 8⅜ × 11¾. 24483-0 Pa. $11.95

GOBLIN MARKET, Christina Rossetti. Best-known work by poet comparable to Emily Dickinson, Alfred Tennyson. With 46 delightfully grotesque illustrations by Laurence Housman. 64pp. 4 × 6¾. 24516-0 Pa. $2.50

THE HEART OF THOREAU'S JOURNALS, edited by Odell Shepard. Selections from *Journal*, ranging over full gamut of interests. 228pp. 5⅜ × 8½. 20741-2 Pa. $4.50

MR. LINCOLN'S CAMERA MAN: MATHEW B. BRADY, Roy Meredith. Over 300 Brady photos reproduced directly from original negatives, photos. Lively commentary. 368pp. 8⅜ × 11¼. 23021-X Pa. $14.95

PHOTOGRAPHIC VIEWS OF SHERMAN'S CAMPAIGN, George N. Barnard. Reprint of landmark 1866 volume with 61 plates: battlefield of New Hope Church, the Etawah Bridge, the capture of Atlanta, etc. 80pp. 9 × 12. 23445-2 Pa. $6.00

A SHORT HISTORY OF ANATOMY AND PHYSIOLOGY FROM THE GREEKS TO HARVEY, Dr. Charles Singer. Thoroughly engrossing non-technical survey. 270 illustrations. 211pp. 5⅜ × 8½. 20389-1 Pa. $4.95

REDOUTE ROSES IRON-ON TRANSFER PATTERNS, Barbara Christopher. Redouté was botanical painter to the Empress Josephine; transfer his famous roses onto fabric with these 24 transfer patterns. 80pp. 8¼ × 10⅞. 24292-7 Pa. $3.50

THE FIVE BOOKS OF ARCHITECTURE, Sebastiano Serlio. Architectural milestone, first (1611) English translation of Renaissance classic. Unabridged reproduction of original edition includes over 300 woodcut illustrations. 416pp. 9⅜ × 12¼. 24349-4 Pa. $14.95

CARLSON'S GUIDE TO LANDSCAPE PAINTING, John F. Carlson. Authoritative, comprehensive guide covers, every aspect of landscape painting. 34 reproductions of paintings by author; 58 explanatory diagrams. 144pp. 8⅜ × 11. 22927-0 Pa. $5.95

101 PUZZLES IN THOUGHT AND LOGIC, C.R. Wylie, Jr. Solve murders, robberies, see which fishermen are liars—purely by reasoning! 107pp. 5⅜ × 8½. 20367-0 Pa. $2.00

TEST YOUR LOGIC, George J. Summers. 50 more truly new puzzles with new turns of thought, new subtleties of inference. 100pp. 5⅜ × 8½. 22877-0 Pa. $2.50

THE MURDER BOOK OF J.G. REEDER, Edgar Wallace. Eight suspenseful stories by bestselling mystery writer of 20s and 30s. Features the donnish Mr. J.G. Reeder of Public Prosecutor's Office. 128pp. 5⅜ × 8½.

24374-5 Pa. $3.95

ANNE ORR'S CHARTED DESIGNS, Anne Orr. Best designs by premier needlework designer, all on charts: flowers, borders, birds, children, alphabets, etc. Over 100 charts, 10 in color. Total of 40pp. 8¼ × 11. 23704-4 Pa. $2.50

BASIC CONSTRUCTION TECHNIQUES FOR HOUSES AND SMALL BUILDINGS SIMPLY EXPLAINED, U.S. Bureau of Naval Personnel. Grading, masonry, woodworking, floor and wall framing, roof framing, plastering, tile setting, much more. Over 675 illustrations. 568pp. 6½ × 9¼. 20242-9 Pa. $9.95

MATISSE LINE DRAWINGS AND PRINTS, Henri Matisse. Representative collection of female nudes, faces, still lifes, experimental works, etc., from 1898 to 1948. 50 illustrations. 48pp. 8⅜ × 11¼. 23877-6 Pa. $3.50

HOW TO PLAY THE CHESS OPENINGS, Eugene Znosko-Borovsky. Clear, profound examinations of just what each opening is intended to do and how opponent can counter. Many sample games. 147pp. 5⅜ × 8½. 22795-2 Pa. $3.50

DUPLICATE BRIDGE, Alfred Sheinwold. Clear, thorough, easily followed account: rules, etiquette, scoring, strategy, bidding; Goren's point-count system, Blackwood and Gerber conventions, etc. 158pp. 5⅜ × 8½. 22741-3 Pa. $3.50

SARGENT PORTRAIT DRAWINGS, J.S. Sargent. Collection of 42 portraits reveals technical skill and intuitive eye of noted American portrait painter, John Singer Sargent. 48pp. 8¼ × 11⅛. 24524-1 Pa. $3.50

ENTERTAINING SCIENCE EXPERIMENTS WITH EVERYDAY OBJECTS, Martin Gardner. Over 100 experiments for youngsters. Will amuse, astonish, teach, and entertain. Over 100 illustrations. 127pp. 5⅜ × 8½. 24201-3 Pa. $2.50

TEDDY BEAR PAPER DOLLS IN FULL COLOR: A Family of Four Bears and Their Costumes, Crystal Collins. A family of four Teddy Bear paper dolls and nearly 60 cut-out costumes. Full color, printed one side only. 32pp. 9¼ × 12¼.

24550-0 Pa. $3.50

NEW CALLIGRAPHIC ORNAMENTS AND FLOURISHES, Arthur Baker. Unusual, multi-useable material: arrows, pointing hands, brackets and frames, ovals, swirls, birds, etc. Nearly 700 illustrations. 80pp. 8⅜ × 11¼.

24095-9 Pa. $3.75

DINOSAUR DIORAMAS TO CUT & ASSEMBLE, M. Kalmenoff. Two complete three-dimensional scenes in full color, with 31 cut-out animals and plants. Excellent educational toy for youngsters. Instructions; 2 assembly diagrams. 32pp. 9¼ × 12¼. 24541-1 Pa. $4.50

SILHOUETTES: A PICTORIAL ARCHIVE OF VARIED ILLUSTRATIONS, edited by Carol Belanger Grafton. Over 600 silhouettes from the 18th to 20th centuries. Profiles and full figures of men, women, children, birds, animals, groups and scenes, nature, ships, an alphabet. 144pp. 8⅜ × 11¼. 23781-8 Pa. $5.95

25 KITES THAT FLY, Leslie Hunt. Full, easy-to-follow instructions for kites made from inexpensive materials. Many novelties. 70 illustrations. 110pp. 5⅜ × 8½.
22550-X Pa. $2.50

PIANO TUNING, J. Cree Fischer. Clearest, best book for beginner, amateur. Simple repairs, raising dropped notes, tuning by easy method of flattened fifths. No previous skills needed. 4 illustrations. 201pp. 5⅜ × 8½. 23267-0 Pa. $3.50

EARLY AMERICAN IRON-ON TRANSFER PATTERNS, edited by Rita Weiss. 75 designs, borders, alphabets, from traditional American sources. 48pp. 8¼ × 11.
23162-3 Pa. $1.95

CROCHETING EDGINGS, edited by Rita Weiss. Over 100 of the best designs for these lovely trims for a host of household items. Complete instructions, illustrations. 48pp. 8¼ × 11. 24031-2 Pa. $2.95

FINGER PLAYS FOR NURSERY AND KINDERGARTEN, Emilie Poulsson. 18 finger plays with music (voice and piano); entertaining, instructive. Counting, nature lore, etc. Victorian classic. 53 illustrations. 80pp. 6½ × 9¼. 22588-7 Pa. $2.25

BOSTON THEN AND NOW, Peter Vanderwarker. Here in 59 side-by-side views are photographic documentations of the city's past and present. 119 photographs. Full captions. 122pp. 8¼ × 11. 24312-5 Pa. $7.95

CROCHETING BEDSPREADS, edited by Rita Weiss. 22 patterns, originally published in three instruction books 1939-41. 39 photos, 8 charts. Instructions. 48pp. 8¼ × 11. 23610-2 Pa. $2.00

HAWTHORNE ON PAINTING, Charles W. Hawthorne. Collected from notes taken by students at famous Cape Cod School; hundreds of direct, personal *apercus*, ideas, suggestions. 91pp. 5⅜ × 8½. 20653-X Pa. $2.95

THERMODYNAMICS, Enrico Fermi. A classic of modern science. Clear, organized treatment of systems, first and second laws, entropy, thermodynamic potentials, etc. Calculus required. 160pp. 5⅜ × 8½. 60361-X Pa. $4.50

TEN BOOKS ON ARCHITECTURE, Vitruvius. The most important book ever written on architecture. Early Roman aesthetics, technology, classical orders, site selection, all other aspects. Morgan translation. 331pp. 5⅜ × 8½. 20645-9 Pa. $6.95

THE CORNELL BREAD BOOK, Clive M. McCay and Jeanette B. McCay. Famed high-protein recipe incorporated into breads, rolls, buns, coffee cakes, pizza, pie crusts, more. Nearly 50 illustrations. 48pp. 8¼ × 11. 23995-0 Pa. $2.00

THE CRAFTSMAN'S HANDBOOK, Cennino Cennini. 15th-century handbook, school of Giotto, explains applying gold, silver leaf; gesso; fresco painting, grinding pigments, etc. 142pp. 6⅛ × 9¼. 20054-X Pa. $3.95

FRANK LLOYD WRIGHT'S FALLINGWATER, Donald Hoffmann. Full story of Wright's masterwork at Bear Run, Pa. 100 photographs of site, construction, and details of completed structure. 112pp. 9¼ × 10. 23671-4 Pa. $7.95

OVAL STAINED GLASS PATTERN BOOK, C. Eaton. 60 new designs framed in shape of an oval. Greater complexity, challenge with sinuous cats, birds, mandalas framed in antique shape. 64pp. 8¼ × 11. 24519-5 Pa. $3.95

THE BOOK OF WOOD CARVING, Charles Marshall Sayers. Still finest book for beginning student. Fundamentals, technique; gives 34 designs, over 34 projects for panels, bookends, mirrors, etc. 33 photos. 118pp. 7¾ × 10⅝.   23654-4 Pa. $3.95

CARVING COUNTRY CHARACTERS, Bill Higginbotham. Expert advice for beginning, advanced carvers on materials, techniques for creating 18 projects— mirthful panorama of American characters. 105 illustrations. 80pp. 8⅜ × 11.
24135-1 Pa. $2.95

300 ART NOUVEAU DESIGNS AND MOTIFS IN FULL COLOR, C.B. Grafton. 44 full-page plates display swirling lines and muted colors typical of Art Nouveau. Borders, frames, panels, cartouches, dingbats, etc. 48pp. 9⅜ × 12¼.
24354-0 Pa. $6.95

SELF-WORKING CARD TRICKS, Karl Fulves. Editor of *Pallbearer* offers 72 tricks that work automatically through nature of card deck. No sleight of hand needed. Often spectacular. 42 illustrations. 113pp. 5⅜ × 8½.   23334-0 Pa. $3.50

CUT AND ASSEMBLE A WESTERN FRONTIER TOWN, Edmund V. Gillon, Jr. Ten authentic full-color buildings on heavy cardboard stock in H-O scale. Sheriff's Office and Jail, Saloon, Wells Fargo, Opera House, others. 48pp. 9¼ × 12¼.
23736-2 Pa. $4.95

CUT AND ASSEMBLE AN EARLY NEW ENGLAND VILLAGE, Edmund V. Gillon, Jr. Printed in full color on heavy cardboard stock. 12 authentic buildings in H-O scale: Adams home in Quincy, Mass., Oliver Wight house in Sturbridge, smithy, store, church, others. 48pp. 9¼ × 12¼.   23536-X Pa. $4.95

THE TALE OF TWO BAD MICE, Beatrix Potter. Tom Thumb and Hunca Munca squeeze out of their hole and go exploring. 27 full-color Potter illustrations. 59pp. 4¼ × 5½. (Available in U.S. only)   23065-1 Pa. $1.75

CARVING FIGURE CARICATURES IN THE OZARK STYLE, Harold L. Enlow. Instructions and illustrations for ten delightful projects, plus general carving instructions. 22 drawings and 47 photographs altogether. 39pp. 8⅜ × 11.
23151-8 Pa. $2.95

A TREASURY OF FLOWER DESIGNS FOR ARTISTS, EMBROIDERERS AND CRAFTSMEN, Susan Gaber. 100 garden favorites lushly rendered by artist for artists, craftsmen, needleworkers. Many form frames, borders. 80pp. 8¼ × 11.
24096-7 Pa. $3.95

CUT & ASSEMBLE A TOY THEATER/THE NUTCRACKER BALLET, Tom Tierney. Model of a complete, full-color production of Tchaikovsky's classic. 6 backdrops, dozens of characters, familiar dance sequences. 32pp. 9⅜ × 12¼.
24194-7 Pa. $4.50

ANIMALS: 1,419 COPYRIGHT-FREE ILLUSTRATIONS OF MAMMALS, BIRDS, FISH, INSECTS, ETC., edited by Jim Harter. Clear wood engravings present, in extremely lifelike poses, over 1,000 species of animals. 284pp. 9 × 12.
23766-4 Pa. $9.95

MORE HAND SHADOWS, Henry Bursill. For those at their 'finger ends," 16 more effects—Shakespeare, a hare, a squirrel, Mr. Punch, and twelve more—each explained by a full-page illustration. Considerable period charm. 30pp. 6½ × 9¼.
21384-6 Pa. $1.95

SURREAL STICKERS AND UNREAL STAMPS, William Rowe. 224 haunting, hilarious stamps on gummed, perforated stock, with images of elephants, geisha girls, George Washington, etc. 16pp. one side. 8¼ × 11. 24371-0 Pa. $3.50

GOURMET KITCHEN LABELS, Ed Sibbett, Jr. 112 full-color labels (4 copies each of 28 designs). Fruit, bread, other culinary motifs. Gummed and perforated. 16pp. 8¼ × 11. 24087-8 Pa. $2.95

PATTERNS AND INSTRUCTIONS FOR CARVING AUTHENTIC BIRDS, H.D. Green. Detailed instructions, 27 diagrams, 85 photographs for carving 15 species of birds so life-like, they'll seem ready to fly! 8¼ × 11. 24222-6 Pa. $3.00

FLATLAND, E.A. Abbott. Science-fiction classic explores life of 2-D being in 3-D world. 16 illustrations. 103pp. 5⅜ × 8. 20001-9 Pa. $2.00

DRIED FLOWERS, Sarah Whitlock and Martha Rankin. Concise, clear, practical guide to dehydration, glycerinizing, pressing plant material, and more. Covers use of silica gel. 12 drawings. 32pp. 5⅜ × 8½. 21802-3 Pa. $1.00

EASY-TO-MAKE CANDLES, Gary V. Guy. Learn how easy it is to make all kinds of decorative candles. Step-by-step instructions. 82 illustrations. 48pp. 8¼ × 11. 23881-4 Pa. $2.95

SUPER STICKERS FOR KIDS, Carolyn Bracken. 128 gummed and perforated full-color stickers: GIRL WANTED, KEEP OUT, BORED OF EDUCATION, X-RATED, COMBAT ZONE, many others. 16pp. 8¼ × 11. 24092-4 Pa. $3.50

CUT AND COLOR PAPER MASKS, Michael Grater. Clowns, animals, funny faces...simply color them in, cut them out, and put them together, and you have 9 paper masks to play with and enjoy. 32pp. 8¼ × 11. 23171-2 Pa. $2.95

A CHRISTMAS CAROL: THE ORIGINAL MANUSCRIPT, Charles Dickens. Clear facsimile of Dickens manuscript, on facing pages with final printed text. 8 illustrations by John Leech, 4 in color on covers. 144pp. 8⅜ × 11¼. 20980-6 Pa. $5.95

CARVING SHOREBIRDS, Harry V. Shourds & Anthony Hillman. 16 full-size patterns (all double-page spreads) for 19 North American shorebirds with step-by-step instructions. 72pp. 9¼ × 12¼. 24287-0 Pa. $5.95

THE GENTLE ART OF MATHEMATICS, Dan Pedoe. Mathematical games, probability, the question of infinity, topology, how the laws of algebra work, problems of irrational numbers, and more. 42 figures. 143pp. 5⅜ × 8½. 22949-1 Pa. $3.50

READY-TO-USE DOLLHOUSE WALLPAPER, Katzenbach & Warren, Inc. Stripe, 2 floral stripes, 2 allover florals, polka dot; all in full color. 4 sheets (350 sq. in.) of each, enough for average room. 48pp. 8¼ × 11. 23495-9 Pa. $2.95

MINIATURE IRON-ON TRANSFER PATTERNS FOR DOLLHOUSES, DOLLS, AND SMALL PROJECTS, Rita Weiss and Frank Fontana. Over 100 miniature patterns: rugs, bedspreads, quilts, chair seats, etc. In standard dollhouse size. 48pp. 8¼ × 11. 23741-9 Pa. $1.95

THE DINOSAUR COLORING BOOK, Anthony Rao. 45 renderings of dinosaurs, fossil birds, turtles, other creatures of Mesozoic Era. Scientifically accurate. Captions. 48pp. 8¼ × 11. 24022-3 Pa. $2.50

JAPANESE DESIGN MOTIFS, Matsuya Co. Mon, or heraldic designs. Over 4000 typical, beautiful designs: birds, animals, flowers, swords, fans, geometrics; all beautifully stylized. 213pp. 11⅜ × 8¼. 22874-6 Pa. $7.95

THE TALE OF BENJAMIN BUNNY, Beatrix Potter. Peter Rabbit's cousin coaxes him back into Mr. McGregor's garden for a whole new set of adventures. All 27 full-color illustrations. 59pp. 4¼ × 5½. (Available in U.S. only) 21102-9 Pa. $1.75

THE TALE OF PETER RABBIT AND OTHER FAVORITE STORIES BOXED SET, Beatrix Potter. Seven of Beatrix Potter's best-loved tales including Peter Rabbit in a specially designed, durable boxed set. 4¼ × 5½. Total of 447pp. 158 color illustrations. (Available in U.S. only) 23903-9 Pa. $12.25

PRACTICAL MENTAL MAGIC, Theodore Annemann. Nearly 200 astonishing feats of mental magic revealed in step-by-step detail. Complete advice on staging, patter, etc. Illustrated. 320pp. 5⅜ × 8½. 24426-1 Pa. $5.95

CELEBRATED CASES OF JUDGE DEE (DEE GOONG AN), translated by Robert Van Gulik. Authentic 18th-century Chinese detective novel; Dee and associates solve three interlocked cases. Led to van Gulik's own stories with same characters. Extensive introduction. 9 illustrations. 237pp. 5⅜ × 8½. 23337-5 Pa. $4.95

CUT & FOLD EXTRATERRESTRIAL INVADERS THAT FLY, M. Grater. Stage your own lilliputian space battles.By following the step-by-step instructions and explanatory diagrams you can launch 22 full-color fliers into space. 36pp. 8¼ × 11. 24478-4 Pa. $2.95

CUT & ASSEMBLE VICTORIAN HOUSES, Edmund V. Gillon, Jr. Printed in full color on heavy cardboard stock, 4 authentic Victorian houses in H-O scale: Italian-style Villa, Octagon, Second Empire, Stick Style. 48pp. 9¼ × 12¼. 23849-0 Pa. $4.95

BEST SCIENCE FICTION STORIES OF H.G. WELLS, H.G. Wells. Full novel *The Invisible Man,* plus 17 short stories: "The Crystal Egg," "Aepyornis Island," "The Strange Orchid," etc. 303pp. 5⅜ × 8½. (Available in U.S. only) 21531-8 Pa. $4.95

TRADEMARK DESIGNS OF THE WORLD, Yusaku Kamekura. A lavish collection of nearly 700 trademarks, the work of Wright, Loewy, Klee, Binder, hundreds of others. 160pp. 8¾ × 8. (EJ) 24191-2 Pa. $5.95

THE ARTIST'S AND CRAFTSMAN'S GUIDE TO REDUCING, ENLARGING AND TRANSFERRING DESIGNS, Rita Weiss. Discover, reduce, enlarge, transfer designs from any objects to any craft project. 12pp. plus 16 sheets special graph paper. 8¼ × 11. 24142-4 Pa. $3.95

TREASURY OF JAPANESE DESIGNS AND MOTIFS FOR ARTISTS AND CRAFTSMEN, edited by Carol Belanger Grafton. Indispensable collection of 360 traditional Japanese designs and motifs redrawn in clean, crisp black-and-white, copyright-free illustrations. 96pp. 8¼ × 11. 24435-0 Pa. $4.50

CHANCERY CURSIVE STROKE BY STROKE, Arthur Baker. Instructions and illustrations for each stroke of each letter (upper and lower case) and numerals. 54 full-page plates. 64pp. 8¼ × 11. 24278-1 Pa. $2.50

THE ENJOYMENT AND USE OF COLOR, Walter Sargent. Color relationships, values, intensities; complementary colors, illumination, similar topics. Color in nature and art. 7 color plates, 29 illustrations. 274pp. 5⅜ × 8½. 20944-X Pa. $4.95

SCULPTURE PRINCIPLES AND PRACTICE, Louis Slobodkin. Step-by-step approach to clay, plaster, metals, stone; classical and modern. 253 drawings, photos. 255pp. 8⅛ × 11. 22960-2 Pa. $7.50

VICTORIAN FASHION PAPER DOLLS FROM HARPER'S BAZAR, 1867-1898, Theodore Menten. Four female dolls with 28 elegant high fashion costumes, printed in full color. 32pp. 9¼ × 12¼. 23453-3 Pa. $3.95

FLOPSY, MOPSY AND COTTONTAIL: A Little Book of Paper Dolls in Full Color, Susan LaBelle. Three dolls and 21 costumes (7 for each doll) show Peter Rabbit's siblings dressed for holidays, gardening, hiking, etc. Charming borders, captions. 48pp. 4¼ × 5½. (USCO) 24376-1 Pa. $2.50

NATIONAL LEAGUE BASEBALL CARD CLASSICS, Bert Randolph Sugar. 83 big-leaguers from 1909-69 on facsimile cards. Hubbell, Dean, Spahn, Brock plus advertising, info, no duplications. Perforated, detachable. 16pp. 8¼ × 11. 24308-7 Pa. $3.50

THE LOGICAL APPROACH TO CHESS, Dr. Max Euwe, et al. First-rate text of comprehensive strategy, tactics, theory for the amateur. No gambits to memorize, just a clear, logical approach. 224pp. 5⅜ × 8½. 24353-2 Pa. $4.50

MAGICK IN THEORY AND PRACTICE, Aleister Crowley. The summation of the thought and practice of the century's most famous necromancer, long hard to find. Crowley's best book. 436pp. 5⅜ × 8½. (Available in U.S. only) 23295-6 Pa. $6.95

THE HAUNTED HOTEL, Wilkie Collins. Collins' last great tale; doom and destiny in a Venetian palace. Praised by T.S. Eliot. 127pp. 5⅜ × 8½. 24333-8 Pa. $3.00

ART DECO DISPLAY ALPHABETS, Dan X. Solo. Wide variety of bold yet elegant lettering in handsome Art Deco styles. 100 complete fonts, with numerals, punctuation, more. 104pp. 8⅛ × 11. 24372-9 Pa. $4.50

CALLIGRAPHIC ALPHABETS, Arthur Baker. Nearly 150 complete alphabets by outstanding contemporary. Stimulating ideas; useful source for unique effects. 154 plates. 157pp. 8⅜ × 11¼. 21045-6 Pa. $5.95

ARTHUR BAKER'S HISTORIC CALLIGRAPHIC ALPHABETS, Arthur Baker. From monumental capitals of first-century Rome to humanistic cursive of 16th century, 33 alphabets in fresh interpretations. 88 plates. 96pp. 9 × 12. 24054-1 Pa. $4.50

LETTIE LANE PAPER DOLLS, Sheila Young. Genteel turn-of-the-century family very popular then and now. 24 paper dolls. 16 plates in full color. 32pp. 9¼ × 12¼. 24089-4 Pa. $3.95

CATALOG OF DOVER BOOKS

KEYBOARD WORKS FOR SOLO INSTRUMENTS, G.F. Handel. 35 neglected works from Handel's vast oeuvre, originally jotted down as improvisations. Includes Eight Great Suites, others. New sequence. 174pp. 9⅜ × 12¼.
24338-9 Pa. $7.50

AMERICAN LEAGUE BASEBALL CARD CLASSICS, Bert Randolph Sugar. 82 stars from 1900s to 60s on facsimile cards. Ruth, Cobb, Mantle, Williams, plus advertising, info, no duplications. Perforated, detachable. 16pp. 8¼ × 11.
24286-2 Pa. $3.50

A TREASURY OF CHARTED DESIGNS FOR NEEDLEWORKERS, Georgia Gorham and Jeanne Warth. 141 charted designs: owl, cat with yarn, tulips, piano, spinning wheel, covered bridge, Victorian house and many others. 48pp. 8¼ × 11.
23558-0 Pa. $1.95

DANISH FLORAL CHARTED DESIGNS, Gerda Bengtsson. Exquisite collection of over 40 different florals: anemone, Iceland poppy, wild fruit, pansies, many others. 45 illustrations. 48pp. 8¼ × 11.
23957-8 Pa. $2.50

OLD PHILADELPHIA IN EARLY PHOTOGRAPHS 1839-1914, Robert F. Looney. 215 photographs: panoramas, street scenes, landmarks, President-elect Lincoln's visit, 1876 Centennial Exposition, much more. 230pp. 8⅞ × 11¾.
23345-6 Pa. $9.95

PRELUDE TO MATHEMATICS, W.W. Sawyer. Noted mathematician's lively, stimulating account of non-Euclidean geometry, matrices, determinants, group theory, other topics. Emphasis on novel, striking aspects. 224pp. 5⅜ × 8½.
24401-6 Pa. $4.50

ADVENTURES WITH A MICROSCOPE, Richard Headstrom. 59 adventures with clothing fibers, protozoa, ferns and lichens, roots and leaves, much more. 142 illustrations. 232pp. 5⅜ × 8½.
23471-1 Pa. $3.95

IDENTIFYING ANIMAL TRACKS: MAMMALS, BIRDS, AND OTHER ANIMALS OF THE EASTERN UNITED STATES, Richard Headstrom. For hunters, naturalists, scouts, nature-lovers. Diagrams of tracks, tips on identification. 128pp. 5⅜ × 8.
24442-3 Pa. $3.50

VICTORIAN FASHIONS AND COSTUMES FROM HARPER'S BAZAR, 1867-1898, edited by Stella Blum. Day costumes, evening wear, sports clothes, shoes, hats, other accessories in over 1,000 detailed engravings. 320pp. 9⅜ × 12¼.
22990-4 Pa. $10.95

EVERYDAY FASHIONS OF THE TWENTIES AS PICTURED IN SEARS AND OTHER CATALOGS, edited by Stella Blum. Actual dress of the Roaring Twenties, with text by Stella Blum. Over 750 illustrations, captions. 156pp. 9 × 12.
24134-3 Pa. $8.95

HALL OF FAME BASEBALL CARDS, edited by Bert Randolph Sugar. Cy Young, Ted Williams, Lou Gehrig, and many other Hall of Fame greats on 92 full-color, detachable reprints of early baseball cards. No duplication of cards with *Classic Baseball Cards*. 16pp. 8¼ × 11.
23624-2 Pa. $3.50

THE ART OF HAND LETTERING, Helm Wotzkow. Course in hand lettering, Roman, Gothic, Italic, Block, Script. Tools, proportions, optical aspects, individual variation. Very quality conscious. Hundreds of specimens. 320pp. 5⅜ × 8½.
21797-3 Pa. $5.95

HOW THE OTHER HALF LIVES, Jacob A. Riis. Journalistic record of filth, degradation, upward drive in New York immigrant slums, shops, around 1900. New edition includes 100 original Riis photos, monuments of early photography. 233pp. 10 × 7⅞. 22012-5 Pa. $9.95

CHINA AND ITS PEOPLE IN EARLY PHOTOGRAPHS, John Thomson. In 200 black-and-white photographs of exceptional quality photographic pioneer Thomson captures the mountains, dwellings, monuments and people of 19th-century China. 272pp. 9⅜ × 12¼. 24393-1 Pa. $13.95

GODEY COSTUME PLATES IN COLOR FOR DECOUPAGE AND FRAMING, edited by Eleanor Hasbrouk Rawlings. 24 full-color engravings depicting 19th-century Parisian haute couture. Printed on one side only. 56pp. 8¼ × 11. 23879-2 Pa. $3.95

ART NOUVEAU STAINED GLASS PATTERN BOOK, Ed Sibbett, Jr. 104 projects using well-known themes of Art Nouveau: swirling forms, florals, peacocks, and sensuous women. 60pp. 8¼ × 11. 23577-7 Pa. $3.95

QUICK AND EASY PATCHWORK ON THE SEWING MACHINE: Susan Aylsworth Murwin and Suzzy Payne. Instructions, diagrams show exactly how to machine sew 12 quilts. 48pp. of templates. 50 figures. 80pp. 8¼ × 11. 23770-2 Pa. $3.95

THE STANDARD BOOK OF QUILT MAKING AND COLLECTING, Marguerite Ickis. Full information, full-sized patterns for making 46 traditional quilts, also 150 other patterns. 483 illustrations. 273pp. 6⅞ × 9⅝. 20582-7 Pa. $5.95

LETTERING AND ALPHABETS, J. Albert Cavanagh. 85 complete alphabets lettered in various styles; instructions for spacing, roughs, brushwork. 121pp. 8¾ × 8. 20053-1 Pa. $3.95

LETTER FORMS: 110 COMPLETE ALPHABETS, Frederick Lambert. 110 sets of capital letters; 16 lower case alphabets; 70 sets of numbers and other symbols. 110pp. 8⅛ × 11. 22872-X Pa. $4.50

ORCHIDS AS HOUSE PLANTS, Rebecca Tyson Northen. Grow cattleyas and many other kinds of orchids—in a window, in a case, or under artificial light. 63 illustrations. 148pp. 5⅜ × 8½. 23261-1 Pa. $2.95

THE MUSHROOM HANDBOOK, Louis C.C. Krieger. Still the best popular handbook. Full descriptions of 259 species, extremely thorough text, poisons, folklore, etc. 32 color plates; 126 other illustrations. 560pp. 5⅜ × 8½. 21861-9 Pa. $8.50

THE DORÉ BIBLE ILLUSTRATIONS, Gustave Doré. All wonderful, detailed plates: Adam and Eve, Flood, Babylon, life of Jesus, etc. Brief King James text with each plate. 241 plates. 241pp. 9 × 12. 23004-X Pa. $8.95

THE BOOK OF KELLS: Selected Plates in Full Color, edited by Blanche Cirker. 32 full-page plates from greatest manuscript-icon of early Middle Ages. Fantastic, mysterious. Publisher's Note. Captions. 32pp. 9¾ × 12¼. 24345-1 Pa. $4.50

THE PERFECT WAGNERITE, George Bernard Shaw. Brilliant criticism of the Ring Cycle, with provocative interpretation of politics, economic theories behind the Ring. 136pp. 5⅜ × 8½. (EUK) 21707-8 Pa. $3.95

THE RIME OF THE ANCIENT MARINER, Gustave Doré, S.T. Coleridge. Doré's finest work, 34 plates capture moods, subtleties of poem. Full text. 77pp. 9¼ × 12. 22305-1 Pa. $4.95

SONGS OF INNOCENCE, William Blake. The first and most popular of Blake's famous "Illuminated Books," in a facsimile edition reproducing all 31 brightly colored plates. Additional printed text of each poem. 64pp. 5¼ × 7.
22764-2 Pa. $3.50

AN INTRODUCTION TO INFORMATION THEORY, J.R. Pierce. Second (1980) edition of most impressive non-technical account available. Encoding, entropy, noisy channel, related areas, etc. 320pp. 5⅜ × 8½. 24061-4 Pa. $5.95

THE DIVINE PROPORTION: A STUDY IN MATHEMATICAL BEAUTY, H.E. Huntley. "Divine proportion" or "golden ratio" in poetry, Pascal's triangle, philosophy, psychology, music, mathematical figures, etc. Excellent bridge between science and art. 58 figures. 185pp. 5⅜ × 8½. 22254-3 Pa. $4.50

THE DOVER NEW YORK WALKING GUIDE: From the Battery to Wall Street, Mary J. Shapiro. Superb inexpensive guide to historic buildings and locales in lower Manhattan: Trinity Church, Bowling Green, more. Complete Text; maps. 36 illustrations. 48pp. 3⅞ × 9¼. 24225-0 Pa. $2.50

NEW YORK THEN AND NOW, Edward B. Watson, Edmund V. Gillon, Jr. 83 important Manhattan sites: on facing pages early photographs (1875-1925) and 1976 photos by Gillon. 172 illustrations. 171pp. 9¼ × 10. 23361-8 Pa. $9.95

HISTORIC COSTUME IN PICTURES, Braun & Schneider. Over 1450 costumed figures from dawn of civilization to end of 19th century. English captions. 125 plates. 256pp. 8⅜ × 11¼. 23150-X Pa. $7.95

VICTORIAN AND EDWARDIAN FASHION: A Photographic Survey, Alison Gernsheim. First fashion history completely illustrated by contemporary photographs. Full text plus 235 photos, 1840-1914, in which many celebrities appear. 240pp. 6½ × 9¼. 24205-6 Pa. $6.00

CHARTED CHRISTMAS DESIGNS FOR COUNTED CROSS-STITCH AND OTHER NEEDLECRAFTS, Lindberg Press. Charted designs for 45 beautiful needlecraft projects with many yuletide and wintertime motifs. 48pp. 8¼ × 11.
(EDNS) 24356-7 Pa. $2.50

101 FOLK DESIGNS FOR COUNTED CROSS-STITCH AND OTHER NEEDLE-CRAFTS, Carter Houck. 101 authentic charted folk designs in a wide array of lovely representations with many suggestions for effective use. 48pp. 8¼ × 11.
24369-9 Pa. $2.25

FIVE ACRES AND INDEPENDENCE, Maurice G. Kains. Great back-to-the-land classic explains basics of self-sufficient farming. The one book to get. 95 illustrations. 397pp. 5⅜ × 8½. 20974-1 Pa. $6.50

A MODERN HERBAL, Margaret Grieve. Much the fullest, most exact, most useful compilation of herbal material. Gigantic alphabetical encyclopedia, from aconite to zedoary, gives botanical information, medical properties, folklore, economic uses, and much else. Indispensable to serious reader. 161 illustrations. 888pp. 6½ × 9¼. (Available in U.S. only) 22798-7, 22799-5 Pa., Two-vol. set $17.00

DECORATIVE NAPKIN FOLDING FOR BEGINNERS, Lillian Oppenheimer and Natalie Epstein. 22 different napkin folds in the shape of a heart, clown's hat, love knot, etc. 63 drawings. 48pp. 8¼ × 11. 23797-4 Pa. $2.25

DECORATIVE LABELS FOR HOME CANNING, PRESERVING, AND OTHER HOUSEHOLD AND GIFT USES, Theodore Menten. 128 gummed, perforated labels, beautifully printed in 2 colors. 12 versions. Adhere to metal, glass, wood, ceramics. 24pp. 8¼ × 11. 23219-0 Pa. $3.50

EARLY AMERICAN STENCILS ON WALLS AND FURNITURE, Janet Waring. Thorough coverage of 19th-century folk art: techniques, artifacts, surviving specimens. 166 illustrations, 7 in color. 147pp. of text. 7⅞ × 10¾. 21906-2 Pa. $9.95

AMERICAN ANTIQUE WEATHERVANES, A.B. & W.T. Westervelt. Extensively illustrated 1883 catalog exhibiting over 550 copper weathervanes and finials. Excellent primary source by one of the principal manufacturers. 104pp. 6⅛ × 9¼. 24396-6 Pa. $3.95

ART STUDENTS' ANATOMY, Edmond J. Farris. Long favorite in art schools. Basic elements, common positions, actions. Full text, 158 illustrations. 159pp. 5⅜ × 8½. 20744-7 Pa. $3.95

BRIDGMAN'S LIFE DRAWING, George B. Bridgman. More than 500 drawings and text teach you to abstract the body into its major masses. Also specific areas of anatomy. 192pp. 6½ × 9¼. 22710-3 Pa. $4.50

COMPLETE PRELUDES AND ETUDES FOR SOLO PIANO, Frederic Chopin. All 26 Preludes, all 27 Etudes by greatest composer of piano music. Authoritative Paderewski edition. 224pp. 9 × 12. (Available in U.S. only) 24052-5 Pa. $7.50

PIANO MUSIC 1888-1905, Claude Debussy. Deux Arabesques, Suite Bergamesque, Masques, 1st series of Images, etc. 9 others, in corrected editions. 175pp. 9⅜ × 12¼. 22771-5 Pa. $6.95

TEDDY BEAR IRON-ON TRANSFER PATTERNS, Ted Menten. 80 iron-on transfer patterns of male and female Teddys in a wide variety of activities, poses, sizes. 48pp. 8¼ × 11. 24596-9 Pa. $2.25

A PICTURE HISTORY OF THE BROOKLYN BRIDGE, M.J. Shapiro. Profusely illustrated account of greatest engineering achievement of 19th century. 167 rare photos & engravings recall construction, human drama. Extensive, detailed text. 122pp. 8¼ × 11. 24403-2 Pa. $7.95

NEW YORK IN THE THIRTIES, Berenice Abbott. Noted photographer's fascinating study shows new buildings that have become famous and old sights that have disappeared forever. 97 photographs. 97pp. 11⅜ × 10. 22967-X Pa. $7.50

MATHEMATICAL TABLES AND FORMULAS, Robert D. Carmichael and Edwin R. Smith. Logarithms, sines, tangents, trig functions, powers, roots, reciprocals, exponential and hyperbolic functions, formulas and theorems. 269pp. 5⅜ × 8½. 60111-0 Pa. $4.95

HANDBOOK OF MATHEMATICAL FUNCTIONS WITH FORMULAS, GRAPHS, AND MATHEMATICAL TABLES, edited by Milton Abramowitz and Irene A. Stegun. Vast compendium: 29 sets of tables, some to as high as 20 places. 1,046pp. 8 × 10½. 61272-4 Pa. $21.95

REASON IN ART, George Santayana. Renowned philosopher's provocative, seminal treatment of basis of art in instinct and experience. Volume Four of *The Life of Reason*. 230pp. 5⅜ × 8. 24358-3 Pa. $4.50

LANGUAGE, TRUTH AND LOGIC, Alfred J. Ayer. Famous, clear introduction to Vienna, Cambridge schools of Logical Positivism. Role of philosophy, elimination of metaphysics, nature of analysis, etc. 160pp. 5⅜ × 8½. (USCO) 20010-8 Pa. $2.95

BASIC ELECTRONICS, U.S. Bureau of Naval Personnel. Electron tubes, circuits, antennas, AM, FM, and CW transmission and receiving, etc. 560 illustrations. 567pp. 6½ × 9¼. 21076-6 Pa. $9.95

THE ART DECO STYLE, edited by Theodore Menten. Furniture, jewelry, metalwork, ceramics, fabrics, lighting fixtures, interior decors, exteriors, graphics from pure French sources. Over 400 photographs. 183pp. 8⅜ × 11¼. 22824-X Pa. $7.95

THE FOUR BOOKS OF ARCHITECTURE, Andrea Palladio. 16th-century classic covers classical architectural remains, Renaissance revivals, classical orders, etc. 1738 Ware English edition. 216 plates. 110pp. of text. 9½ × 12¾. 21308-0 Pa. $11.95

THE WIT AND HUMOR OF OSCAR WILDE, edited by Alvin Redman. More than 1000 ripostes, paradoxes, wisecracks: Work is the curse of the drinking classes, I can resist everything except temptations, etc. 258pp. 5⅜ × 8½. 20602-5 Pa. $4.50

THE DEVIL'S DICTIONARY, Ambrose Bierce. Barbed, bitter, brilliant witticisms in the form of a dictionary. Best, most ferocious satire America has produced. 145pp. 5⅜ × 8½. 20487-1 Pa. $2.95

ERTÉ'S FASHION DESIGNS, Erté. 210 black-and-white inventions from *Harper's Bazar*, 1918-32, plus 8pp. full-color covers. Captions. 88pp. 9 × 12. 24203-X Pa. $7.95

ERTÉ GRAPHICS, Erté. Collection of striking color graphics: *Seasons, Alphabet, Numerals, Aces* and *Precious Stones*. 50 plates, including 4 on covers. 48pp. 9⅜ × 12¼. 23580-7 Pa. $6.95

PAPER FOLDING FOR BEGINNERS, William D. Murray and Francis J. Rigney. Clearest book for making origami sail boats, roosters, frogs that move legs, etc. 40 projects. More than 275 illustrations. 94pp. 5⅜ × 8½. 20713-7 Pa. $2.50

ORIGAMI FOR THE ENTHUSIAST, John Montroll. Fish, ostrich, peacock, squirrel, rhinoceros, Pegasus, 19 other intricate subjects. Instructions. Diagrams. 128pp. 9 × 12. 23799-0 Pa. $5.95

CROCHETING NOVELTY POT HOLDERS, edited by Linda Macho. 64 useful, whimsical pot holders feature kitchen themes, animals, flowers, other novelties. Surprisingly easy to crochet. Complete instructions. 48pp. 8¼ × 11. 24296-X Pa. $1.95

CROCHETING DOILIES, edited by Rita Weiss. Irish Crochet, Jewel, Star Wheel, Vanity Fair and more. Also luncheon and console sets, runners and centerpieces. 51 illustrations. 48pp. 8¼ × 11. 23424-X Pa. $2.75

YUCATAN BEFORE AND AFTER THE CONQUEST, Diego de Landa. Only significant account of Yucatan written in the early post-Conquest era. Translated by William Gates. Over 120 illustrations. 162pp. 5⅜ × 8½.   23622-6 Pa. $3.95

ORNATE PICTORIAL CALLIGRAPHY, E.A. Lupfer. Complete instructions, over 150 examples help you create magnificent "flourishes" from which beautiful animals and objects gracefully emerge. 8⅛ × 11.   21957-7 Pa. $3.50

DOLLY DINGLE PAPER DOLLS, Grace Drayton. Cute chubby children by same artist who did Campbell Kids. Rare plates from 1910s. 30 paper dolls and over 100 outfits reproduced in full color. 32pp. 9¼ × 12¼.   23711-7 Pa. $3.50

CURIOUS GEORGE PAPER DOLLS IN FULL COLOR, H. A. Rey, Kathy Allert. Naughty little monkey-hero of children's books in two doll figures, plus 48 full-color costumes: pirate, Indian chief, fireman, more. 32pp. 9¼ × 12¼.
24386-9 Pa. $3.50

GERMAN: HOW TO SPEAK AND WRITE IT, Joseph Rosenberg. Like *French, How to Speak and Write It.* Very rich modern course, with a wealth of pictorial material. 330 illustrations. 384pp. 5⅜ × 8½.   20271-2 Pa. $4.95

CATS AND KITTENS: 24 Ready-to-Mail Color Photo Postcards, D. Holby. Handsome collection; feline in a variety of adorable poses. Identifications. 12pp. on postcard stock. 8¼ × 11.   24469-5 Pa. $2.95

MARILYN MONROE PAPER DOLLS, Tom Tierney. 31 full-color designs on heavy stock, from *The Asphalt Jungle,Gentlemen Prefer Blondes,* 22 others.1 doll. 16 plates. 32pp. 9⅜ × 12¼.   23769-9 Pa. $3.95

FUNDAMENTALS OF LAYOUT, F.H. Wills. All phases of layout design discussed and illustrated in 121 illustrations. Indispensable as student's text or handbook for professional. 124pp. 8⅛.× 11.   21279-3 Pa. $4.50

FANTASTIC SUPER STICKERS, Ed Sibbett, Jr. 75 colorful pressure-sensitive stickers. Peel off and place for a touch of pizzazz: clowns, penguins, teddy bears, etc. Full color. 16pp. 8¼ × 11.   24471-7 Pa. $3.50

LABELS FOR ALL OCCASIONS, Ed Sibbett, Jr. 6 labels each of 16 different designs—baroque, art nouveau, art deco, Pennsylvania Dutch, etc.—in full color. 24pp. 8¼ × 11.   23688-9 Pa. $3.95

HOW TO CALCULATE QUICKLY: RAPID METHODS IN BASIC MATHE-MATICS, Henry Sticker. Addition, subtraction, multiplication, division, checks, etc. More than 8000 problems, solutions. 185pp. 5 × 7¼.   20295-X Pa. $2.95

THE CAT COLORING BOOK, Karen Baldauski. Handsome, realistic renderings of 40 splendid felines, from American shorthair to exotic types. 44 plates. Captions. 48pp. 8¼ × 11.   24011-8 Pa. $2.50

THE TALE OF PETER RABBIT, Beatrix Potter. The inimitable Peter's terrifying adventure in Mr. McGregor's garden, with all 27 wonderful, full-color Potter illustrations. 55pp. 4¼ × 5½. (Available in U.S. only)   22827-4 Pa. $1.75

BASIC ELECTRICITY, U.S. Bureau of Naval Personnel. Batteries, circuits, conductors, AC and DC, inductance and capacitance, generators, motors, trans-formers, amplifiers, etc. 349 illustrations. 448pp. 6½ × 9¼.   20973-3 Pa. $7.95